GENERAL CHEMISTRY EXAMINATION QUESTIONS

Fourth Edition

K. J. Johnson

L. M. Epstein

University of Pittsburgh
Pittsburgh, Pennsylvania

Burgess Publishing Company
Minneapolis, Minnesota

CONTENTS

i

CHAPTER 1.   INTRODUCTON

This book is a gold mine both for students in general chemistry courses and for those preparing for standardized examinations. There are 505 multiple-choice exam questions, covering the topics emphasized in general chemistry. The questions in each chapter are ordered according to increasing difficulty. Detailed solutions to the 189 questions marked with an asterisk (*) are found in Part II (pages 279-336). There are 140 sample problems with solutions in the textual material in each chapter. Five 20-question practice exams are included as Appendix A. Appendix B contains a set of constants, formulas, and conversion factors. The answers to the 505 questions and the sample exams are given in Appendix C.

Often the most effective way to find out if one is adequately prepared for a chemistry examination is to obtain a copy of an exam covering the same material and take the test in the time alloted. This book offers a wide selection of sample exam questions that range in difficulty from easy to challenging. Students are advised to resist the temptation to simply learn the procedures for solving these problems and not concentrate on the chemical concepts being tested. It is not difficult for an instructor to change the question to test understanding of the concept rather than memorization of the procedures for solving chemistry problems.

Enough space is provided between questions so that students can use this book as a workbook. One of the most important aspects of successful problem solving in chemistry is setting up the problem correctly. Students are encouraged to use the space provided to set the problem up. In the unhappy event that this set-up leads to an incorrect answer, the student can flag the question and be reminded of this error in subsequent review of the question. Students are also urged to use dimensional analysis in chemical problem solving. As the following example illustrates this approach not only can save time but also reduces the chance for error.

<u>Example</u> What is the volume in quarts of 25.0 pounds of liquid mercury? The density of liquid mercury is 13.6 g/ml. There are 454 grams per pound, and 1 quart = 946 ml.

<u>Solution</u> The solution requires two conversion factors:

$$1 \text{ qt} = 946 \text{ ml, and } 1 \text{ lb} = 454 \text{ g,}$$

A stepwise solution could proceed as follows:

(1) Convert pounds to grams

$$\# \text{ g} = \frac{454 \text{ g}}{\cancel{\text{lb}}} \times 25 \cancel{\text{ lb}} = 11350 \text{ g}$$

(2) Use the density to calculate the volume of mercury

$$\# \text{ ml} = \frac{1 \text{ ml}}{13.6 \cancel{\text{g}}} \times 11350 \cancel{\text{g}} = 834.6 \text{ ml}$$

(3) Convert milliliters to quarts

$$\# \text{ qt} = \frac{1 \text{ qt}}{940 \cancel{\text{ ml}}} \times 834.6 \cancel{\text{ ml}} = 0.882 \text{ qt}$$

Alternatively, the problem can be solved in one step by carefully combining the four factors so that the units cancel to give quarts.

$$\# \text{ qt} = \frac{1 \text{ qt}}{946 \cancel{\text{ ml}}} \times \frac{1 \cancel{\text{ ml}}}{13.6 \cancel{\text{g}}} \times \frac{454 \cancel{\text{g}}}{\cancel{\text{lb}}} \times 25 \cancel{\text{ lb}} = 0.882 \text{ qt}$$

The one-step solution is an example of the use of dimensional analysis. The procedure is to write down the information given in the problem and the conversion factors in such a way that all units save the desired unit cancels out.

Success in chemistry, like anything else, requires practice. The sample examination questions in this book will provide ample practice for students to acquire the skills and concepts needed to master general chemistry.

CHAPTER 2. UNITS AND CONVERSION FACTORS

Scientific Notation

Chemical and physical quantities are often very large or very small numbers. The following scheme is used to avoid the tedium and space required to write these numbers.

Number	Scientific Notation
63,500,000	6.35×10^7
0.000129	1.29×10^{-4}

The procedure is to change the very large (or small) number to a leading number (between one and ten) times ten raised to an appropriate power. For large numbers the power is a positive integer, and is one less than the total number of digits. For small numbers it is a negative integer, and is one more than the number of zeros following the decimal point and preceding the first non-zero digit.

Example 2.1

Write the following numbers using scientific notation.

a) 107   b) 0.00623   c) 12,400,000   d) 7.23 e) 1/4

Solution

a) 1.07×10^2 b) 6.23×10^{-3} c) 1.24×10^7

d) 7.23   e) 2.5×10^{-1}

Notice that for case (e) it is necessary to convert the number to a decimal first.

When numbers in scientific notation are multiplied the leading numbers are multiplied, and the power of ten of the product is the sum of the powers. In division the leading numbers are divided and the powers of ten subtracted. For addition and subtraction it is first necessary to convert all numbers to the same power of ten. Some examples are given below.

$$(1.25 \times 10^2) \times (2.00 \times 10^3) = 2.50 \times 10^5$$

$$(2.00 \times 10^7) \times (6.50 \times 10^{-3}) = 13.0 \times 10^4 = (1.30 \times 10^1) \times 10^4$$

$$= 1.30 \times 10^5$$

$$\frac{6.50 \times 10^8}{2.60 \times 10^3} = \frac{6.50}{2.60} \times 10^{8-3} = 2.50 \times 10^5$$

$$(1.671 \times 10^4) + (3.93 \times 10^3) = (1.671 \times 10^4) + (0.393 \times 10^4) = 2.064 \times 10^4$$

Significant Figures

A rough idea of the precision of a number is given by the number of digits used, not counting zeros used as spacers to indicate magnitude. When scientific notation is used all the digits in the leading number are significant.

When reporting the results of a measurement or calculation the number of digits used must not exceed the number which are significant. As a rule it is accepted that there is some error in the last significant figure reported. It is not only unnecessary to report more, but it is deceptive since it implies there is no error in the preceding digit.

In large numbers not expressed in scientific notation there is no way to tell how many spacer zeros, if any, are significant. It will be assumed that none are. If a wavelength is known to be 290 nanometers precisely to the nearest nanometer it should be expressed as 2.90×10^2 nanometers; otherwise it would be inferred that the 9 is doubtful. When numbers are multiplied and divided the number of significant figures in the answer cannot exceed that in the factor containing the least number of significant figures.

Example 2.2

How many significant figures are there in the following numbers?

a) 360   b) 3.60×10^2 c) 0.00360   d) 3.6×10^{-4}

e) 36,060

Solution:

    a) 2   b) 3   c) 3   d) 2   e) 4

Temperature Conversions

The common scientific temperature scale is the Celsius scale. The unit "degrees Celsius" is abbreviated °C. The Celsius scale is also called the centigrade scale. On this scale water freezes at 0°C and boils at 100°C. Absolute zero is −273°C. The absolute temperature is measured in degrees the same size as centigrade degrees but starting at the absolute zero, and is expressed as degrees Kelvin, abbreviated K. On the absolute scale water freezes at 273K and boils at 373K. If t_c is the temperature in degrees centigrade and T is the absolute temperature in degrees Kelvin, then

$$T = t_c + 273$$

The temperature scale widely used in the U. S. in non-scientific work is the Fahrenheit (°F) on which water freezes at 32°F and boils at 212°F. A degree on the centigrade scale is 1.8 times a degree on the Fahrenheit scale. If t_F is in °F and t_c is the same temperature in °C, then

$$t_c = (t_F - 32)/1.8$$

$$\text{or} \quad t_F = 1.8\, t_c + 32$$

Example 2.3

Convert 80°F to centigrade and absolute temperatures.

Solution:

    °C = (°F − 32)/1.8 = (80−32)/1.8 = 27°C

    K = °C + 273 = 27 + 273 = 300K

Example 2.4

Convert −20°C to Fahrenheit and absolute temperatures.

6

Solution:

$$°F = 32 + 1.8°C = 32 + 1.8(-20) = -4°F$$

$$K = °C + 273 = -20 + 273 = 253K$$

The Metric System

The fundamental units of the metric system are the second (s) for time; the meter (m) for distance; the kilogram (kg), equal to 1000 grams for mass; and the liter, equal to 1000 cm^3, for volume. All other units are derived from these by multiplying by a power of ten. The names of the units are derived by using a prefix that indicates the power of ten. The prefixes and their abbreviations are as follows:

X 10^1	deka	da		X 10^{-1}	deci	d
X 10^2	hecto	h		X 10^{-2}	centi	c
X 10^3	kilo	k		X 10^{-3}	milli	m
X 10^6	mega	M		X 10^{-6}	micro	μ
X 10^9	giga	G		X 10^{-9}	nano	n
X 10^{12}	tera	T		X 10^{-12}	pico	p

Some units used in the "English system" and their metric equivalents are listed below.

Dimension	English	Metric
time	1 second	1 second
mass	1 lb.	454 grams
distance	1 inch	2.54 centimeters
volume	1 quart	0.946 liter

Example 2.5

Convert the following lengths to meters:

a) 174 km   b) 0.247 mm   c) 8.62 cm   d) 9.6 x 10^{-3} dm

e) 8.72 nm

Solution:

a) 1.74×10^5 m   b) 2.47×10^{-4} m   c) 8.62×10^{-2} m

d) 9.6×10^{-4} m   e) 8.72×10^{-9} m

When making calculations involving physical units you should write in the units along with the numbers. They may be multiplied and divided like algebraic variables to yield the units of the answer. This procedure assures that all necessary conversion factors have been used and that none is accidentally inverted.

Example 2.6

A person weighs 180 lbs. and is 5'10" tall. Convert these measurements to the corresponding values in kilograms and meters.

Solution:

$$\#kg = \frac{1 \ kg}{1000 \ g} \times \frac{454 \ g}{lb} \times 180 \ lbs. = 82 \ kg$$

$$\#m = \frac{1 \ m}{100 \ cm} \times \frac{2.54 \ cm}{1 \ in.} \times [(5 \times 12) + 10] \ in. = 1.8 \ m$$

Density

The density, a fundamental property of matter, is simply the mass per unit volume. For pure solids and liquids the density is measured in grams per cubic centimeter (g/ml). One cubic centimeter is equal to one milliliter. For gases the density is measured in grams per liter (g/ℓ). The density is often abbreviated with the Greek symbol rho (ρ).

Example 2.7

Calculate the density in g/ml of a block of styrofoam 1.00 foot x 1.50 feet x 3.00 feet which weighs 3.50 pounds.

Solution:

$$Density = \frac{mass \ (g)}{volume \ (ml)} = \rho$$

$$Mass = \frac{454 \ g}{lb} \times 3.50 \ lb. = 1590 \ g$$

$$\text{Volume} = \left(\frac{2.54 \text{ cm}}{1 \text{ in}}\right)^3 \times \left(\frac{12 \text{ in}}{1 \text{ ft}}\right)^3 \times (1.00)(1.50)(3.00) \text{ ft}^3$$

$$= 1.27 \times 10^5 \text{ ml}$$

$$\rho = \frac{1590 \text{ g}}{1.27 \times 10^5 \text{ ml}} = 1.25 \times 10^{-2} \text{ g/ml}$$

Note that the answer is rounded to three significant figures in accordance with the precision of the original data.

Example 2.8

A flask weighs 10.473 grams empty and 38.769 grams when filled with water. The same flask weighs 34.217 grams when filled with an organic solvent. Assuming the density of water is 1.00 g/ml, what is the density of the organic solvent?

Solution:

The mass of water in the flask is 38.769–10.473 or 28.296 grams. Since the density of water is 1.00 g/ml, then the volume of the flask is 28.296 mls. The density of the organic solvent is the mass of the solvent divided by the volume of the solvent,

$$\rho = \frac{(34.217 - 10.473) \text{ g}}{28.296 \text{ ml}} = 0.839 \text{ g/ml}$$

Weight and Volume Percentage

When substances are mixed, the relative proportions may be indicated by the percentage of each component of the mixture. The weight of each substance, divided by the total weight of the mixture (in the same units), times 100, is its percent by weight. It follows that the percent by weight is equal to the number of grams of a component present in 100 grams of the mixture.

Example 2.9

How many grams of $AgNO_3$ are in 250 ml of a solution which is 5.0% $AgNO_3$ by weight and has a density of 1.10 g/ml?

<u>Solution:</u>

The mass of the solution can be calculated from its volume and density,

$$\text{mass} = \frac{1.10 \text{ g}}{\text{ml}} \times 250 \text{ ml} = 275 \text{ grams}$$

Then the mass of $AgNO_3$ is 5% of 275, or

$$\#\text{g } AgNO_3 = \frac{5 \text{ grams } AgNO_3}{100 \text{ grams solution}} \times 275 \text{ grams solution} = 14 \text{ grams}$$

The percent by volume, similarly, is 100 times the volume of each component divided by the volume of the mixture. The percent by volume is a useful measure of composition only if the volumes are additive, that is, the total volume is simply the sum of the constituent volumes.

<u>Example 2.10</u>

Calculate the volume percent methanol in a mixture consisting of 175 grams of ethanol and 225 grams of methanol. The densities of ethanol and methanol are 0.789 and 0.793 g/ml respectively. Assume the volumes are additive.

<u>Solution:</u>

The volumes are

$$\#\text{ml ethanol} = \frac{1 \text{ml}}{0.789 \text{g}} \times 175 \text{ g} = 222 \text{ ml}$$

$$\#\text{ml methanol} = \frac{1 \text{ml}}{0.793 \text{g}} \times 225 \text{ g} = 284 \text{ ml}$$

$$\text{Volume \% of methanol} = \frac{284 \times 100}{222 + 284} = 56.1\%$$

Questions

2.1 Which of the following numbers is(are) equivalent?

   1) 1,470   2) 1.470×10^3 3) 147000×10^{-3}

   a) 1 and 2
   b) 2 and 3
   c) 1 and 3
   d) All of them
   e) None of them

2.2 Which of the following is(are) correctly represented?

1) $12,000 = 1.2 \times 10^4$ 2) $0.0014 = 1.4 \times 10^{-4}$ 3) $62.4 = 6.24 \times 10^1$

a)   1
b)   1 and 2
c)   1 and 3
d)   2 and 3
e)   All of them

2.3 Which one of the following numbers has 4 significant figures?

a)   0.004398
b)   0.043
c)   0.0439
d)   0.43980
e)   None of these

2.4 Which of the following measurements of mass is(are) equivalent?

1) 2.5 pg   2) 2.5×10^6 mg   3) 2.5×10^{-3} kg

a)   1 and 2
b)   2 and 3
c)   1 and 3
d)   All of them
e)   None of them

2.5 A scientist measured a certain length and found it to be 1.01 m.
Which of the following is equivalent?

a)   101 mm
b)   .101 km
c) 1.01×10^3 cm
d)   1.01 cm
e) 1.01×10^3 mm

2.6 Which of the following measurements of length is(are) equivalent?

1) 5 km   2) 5000 m   3) 5×10^6 cm

a)   1 and 2
b)   2 and 3
c)   1 and 3
d)   All of them
e)   None of them

2.7 In most countries of the world, gasoline is sold by the liter. If you
wanted 20.0 gallons of gas, how many liters would that be?

$$1.00 \text{ liter} = 1.06 \text{ quarts}$$

$$4.00 \text{ quarts} = 1.00 \text{ gallon}$$

(see next)

a)   4.72
b)   84.8
c)   75.5
d)   21.2
e)   5.30

2.8* On a recent trip to Europe, a traveler was able to drive 427 kilometers
on a tank filled with 37.9 liters of gasoline. What is his mileage in
miles per gallon?

$$1.00 \text{ liter} = 0.264 \text{ gallon}$$

$$1.00 \text{ kilometer} = 0.621 \text{ mile}$$

a)   23.8 miles per gallon
b)   26.5 miles per gallon
c)   18.5 miles per gallon
d)   15.9 miles per gallon
e)   21.2 miles per gallon

2.9 Calculate the time in minutes that it takes for light to travel from
the sun to the earth. Assume the distance is 9.2×10^7 miles and
that the speed of light is 3.0×10^{10} cm/sec. 1 km = 0.621 mile.

a)   0.68
b)   3.2
c)   8.2
d)   16
e)   21

2.10 The market value of a certain gem is \$50.00 per carat. What is the
value of a sample of this gem that weighs 1.24 oz?
(1 oz = 28.4 g; 1 carat = 200 mg)

a)   \$310
b)   \$352
c)   \$1760
d)   \$7100
e)   \$8800

2.11 Assume that the standard established by the United States government for
carbon monoxide emission for automobiles limits exhaust to 23.0 grams
CO per vehicle-mile. Assume also that in a typical metropolitan area
of 517000 people there are 82700 automobiles, driven an average of 13.5
miles per 24 hour period. How much CO (in tons) can legally be discharged into the area's atmosphere per day (1 ton = 2000 x 454 grams.)?

a)   270 tons/day
b)   .155 tons/day
c)   28.3 tons/day
d)   .0536 tons/day
e)   36.2 tons/day

2.12 Convert 250K to centigrade and Fahrenheit temperatures.

	°C	°F
a)	23	-9.4
b)	-23	-9.4
c)	23	-19.2
d)	-23	44.8
e)	523	973

2.13 Which of the following temperatures are equivalent?

1)   77°F   2)   25°C   3)   298K

a)   1 and 2
b)   2 and 3
c)   1 and 3
d)   All of them
e)   None of them

2.14*At what temperature are the Fahrenheit and centigrade temperatures equal in magnitude but opposite in sign?

a)   -11.4°F
b)   11.4°F
c)   -40°F
d)   40°F
e)   32°F

2.15 The Rankine (°R) temperature scale is used in engineering. The degree interval on the Rankine and Fahrenheit scales is the same. The zero or the Rankine scale corresponds to absolute zero, or

$$0°R = -273°C$$

What is the boiling point of water on the Rankine scale?

Hint: Determine the constant A in the formula °R = A + °F.

a)   0
b)   173
c)   212
d)   492
e)   671

2.16 The following data were collected from measurements on a rectangular block of pure tin:

$$\text{Dimensions:} \quad 2.93 \text{ in.}$$
$$.0611 \text{ meter}$$
$$.261 \text{ ft}$$

$$\text{Mass:} \quad 2.08 \times 10^5 \text{ grams}$$

What is the density of tin in g/cm^3?

$$2.54 \text{ cm.} = 1.00 \text{ in.}$$
$$30.5 \text{ cm.} = 1.00 \text{ ft.}$$

a) $\dfrac{2.08 \times 10^5}{(2.93 \times 2.54)(.0611 \times 100)(.261 \times 30.5)}$

b) $\dfrac{2.08 \times 10^5}{(\frac{2.93}{2.54})(\frac{.0611}{100})(\frac{.261}{30.5})}$

c) $\dfrac{(2.93 \times 2.54)(.0611 \times 100)(.261 \times 30.5)}{2.08 \times 10^5}$

d) $\dfrac{(\frac{2.93}{2.54})(\frac{.0611}{100})(\frac{.261}{30.5})}{2.08 \times 10^5}$

e) $\dfrac{1}{(2.08 \times 10^5)(2.93 \times 2.54)(.0611 \times 100)(.261 \times 30.5)}$

2.17 The density of the liquid compound S_2Cl_2 is 1.68 g/ml. What volume (in ml) should be used if 0.185 pound is required? (1 lb = 454 grams)

a) 31.1
b) 50.0
c) 84.0
d) 110
e) 141

2.18* The density of liquid mercury is 13.6 g/ml. How much does 100 gallons of mercury weigh? (1 gallon = 3.7854 liters; 1 ton = 2000 lbs, 454 g = 1 lb.)

a) 5.67 tons
b) 16.3 tons
c) 22.6 tons
d) 55.6 tons
e) 441 tons

2.19 An object weighs 10.17 grams in air and 8.54 grams when submerged
in water. What is the density of the object?

a)   3.87
b)   5.24
c)   6.24
d)   11.5
e)   13.9

2.20 A solid object which weighs 47.6 grams in air when placed in a
beaker of water sank to the bottom and displaced 37.8 ml of
water. In a similar experiment, it displaced 32.7 grams of
an oil. Calculate both the density of the object and the density
of the oil.

	Object	Oil
a)	1.26 g/ml	0.865 g/ml
b)	1.46 g/ml	1.26 g/ml
c)	1.26 g/ml	1.16 g/ml
d)	3.86 g/ml	2.19 g/ml
e)	0.865 g/ml	1.76 g/ml

2.21 A weighing bottle has a mass of 17.66 grams when empty and 35.66
grams when filled with H_2O. The same weighing bottle weighs
42.15 grams when filled with a liquid of unknown composition.
What is the density of grams per milliliter of the unknown
liquid? (The density of water is 1.00 grams per milliliter).

a)   2.39
b)   2.34
c)   0.69
d)   1.36
e)   1.18

2.22 Let two different solid substances (for example, a block of
wood and a piece of metal) be denoted S1 and S2. A chunk of
each substance is added to a pail of water. S1 floats on the
water and S2 sinks. Which one of the following statements is
not correct?

a)   The density of S1 is less than the density of water.
b)   The density of S2 is greater than the density of S1.
c)   The mass of a fixed volume of S1 is less than the mass of
     a fixed volume of S2.
d)   The volume of a fixed mass of water is less than the volume
     of a fixed mass of S2.
e)   The mass of a fixed volume of water is intermediate between
     that of S1 and S2.

2.23   A shipment of coal contains 3.27% sulfur by weight. How many
        kilograms of sulfur can theoretically be separated from every
        ton of coal?   (1 ton = 2000 lb;   454 grams = 1 lb.)

        a)   3.27
        b)   14.8
        c)   29.7
        d)   32.7
        e)   65.4

2.24   What is the weight percentage of salt in a sample of sea water
        if 275 grams of the sea water left a solid residue of 10.6 grams
        when evaporated to dryness?

        a)   3.71%
        b)   3.85%
        c)   4.01%
        d)   25.9%
        e)   26.9%

2.25*   How many grams of KCl are in 775 ml of a solution which is 35.5%
        KCl by weight and has a density of 1.28 g/ml?

        a)   215
        b)   275
        c)   352
        d)   640
        e)   972

2.26   15.614 grams of a compound containing carbon, hydrogen, and oxygen
        was analyzed and found to contain 4.164 grams of carbon and 10.177
        grams of oxygen. Determine the percentage by weight of carbon,
        hydrogen, and oxygen.

	% Carbon	% Hydrogen	% Oxygen
a)	26.67	0	65.18
b)	26.67	8.15	65.18
c)	26.67	65.18	8.15
d)	4.164	1.273	10.18
e)	4.164	0	10.18

2.27.   A certain liquor is 110 proof, or 55% ethyl
        alcohol by volume. How many milliliters of alcohol are there
        in a "fifth" of this liquor? (A "fifth" is 1/5 of a gallon;
        1 gallon = 3.7854 liters).

        a)   260
        b)   340
        c)   420
        d)   550
        e)   2100

2.28   Estimate the number of tons of silver in the oceans given that there are approximately 320,000,000 cubic miles of seawater, each cubic mile contains 4 billion (4×10^9) tons of water, and the weight percentage silver in seawater is $1.1 \times 10^{-6}\%$ by weight. (11 parts per billion).

a) 8.8×10^7

b) 1.4×10^8

c) 8.8×10^9

d) 1.4×10^{10}

e) 1.3×10^{12}

2.29   Concentrated sodium hydroxide solution contains 50.0% by weight of solid NaOH and has a density 1.52 g/ml. What volume (in ml) is required to obtain 275 grams of sodium hydroxide?

a)   164
b)   128
c)   362
d)   500
e)   720

2.30*   Pure calcium carbonate, when dissolved in an acidic solution at 25°C and one atmosphere, released 245 ml of carbon dioxide gas for every gram of calcium carbonate. What is the percentage calcium carbonate by weight in a sample of calcite if 1.69 grams of the sample released 331 ml of CO_2 under the same conditions of temperature and pressure?

a)   26.0
b)   35.1
c)   43.8
d)   79.9
e)   98.4

CHAPTER 3.   NOMENCLATURE

<u>Elements and Ionic Compounds</u>

Each element has a unique name that does not depend on the
molecular composition. For example Ar is argon, Cl_2 is chlorine, P_4 is
phosphorous, and S_8 is sulfur. The elements which occur in more than
one chemical form are distinguished by special names. Carbon, for
example, occurs both as graphite and diamond.

Salts are hard, brittle solids consisting of a regular array of
positive and negative ions, called cations and anions, respectively.
Elements in groups I, II, and III form cations with positive charge
equal to the group number, for example, Na^+, Ca^{2+}, and Al^{3+}. Most other
metals are capable of forming positive ions of more than one charge.
These are named by writing the charge in parentheses using Roman numerals
to indicate the charge, for example, Pb^{2+} - lead(II), and Fe^{3+} - iron(III).
The only polyatomic positive ion commonly encountered is the ammonium
ion, NH_4^+.

Monatomic anions are named by adding the suffix "ide" to the stem
of the element's name, for example, F^- - fluoride, S^{2-} - sulfide, and
P^{3-} - phosphide. The following table contains the names and formulas of
several polyatomic anions.

Table 3.1 Polyatomic Anions

<u>Anion</u>	<u>Name</u>	<u>Anion</u>	<u>Name</u>
NO_3^-	nitrate	ClO_4^-	perchlorate
NO_2^-	nitrite	ClO_3^-	chlorate
PO_4^{3-}	phosphate	ClO_2^-	chlorite
HPO_3^{2-}	phosphite	ClO^-	hypochlorite
SO_4^{2-}	sulfate	O_2^{2-}	peroxide
HSO_4^-	hydrogen sulfate	N_3^-	azide
SO_3^{2-}	sulfite	MnO_4^-	permanganate
HSO_3^-	hydrogen sulfite	$Cr_2O_7^{2-}$	dichromate

18

Table 3.1 continued

Anion	Name		Anion	Name
CO_3^{2-}	carbonate		CrO_4^{2-}	chromate
HCO_3^-	hydrogen carbonate		$S_2O_3^{2-}$	thiosulfate
$CH_3CO_2^-$	acetate		CN^-	cyanide
$C_2O_4^{2-}$	oxalate		OCN^-	cyanate
			SCN^-	thiocyanate

The prefixes "per" and "hypo" and the suffixes "ate" and "ite" are used to name the anions according to the "oxidation number" of the central element in the anion. Oxidation numbers are defined in Chapter 6.

Example 3.1 Write the formulas for sodium permanganate, potassium oxalate, and lead(II) phosphate.

Solution $NaMnO_4$, $K_2C_2O_4$ and $Pb_3(PO_4)_2$

Note that two K^+ ions are required to balance the two negative charges in $C_2O_4^{2-}$ and that in lead(II) phosphate the numbers of Pb^{2+} and PO_4^{3-} ions must be chosen so that the compound is electrically neutral.

Example 3.2 What are the names of K_2SO_3, $Sn(CH_3COO)_2$ and $Ba(SCN)_2$?

Solution   Potassium sulfite, tin(II) acetate and barium thiocyanate.

Acids

Binary acids with hydrogen and one other element use the prefix "hydro", followed by the name of the element ending with the suffix "ic", for example HBr – hydrobromic acid, and H_2S – hydrosulfuric acid. This nomenclature convention applies only to elements in groups VI and VII.

For oxyacids, that is, acids containing one or more oxygen atoms per molecule, the prefix "hydro" is omitted and the suffixes "ic" and "ous" are used:

HNO_3	nitric acid		H_3PO_4	phosphoric acid
HNO_2	nitrous acid		H_3PO_3	phosphorous acid
H_2SO_4	sulfuric acid		$HClO_3$	chloric acid
H_2SO_3	sulfurous acid		$HClO_2$	chlorous acid

Some acids are formed by joining simple acid molecules together, and the prefixes ortho, pyro, and meta are used to indicate one, two, and many respectively. For example:

H_3PO_4 — orthophosphoric acid

$H_4P_2O_7$ — pyrophosphoric acid (also diphosphoric acid)

$(HPO_3)_n$ — metaphosphoric acid (n is a large integer)

The last formula is often written simply HPO_3. It is understood to be polymeric (many parts) in structure. Other examples of polymeric acids are metaboric, $(HBO_2)_n$, and metasilicic, $(H_2SiO_3)_n$. Table 3.2 provides the names for several other acids.

Table 3.2 Acids

Acid	Name	Acid	Name
HOCl	hypochlorous	HCOOH	formic
$HClO_4$	perchloric	CH_3COOH	acetic
H_2CO_3	carbonic	$H_2C_2O_4$	oxalic
HCN	hydrocyanic	H_2O_2	hydrogen peroxide

Common Substances

Although a systematic nomenclature has been developed, many common substances are known by non-systematic names. Some of these are included in Table 3.3.

Table 3.3 Names and Formulas of Some Common Substances

Common Name	Formula	Systematic Name
water	H_2O	hydrogen oxide
ammonia	NH_3	hydrogen nitride
muriatic acid	HCl	hydrochloric acid
lye, caustic soda	NaOH	sodium hydroxide
lime	CaO	calcium oxide
slaked lime	$Ca(OH)_2$	calcium hydroxide
baking soda	$NaHCO_3$	sodium hydrogen carbonate (sodium bicarbonate)
washing soda	$Na_2CO_3 \cdot 10H_2O$	sodium carbonate decahydrate
blue vitriol	$CuSO_4 \cdot 5H_2O$	copper(II) sulfate pentahydrate
brimstone	S	sulfur

(over)

Table 3.3 continued

Common Name	Formula	Systematic Name
limestone, marble, calcite	$CaCO_3$	calcium carbonate
Epsom salts	$MgSO_4 \cdot 7H_2O$	magnesium sulfate heptahydrate
gypsum	$CaSO_4 \cdot 2H_2O$	calcium sulfate dihydrate
plaster of Paris	$CaSO_4 \cdot 1/2H_2O$	calcium sulfate hemihydrate
hypo	$Na_2S_2O_3$	sodium thiosulfate
laughing gas	N_2O	nitrogen(I) oxide
milk of magnesia	$Mg(OH)_2$	magnesium hydroxide
oil of vitriol	H_2SO_4	sulfuric acid
potash	K_2CO_3	potassium carbonate
pyrite, fool's gold	FeS_2	iron(II) disulfide
quicksilver	Hg	mercury
saltpeter	$NaNO_3$	sodium nitrate
table salt	$NaCl$	sodium chloride

Also, anions with the general formula HXO_n^- are often given the prefix "bi", for example HCO_3^- – bicarbonate, HSO_4^- – bisulfate, and HSO_3^- – bisulfite.

Questions

3.1 Which of the following cations is(are) correctly named?

1. NH_4^+ – ammonium   2. Sn^{4+} – tin(IV)   3. Sb^{5+} – antimony(V)

a) 1 and 2
b) 2 and 3
c) 1 and 3
d) All of them
e) None of them

3.2 Which of the following anions is(are) correctly named?

1. NO_2^- – nitrite   2. SCN^- – thiosulfide   3. P^{3-} – phosphide

a) 1 and 2
b) 1 and 3
c) 2 and 3
d) All of them
e) None of them

3.3 Which of the following acids is(are) correctly named?

1. H_2SO_3 – sulfurous   2. $HClO_3$ – chloric

3. $H_2S_2O_3$ – thiosulfuric

a) 1 and 2
b) 2 and 3
c) 1 and 3
d) All of them
e) None of them

3.4 Which of the following is(are) correctly named?

1. $Rb_2C_2O_4$ – rubidium oxalate

2. $CaCO_3$ – calcium carbonate

3. $Cr(HCO_3)_3$ – chromium(III) hydrogen carbonate

a) 1
b) 1 and 2
c) 1 and 3
d) 2 and 3
e) All of them

3.5 Which of the following is(are) correctly named?

1. $FePO_4$ – iron(III) phosphate

2. $Pb(CH_3CO_2)_2$ – lead(IV) acetate

3. NiS_2O_3 – nickel(II) thiosulfate

a) 1 and 2
b) 2 and 3
c) 1 and 3
d) All of them
e) None of them

3.6 Which of the following is(are) correctly named?

1. NaN_3 – sodium nitride   2. $Al(NO_2)_3$ – aluminum nitrite

3. $Ba(SCN)_2$ – barium thiocyanate

a) 2
b) 1 and 2
c) 2 and 3
d) All of them
e) None of them

22

3.7 Select the proper set of names for the following four compounds

1. $Ag(N_3)_2$ 2. $Sn_3(PO_4)_2$ 3. $CoCl_3$ 4. $Bi_2(C_2O_4)_3$

a) 1. silver(II) azide   2. tin(II) phosphate
   3. cobalt(III) chloride   4. bismuth(III) oxalate

b) 1. silver(II) azide   2. antimony(II) phosphate
   3. cobalt(III) chloride   4. bismuth(III) oxalate

c) 1. silver(II) azide   2. tin(II) phosphate
   3. copper(III) chloride   4. bismuth(III) acetate

d) 1. silver(II) nitride   2. tin(II) phosphate
   3. copper(III) chloride   4. bismuth(III) acetate

e) None of the above

3.8 Which of the following compounds is __incorrectly__ named?

a) $CoBr_2$ – cobalt(II) bromide

b) $AgCN$ – silver(III) cyanide

c) $V_2(SO_4)_3$ – vanadium(III) sulfate

d) $Mn_3(PO_4)_2$ – manganese(II) phosphate

e) $Co(OCN)_2$ – cobalt(II) cyanate

3.9 For which of the following common substances is the chemical formula
or the name __incorrectly__ indicated?

a) baking soda – $NaHCO_3$ – sodium hydrogen carbonate

b) washing soda – $Na_2CO_3 \cdot 10H_2O$ – sodium carbonate decahydrate

c) lime – CaO – calcium oxide

d) slaked lime – $Ca(OH)_2$ – calcium peroxide

e) limestone – $CaCO_3$ – calcium carbonate

3.10 Which of the following common substances is(are) correctly named?

1. $NaOH$ – caustic soda

2. $NaNO_3$ – saltpeter

3. laughing gas – NO_2

a) 2
b) 1 and 2
c) 2 and 3
d) 1 and 3
e) All of them

CHAPTER 4.   ATOMS AND MOLECULES

Atomic Weight

Modern atomic weights are obtained using a mass spectrometer. The atomic mass unit, abbreviated amu, is defined as exactly 1/12 the mass of the carbon-12 isotope, $^{12}_{6}C$. The isotopic masses of the isotopes of all the elements are obtained by comparison to this standard using a mass spectrometer. The fraction of each isotope in a natural mixture is also measured. The atomic weight of an element is the weighted average of the constituent isotopic masses. It can be expresses as a amu per (average) atom, or grams per mole (of atoms). The "mole" is a measure of quantity. One mole of atoms is the amount 6.02×10^{23}. The atomic weight of an element is the mass in grams of 6.02×10^{23} atoms of the element.

Example 4.1 Carbon is found to consist of 98.892% $^{12}_{6}C$ and 1.108% $^{13}_{6}C$. The masses of these isotopes are exactly 12 by definition, and 13.00335 respectively. Calculate the atomic weight of carbon.

Solution The atomic weight is the weighted average of the naturally occurring isotopes,

$$0.98892 \times 12.00000 + 0.01108 \times 13.00335 = 12.01112$$

Before 1960 atomic weights were based on "combining weights" determined by chemical analysis. For instance the combining weight of aluminum, the weight which combines with 8.00 g of oxygen or 1.008 g of hydrogen, was found to be 9.00 g. The atomic weight is an integral multiple, that is either 9.00 g, or 18.00 g, or 27.00 g, or 36.00 g, etc. Other evidence was required to determine this multiple. Another method of estimating the atomic weight was the law of Dulong and Petit, which stated that at room temperature the heat capacity of an element is approximately 6.0 calories per mole per degree. The heat capacity is related to the specific heat (cal/g-deg) and the atomic weight (g/mole),

$$\text{heat capacity} = \text{specific heat} \times \text{atomic weight}$$

$$\frac{cal}{mole\text{-}deg} = \frac{cal}{g\text{-}deg} \times \frac{g}{mole} \approx 6.0$$

The specific heat of a pure substance, like the density, is a fundamental quantity characteristic of the material.

Example 4.2 The specific heat of aluminum is 0.215 cal/g-deg. Estimate the atomic weight of Al.

Solution $\dfrac{\#g}{mole} \sim \dfrac{6 \text{ cal/mole-deg}}{0.215 \text{ cal/g-deg}} \sim 28$ g/mole. This is closest to

27.00, the accurate value based on the combining weight. See preceding paragraph.

Molecular Weight

The chemical formula states the number of atoms of each element which are combined in one molecule of a compound. For instance, one molecule of methyl alcohol, CH_3OH, contains one carbon, one oxygen, and four (3 plus 1) hydrogen atoms. The mass of a CH_3OH molecule is the sum of the masses of all the atoms present using the isotopic weighted averages. The last page of this book is a periodic table which includes atomic weights. The molecular weight of CH_3OH is calculated as follows:

1 carbon atom	1 x 12.01 =	12.01 amu
1 oxygen atom	1 x 16.00 =	16.00 amu
4 hydrogen atoms	4 x 1.008 =	$\dfrac{4.032 \text{ amu}}{32.04 \text{ amu}}$

One molecule of CH_3OH weighs 32.04 amu. The molecular weight of methyl alcohol is 32.04 gram/mole. 6.02×10^{23} molecules of CH_3OH weighs 32.04 grams.

Ionic substances, NaCl for example, do not exist as separate molecules. The formula simply shows the ratio of the ions present. Nonetheless, it is common practice to call the mass of one formula unit (58.5 amu for NaCl) the "molecular weight", as if there were really a NaCl molecule. Some authors, to be more precise, use the term "formula weight" rather than "molecular weight". In any case, one mole of NaCl is understood to be 58.5 grams of NaCl.

Example 4.3 Calculate the molecular weights of the refrigerant freon, CF_2Cl_2, and slaked lime, $Ca(OH)_2$.

Solution CF_2Cl_2:

	carbon	1 x 12.01 =	12.01
	2 fluorine	2 x 19.00 =	38.00
	2 chlorine	2 x 35.45 =	70.90
		Molecular weight =	120.91 g/mole

$$Ca(OH)_2:$$

calcium	$1 \times 40.08 = 40.08$
oxygen	$2 \times 16.00 = 32.00$
hydrogen	$2 \times 1.008 = \underline{2.016}$
Molecular weight	$= 74.10 \text{ g/mole}$

Percentage Composition

The percentage composition of an element in a compound is the percentage by weight of the element. The percentage composition is easily calculated if the chemical formula of the compound is known.

<u>Example 4.4</u> Calculate the percent composition of C, F, and Cl in freon (see Example 4.3)

<u>Solution</u> The molecular weight of CF_2Cl_2 is 120.91 g/mole. Of this, 12.01 g is contributed by one mole of carbon atoms, 38.00 g by two moles of fluorine atoms, and 70.90 g by two moles of chlorine atoms.

$$\text{Percent carbon} = \frac{12.01 \text{ g}}{120.91 \text{ g}} \times 100\% = 9.93\%$$

$$\text{Percent fluorine} = \frac{38.00}{120.91} \times 100\% = 31.43\%$$

$$\text{Percent chlorine} = \frac{70.90}{120.91} \times 10\% = 58.64\%$$

Empirical Formulas

The empirical formula is the simplest chemical formula of a compound.

Name	Chemical Formula	Empirical Formula
Hydrogen Peroxide	H_2O_2	HO
Water	H_2O	H_2O
Ethane	C_2H_6	CH_3
Sodium oxalate	$Na_2C_2O_4$	$NaCO_2$

The empirical formula can be deduced from the percentage composition. The number of grams of each element in 100 grams of the compound is the percentage composition of that element. The number of moles of each element in 100 grams of the compound is determined by dividing the number of grams of each element by its atomic weight. The empirical formula is the chemical formula with the relative numbers of moles of each element represented as integers. Since one mole of an element contains 6.02×10^{23} atoms, the empirical formulas also gives the relative numbers of atoms in the molecule. The actual chemical formula is an integral multiple of the empirical formula,

$$\text{Chemical formula} = \text{Empirical formula} \times i$$

$$\text{where } i = 1, 2, 3, 4 \ldots.$$

Example 4.5 What is the empirical formula of the compound of sulfur and oxygen which is 49.9% S by weight?

Solution 100 grams of the compound contain 49.9 grams of S and 100–49.9 or 50.1 grams of O.

The numbers of moles of S and O in 100 g of the compound are:

$$\text{\# moles of S} = \frac{1 \text{ mole S}}{32.06 \text{ g}} \times 49.9 \text{ g} = 1.556 \text{ moles S}$$

$$\text{\# moles of O} = \frac{1 \text{ mole O}}{16.00 \text{ g}} \times 50.1 \text{ g} = 3.131 \text{ moles O}$$

The compound could be represented $S_{1.556}O_{3.131}$. However, it is more convenient to use integers in writing chemical formulas. Dividing both numbers by the smaller number (1.556)

$$\frac{S_{1.556}}{1.556} \quad \frac{O_{3.131}}{1.556} = S_1O_{2.01}$$

The empirical formula is SO_2. The small discrepancy of .01 can be attributed to experimental error.

Example 4.6 When a sample of a hydrocarbon (a compound containing only carbon and hydrogen) was burned in oxygen, 0.4732 g of CO_2 and 0.1936 g of H_2O were formed. An independent experiment yielded a molecular weight for the hydrocarbon of 56.1 g/mole. What are the empirical and chemical formulas of the hydrocarbon?

Solution In this case the percentage composition of the elements is not known directly but can be calculated from the masses of the compounds derived from them. The CO_2 contains all the carbon and the H_2O all the hydrogen that was present in the hydrocarbon. The molecular weights of CO_2 and H_2O are 44.0 and 18.0 g/mole respectively.

$$\text{\# moles of C} = \text{\# moles of } CO_2 = \frac{0.4732 \text{ g}}{44.0 \text{ g/mole}} = 1.075 \times 10^{-2} \text{ mole C}$$

$$\text{\# moles of H} = \text{\# moles } H_2O \times \frac{2 \text{ moles H}}{\text{mole } H_2O}$$

$$= \frac{0.1936 \text{ g}}{18.0 \text{ g/mole}} \times 2 = 2.150 \times 10^{-2} \text{ mole H}$$

The empirical formula is $C_{.01075}H_{.0215}$ or CH_2.

The molecula weight of CH_2 is $12.01 + 2 \times 1.008$ or 14.03 g/mole.

The actual molecular weight of the compound is 56.1 g/mole.

Since 56.1 is 4×14.03, the molecular formula is C_4H_8.

Avogadro's Number and the Mole

Avogadro's number is the number 6.02×10^{23}. This is the number of atoms in one mole of an element, and the number of molecules in one mole of a compound. The atomic weight of an element is the weight in grams of Avogadro's number of atoms of that element. The molecular weight of a compound is the weight in grams of Avogadro's number of molecules of that compound. For example, the atomic weight of mercury is 200.6 g/mole. This means that 6.02×10^{23} atoms of mercury weigh 200.6 grams. It also means that the "average" mercury atom weights 200.6 amu. The atomic weight is both the weight in grams of one mole of an element and the weight of one atom of the element in amu. Similarly, the molecular weight is both the weight in grams of one mole of the compound and the weight in amu of one molecule of the compound.

Example 4.7 A semiconductor contains 0.00020% by weight of phosphorus. How many phosphorus atoms are there in a 15 mg chip of the semiconductor?

Solution In three steps:

$$(1) \quad \text{g of P} = \frac{0.00020 \text{ gP}}{100 \text{ g chip}} \times \frac{1 \text{ g chip}}{1000 \text{ mg chip}} \times 15 \text{ mg chip}$$

$$= 3.0 \times 10^{-8} \text{ g P}$$

$$(2) \quad \text{moles of P} = \frac{1 \text{ mole of P}}{31.0 \text{ g of P}} \times 3.0 \times 10^{-8} \text{ g of P}$$

$$= 9.7 \times 10^{-10} \text{ mole of P}$$

$$(3) \quad \text{atoms of P} = \frac{6.02 \times 10^{23} \text{ atoms of P}}{\text{mole of P}} \times 9.7 \times 10^{-10} \text{ moles of P}$$

$$= 5.8 \times 10^{14} \text{ atoms of P}$$

In one step:

$$\text{\# atoms of P} = \frac{6.02 \times 10^{23} \text{ atoms}}{\text{mole}} \times \frac{1 \text{ mole}}{31.0 \text{ g}} \times \frac{1 \text{ g}}{1000 \text{ mg}} \times 15 \text{ mg} \times \frac{.00020}{100}$$

$$= 5.8 \times 10^{14} \text{ atoms of P}$$

Questions

4.1 What is the molecular weight of the hydrocarbon $C_{24}H_{50}$?

    a) 288
    b) 74.0
    c) 13.0
    d) 339
    e) 625

4.2 How many atoms are there in a pure sample of 30.4 grams of neon?
(The atomic weight of neon is 20.2 grams/mole).

    a) $\dfrac{20.2}{30.4}$

    b) $\dfrac{(6.02 \times 10^{23})(20.2)}{2.00}$

    c) $\dfrac{(6.02 \times 10^{23})(30.4)}{2.00}$

    d) 30.4

    e) None of the above.

4.3 How many molecules of acetic acid, CH_3COOH, are there in 1.0 kg
of a vinegar solution that is 5.0% acetic acid by weight? The
molecular weight of CH_3COOH is 60.0 g/mole.

    a) 7.2×10^{24}
    b) 5.0×10^{24}
    c) 7.2×10^{23}
    d) 5.0×10^{23}
    e) 5.0×10^{22}

4.4 How many moles of $PbCrO_4$ are there in a 279 gram sample of pure $PbCrO_4$?
(The molecular weight of $PbCrO_4$ is 323 grams/mole).

    a) $\dfrac{(279)(6.02 \times 10^{23})}{323}$

    b) 279

    c) $\dfrac{279}{323}$

    d) $\dfrac{323}{279}$

    e) None of the above.

4.5 If 18.5 moles of the liquid compound C_2Cl_4 are required for a chemical reaction, what volume (in liters) should be taken? The molecular weight of C_2Cl_4 is 166 grams/mole and the density if 1.63 g/ml.

a) 14.6
b) 0.531
c) 1.88
d) 0.182
e) 11.3

4.6 A hypothetical element has the following isotopic composition:

Isotope	Mass	Percent Abundance
60	59.9153	58.1
63	62.9228	41.9

What is the atomic weight of the element?

a) $\dfrac{59.9153 + 62.9228}{2}$

b) $\dfrac{60 + 63}{2}$

c) $59.9153 + 62.9228$

d) $\dfrac{(58.1)(59.9153)+(41.9)(62.9228)}{100}$

e) None of the above

4.7 * How many atoms of Rb are there in a 389 gram sample of pure RbF? The molecular weight of RbF is 105 grams/mole and the atomic weight of Rb is 85.5 grams/mole.

a) 4.90×10^{23}

b) 1.92×10^{26}

c) 2.23×10^{24}

d) 1.32×10^{23}

e) 3.72

4.8 A mixture contains 47.0% $NiSO_3$ by weight. How many moles of $NiSO_3$ are there in a sample of the mixture that weighs 646 grams? The molecular weight of $NiSO_3$ is 139 grams/mole.

a) 3.39×10^{-3}
b) 219
c) 2.19
d) 4.65
e) 4.65×10^{-2}

4.9   How many molecules of Hg_2O are there in a 411 gram sample of pure Hg_2O? The molecular weight of Hg_2O is 418 grams per mole.

    a) 5.92×10^{23}

    b) 2.47×10^{26}

    c) 6.12×10^{23}

    d) 1.46×10^{21}

    e) 2.52×10^{26}

4.10 How many moles of sulfur are there in 1.00 kilogram of a mixture consisting of 75.0% Na_2SO_4 and 25.0% $Na_2S_2O_3$ by weight? The molecular weights are Na_2SO_4 - 142 g/mole and $Na_2S_2O_3$ - 158 g/mole.

    a) 5.00
    b) 6.51
    c) 6.86
    d) 8.27
    e) 8.45

4.11 How many pounds of fluorine are there in 50.0 lbs. of SF_6? The atomic weights are S - 32.1 g/mole and F - 19.0 g/mole.

    a) 2.23
    b) 6.50
    c) 18.6
    d) 39.0
    e) 50.0

4.12 The principle of DuLong and Petit states that the product of the atomic weight of a solid element and its specific heat, measured at room temperature, is approximately 6.0 cal per mole per degree. A prospector has asked you to identify the most abundant element in a mineral sample he has found. You chemically separate the most abundant element and determine that its specific heat is 5.24×10^{-2} cal per gram per degree. What is the element?

    a) Si
    b) Fe
    c) Ni
    d) As
    e) Sn

4.13 What is the percentage by weight of potassium in the compound KNO_3? Atomic weights: K - 39.1, N - 14.0, O - 16.0 g/mole.

    a) 77.3%
    b) 15.5%
    c) 25.5%
    d) 19.3%
    e) 38.7%

4.14* The following nitrogen compounds could possibly be added to the soil
as fertilizers. Which compound has the greatest percent nitrogen
by weight?

a) iron(II) azide, $Fe(N_3)_2$ MW = 140 g/mole

b) sodium nitride, Na_3N MW = 83.0 g/mole

c) potassium nitrite, KNO_2 MW = 85.1 g/mole

d) potassium azide, KN_3 MW = 81.1 g/mole

e) More information is required.

4.15 Which of the following minerals and compounds contains the <u>least</u>
percentage by weight of maganese? (The atomic weight of manganese is
54.9 grams per mole).

a) rhodochrosite $MnCO_3$ 115 g/mole

b) scacchite $MnCl_2$ 126 g/mole

c) pyrolusite MnO_2 87 g/mole

d) hausmannite Mn_3O_4 229 g/mole

e) manganese sulfate $MnSO_4$ 151 g/mole

4.16* A hypothetical element has atomic weight 47.17 grams/mole. It is
composed of two isotopes of mass 47.92 and 46.94. What is the
percentage of the heavier isotope?

a) 30.7
b) .765
c) 76.5
d) 3.26
e) 23.5

4.17* The first chemical compound of a rare gas element was prepared in
1962. Since then several such compounds have been prepared and
characterized. What is the empirical formula of a compound of Xe
which is 67.2% Xe and 32.8% O by weight?

a) XeO_2

b) XeO_5

c) XeO_3

d) XeO_6

e) XeO_4

32

4.18 An analysis of a compound gave 63.6% carbon, 24.8% nitrogen and 11.6% hydrogen. What is the empirical formula of the compound? (Atomic weights: C - 12.0, N - 14.0, H - 1.01 gram/mole)

a) $C_6H_{13}N_2$

b) $C_2H_{13}N_6$

c) $C_3H_{14}N_5$

d) $C_8H_{12}N_2$

e) $C_5H_{12}N_2$

4.19 When added to chlorine, 9.33 grams of copper formed 14.6 grams of a chloride of copper. Find the empirical formula of the chloride.

a) $CuCl$

b) $CuCl_5$

c) $CuCl_3$

d) $CuCl_2$

e) Cu_2Cl

4.20* A sample containing a mixture of NaCl and $CaCl_2$, weighing 70.3 grams, was reacted with carbonate ion to precipitate all the calcium as $CaCO_3$. The $CaCO_3$ precipitate was then converted to CaO by heating. If the final weight of the CaO was 18.4 grams, what is the percentage (by weight) of $CaCl_2$ in the original sample? (The molecular weights of $CaCl_2$ and CaO are 111 and 56.1 grams/mole, respectively.)

a) 26.2%

b) 73.8%

c) 51.9%

d) 13.3%

e) 48.1%

4.21 An electric utility company received a shipment of 55 tons of coal containing 1.3 percent sulfur. How many tons of H_2S can be prepared assuming 86.0 percent recovery of the sulfur?

> 1.00 ton = 2000 pounds, and 1.00 kilogram = 2.20 pounds
> The atomic weight of sulfur is 32.1 grams/mole.
> The molecular weight of H_2S is 34.1 grams/mole.

a) 0.62

b) 0.65

c) 2.1

d) 0.019

e) 0.60

4.22 * The compound BaO_2 (molecular weight 169 grams per mole) forms a hydrate. A 323 gram sample of the hydrate lost 149 grams of H_2O on heating. How many waters of hydration are associated with this hydrate?

a) 7
b) 8
c) 9
d) 10
e) None of the above.

4.23 A 39.8 gram sample of an iron oxide was heated with a stream of hydrogen gas converting it completely to metallic iron. If the resultant iron weighted 30.9 grams, what is the empirical formula of the iron oxide?

a) Fe_4O_3

b) Fe_3O_4

c) Fe_2O_3

d) FeO

e) Fe_3O_2

4.24* Two chlorides of a metal are 37.4% and 47.2% chlorine by weight. The empirical formulas of the chlorides must be?

a) MCl and MCl_2

b) MCl and MCl_3

c) MCl_2 and MCl_3

d) MCl_2 and MCl_4

e) MCl_2 and MCl_5

4.25 A sample of pure TlBr contains 71.90 percent Tl by weight. The relative abundance of the two bromine isotopes in the sample is $^{79}_{35}Br$ – 50.54%, and $^{81}_{35}Br$ – 49.46%. The atomic masses are 78.92 and 80.92 amu respectively. There are two isotopes of Tl present, $^{203}_{81}Tl$ and $^{205}_{81}Tl$, with atomic masses 202.97 and 204.97 amu respectively. What are the approximate relative abundances of the two thallium isotopes?

	% $^{203}_{81}Tl$	% $^{205}_{81}Tl$
a)	25	75
b)	70	30
c)	65	35
d)	35	65
e)	50	50

CHAPTER 5. CHEMICAL REACTIONS

Balancing Equations

The masses of substances involved in chemical changes can be
calculated on the basis of Dalton's laws: atoms do not change mass
in chemical reactions, nor are any atoms created or destroyed. The
first step in any calculation involving chemical reactions is to write
the balanced equation for the chemical change which shows the relative
numbers of atoms and molecules which must participate in accordance with
Dalton's laws.

If the formulas of all reactants and products are known the equation
for the reaction can be balanced by specifying the number of atoms and
molecules which participate. These numbers are called coefficients.

Example 5.1. The reaction between the two gases hydrogen chloride and
ammonia produces the white solid, ammonium chloride. Write the balanced
chemical equation.

Solution:

$$HCl(g) + NH_3(g) \longrightarrow NH_4Cl(g)$$

This equation is balanced. There are four hydrogen
atoms, one chlorine atom, and one nitrogen atom on
both sides of the reaction sign.

Example 5.2. When a solution of ammonia is added to a solution of sulfuric
acid, a solution of ammonium sulfate is formed. The chemical equation is:

$$NH_3(aq) + H_2SO_4(aq) \longrightarrow (NH_4)_2SO_4(aq)$$

Here "aq" stands for aqueous. Balance this chemical equation.

Solution: In this case there is one N atom on the left and two on the
right, in violation of the law of conservation of atoms. However if $2NH_3$
molecules were used the N atoms would be balanced. Inserting the coefficient 2,

$$2NH_3(aq) + H_2SO_4(aq) \longrightarrow (NH_4)_2SO_4(aq)$$

Notice that now the number of H atoms is also balanced, eight on each
side, whereas they were not balanced originally. The S and O atoms remain

in balance.

Most chemical reactions that do not involve oxidation and reduction can be balanced by inspection. One simply places coefficients where indicated until all atoms are in balance.

Example 5.3. Iron(III) oxide reacts with carbon monoxide gas to form iron metal and carbon dioxide gas. Write the balanced equation.

Solution:

$$Fe_2O_3(s) + 3CO(g) \longrightarrow 2Fe(s) + 3CO_2(g)$$

Weight-Weight Stoichiometry

Stoichiometry is the study of the quantities of reactants and products participating in chemical changes. The quantities of substances may be specified in different ways, for example, mass, moles, volume of standard solution, or volume of gas at a certain temperature and pressure. All calculations are based on the one fundamental principle that the numbers of moles of reactants and products are proportional to the coefficients in the balanced equation. Stoichiometry problems involving masses should be solved by the following procedure:

(1) Given the mass of a reactant or product, convert to moles by dividing by the molecular weight.

(2) Refer to the balanced equation and determine the number of moles of the substance sought by multiplying by the appropriate ratio of coefficients.

(3) Convert from the number of moles to the mass of the substance sought by multiplying by the molecular weight.

Example 5.4. How many grams of calcium fluoride are required to produce 63.7g of hydrogen fluoride? The equation is:

$$CaF_2(s) + H_2SO_4(\ell) \longrightarrow CaSO_4(s) + HF(g)$$

Solution: The balanced equation is

$$CaF_2(s) + H_2SO_4(\ell) \longrightarrow CaSO_4(s) + 2HF(g)$$

$$\text{Moles of HF} = \frac{1 \text{ mole HF}}{20.0g \text{ HF}} \times 63.7g \text{ HF} = 3.18 \text{ moles of HF}$$

$$\text{Moles of CaF}_2 = \frac{1 \text{ mole of CaF}_2}{2 \text{ moles of HF}} \times 3.18 \text{ moles of HF}$$

$$= 1.59 \text{ moles of CaF}_2$$

The molecular weight of CaF_2 is $40.1 + 2 \times 19.0 = 78.1$ g/mole

$$\#g\ CaF_2 = \frac{78.1\ g}{mole\ CaF_2} \times 1.59\ mole\ CaF_2 = 124\ g\ CaF_2.$$

In one step,

$$\#g\ CaF_2 = \frac{78.1g\ CaF_2}{mole\ CaF_2} \times \frac{1\ mole\ CaF_2}{2\ moles\ HF} \times \frac{1\ mole\ HF}{20.0g\ HF} \times 63.7g\ HF$$

$$= 124\ g\ CaF_2$$

In the calculation above it was assumed that as much H_2SO_4 as needed (or more) would be available to insure the complete reaction of the CaF_2. In general, the masses of all reagents are given. Unless the reactants are present in the exact stoichiometric quantities, one reagent, called the "limiting reagent", will determine the amount of product formed. For example, consider the following hypothetical reaction,

$$A + 2B \longrightarrow 3C$$

The equation states that one mole of A requires 2 moles of B to react completely. If one mole of A were mixed with only one mole of B, then there would not be enough B for complete reaction. B is the limiting reagent. After the reaction there would be 1/2 mole of A left and 3/2 mole of C formed.

Example 5.5. How many grams of iron can be made by reacting 1.00 kg of Fe_2O_3 with 1.00 kg of aluminum metal? The equation (unbalanced) is:

$$Fe_2O_3(s) + Al(s) \longrightarrow Al_2O_3(s) + Fe(s)$$

Solution: The balanced equation is

$$Fe_2O_3(s) + 2Al(s) \longrightarrow Al_2O_3(s) + 2Fe(s)$$

The molecular weight of Fe_2O_3 is 160 g/mole

The atomic weights of Al and Fe are 27.0 and 55.8 g/mole, respectively. First, convert the masses to moles:

$$\#\ moles\ Fe_2O_3 = \frac{1\ mole\ Fe_2O_3}{160\ g} \times 1000g = 6.25\ moles\ Fe_2O_3$$

$$\#\ moles\ Al = \frac{1\ mole\ Al}{27.0\ g} \times 1000g = 37.0\ moles\ Al$$

There is more Al than needed, so Fe_2O_3 is the limiting reagent.

$$\#\ g\ Fe = \frac{55.8g\ Fe}{mole\ Fe} \times \frac{2\ mole\ Fe}{1\ mole\ Fe_2O_3} \times 6.25\ moles\ Fe_2O_3 = 698g\ Fe.$$

Example 5.6. Show that mass is conserved in the reaction considered in Example 5.5.

<u>Solution</u>: Initially there were 2000 grams. All the Fe_2O_3 reacted and 698 grams of Fe were produced.

$$\#g\ Al_2O_3 = \frac{102g\ Al_2O_3}{mole\ Al_2O_3} \times \frac{1\ mole\ Al_2O_3}{1\ mole\ Fe_2O_3}\ \times\ 6.25\ moles\ Fe_2O_3$$

$$= 638g\ Al_2O_3$$

$$\#\ moles\ Al\ reacted = \frac{2\ moles\ Al}{1\ mole\ Fe_2O_3} \times\ 6.25\ moles\ Fe_2O_3 = 12.5\ moles\ Al$$

$$\#\ moles\ Al\ left = 37.0 - 12.5 = 24.5\ moles\ Al$$

$$\#g\ Al = \frac{27.0g\ Al}{mole\ Al}\ \times\ 24.5\ moles\ Al = 662g\ Al$$

The total number of grams of products after the reaction is:

$$total\ grams = g\ Fe_2O_3 + g\ Al + g\ Al_2O_3 + g\ Fe$$

$$= 0 + 662 + 638 + 698$$

$$= 1998\ \#grams$$

This, to three significant figures, is 2.00 kg. Mass is conserved.

Percentage Yield

The theoretical amount of product in a chemical reaction is seldom attained in practice. All chemical reactions are reversible to some degree. Also, a set of reactants can often give more than one product, for example

$$Fe(s) + O_2(g) \longrightarrow \begin{matrix} FeO\ (s) \\ Fe_2O_3\ (s) \\ Fe_3O_4\ (s) \end{matrix}$$

The percentage yield of a reaction is the actual amount of product divided by the theoretical amount, times 100.

<u>Example 5.7.</u> The following reaction is used in the production of copper,

$$CuFeS_2(s) + O_2(g) \longrightarrow Cu + FeO + SO_2$$

Calculate the percentage yield if the reaction of 1.00 kg of $CuFeS_2$ yielded 312 grams of Cu.

<u>Solution:</u> The balanced equation is

$$CuFeS_2(s) + \frac{5}{2} O_2(g) \longrightarrow Cu(s) + FeO(g) + 2SO_2(g)$$

The theoretical yield is

$$\#g\ Cu = \frac{63.5g\ Cu}{mole\ Cu} \times \frac{1\ mole\ Cu}{1\ mole\ CuFeS_2} \times \frac{1\ mole\ CuFeS_2}{184g\ CuFeS_2} \times 1000g\ CuFeS_2$$

$$= 346g\ Cu$$

The percentage yield is

$$\%\ yield = \frac{actual\ yield}{theoretical\ yield} \times 100 = \frac{312 \times 100}{346} = 90.2\%$$

<u>Weight-Volume Stoichiometry</u>

Many chemical reactions involve gases. It is often necessary to calculate the volume of a gaseous reactant or product. The following is a very useful approximation: one mole of a gas occupies approximately 22.4 liters if the temperature is 273K (0°C) and the pressure is 1.00 atmosphere (760 mm of Hg). These conditions of temperature and pressure are the "standard temperature and pressure" (STP), by definition. The volume is exactly 22.4 liters if the gas is an ideal gas. For most real gases the approximation is excellent.

<u>Example 5.8.</u> Calculate the volumes at STP of O_2 consumed and SO_2 produced in the reaction of Example 5.7.

<u>Solution:</u>

$$\#\ell O_2 = \frac{22.4\ell O_2}{1\ mole\ O_2} \times \frac{2.5\ mole\ O_2}{1\ mole\ CuFeS_2} \times \frac{1\ mole\ CuFeS_2}{184g\ CuFeS_2} \times 1000\ g\ CuFeS_2$$

$$= 304\ liters\ of\ O_2$$

$$\#\ell SO_2 = \frac{22.4\ell\ SO_2}{mole\ SO_2} \times \frac{2\ moles\ SO_2}{2.5\ moles\ O_2} \times \frac{1\ mole\ O_2}{22.4\ell\ O_2} \times 304\ \ell O_2$$

$$= \frac{2}{2.5} \times 304 = 243\ liters\ SO_2$$

Note that as long as the temperature and pressure are at STP, the volumes of gases are related to the number of moles by the conversion factor 22.4 ℓ/mole.

<u>Questions</u>[*]

5.1 Balance the following equation using the minimum integral coefficients.

$$Na_3PO_4 + Fe_2(SO_4)_3 \longrightarrow FePO_4 + Na_2SO_4$$

What is the sum of the coefficients?

a) 4
b) 5
c) 7
d) 8
e) 11

5.2 Given the following combustion reaction for propene, balance the
     equation and determine the sum of the coefficients using the
     minimum integral coefficients.

$$C_3H_6 + O_2 \longrightarrow CO_2 + H_2O$$

What is the sum of the coefficients?

a) 11
b) 18
c) 20
d) 23
e) 26

5.3 The combustion reaction for butane is

$$C_4H_{10} + O_2 \longrightarrow CO_2 + H_2O$$

How many moles of O_2 are required to react completely with 4.0
moles of C_4H_{10}?

a) 1.0
b) 4.0
c) 6.5
d) 13.
e) 26

---

[*] Unless otherwise indicated, assume all reactions are quantitative,
that is, the percentage yield is 100%.

40

5.4 Ammonia is prepared according to the following reaction in the
     gas phase:

$$N_2 + 3 H_2 \longrightarrow 2 NH_3$$

If the reaction conditions are maintained at STP, which of the
following statements is <u>incorrect</u>?

a)   22.4 liters of N_2 will react with 3 x 22.4 liters of H_2 to form
     2 x 22.4 liters of NH_3.

b)   200 liters of N_2 will react with 600 liters of H_2 to form 400
     liters of NH_3.

c)   28.0 grams of N_2 will react with 6.0 grams of H_2 to form 44.8
     liters of NH_3.

d)   300 liters of NH_3 can be prepared from 200 liters of N_2 given an
     excess of H_2

e)   500 liters of NH_3 can be prepared from 750 liters of H_2 given
     an excess of N_2.

5.5 Which of the statements concerning the following reaction is(are)
     <u>incorrect</u>? The molecular weights are:

     P_4 - 124, Cl_2 - 70.9, and PCl_5 - 208 g/mole.

$$P_4(s) + Cl_2(g) \longrightarrow PCl_5(g)$$

1.   124 grams of P_4 react with 70.9 grams of Cl_2 to form 208 grams
     of PCl_5.

2.   124 grams of P_4 react with 22.4 liters of Cl_2 at STP to form
     22.4 liters of PCl_5 at STP.

3.   124 grams of P_4 react with an excess of Cl_2 to form 832 grams
     of PCl_5.

     a)   3
     b)   1 and 2
     c)   2 and 3
     d)   All of them
     e)   None of them

5.6* In an analysis of an ore sample 16.7 grams of pure metal, M, was separated from the other components. This quantity reacted with 2.48 grams of sulfur to form the compound M_2S. What is the metal?

a)   Rb
b)   Cs
c)   Ag
d)   Th
e)   Cu

5.7 How many grams of antimony must be burned in an atmosphere of oxygen to form 24.4 grams of Sb_2O_5?

a)   22.1
b)   4.60
c)   36.8
d)   .0544
e)   18.4

5.8 The combustion products of octane, C_8H_{18}, in the internal combustion engine are CO_2 and H_2O, assuming complete combustion. How many liters of CO_2 and H_2O at STP are formed per mole of C_8H_{18} burned?

	CO_2	H_2O
a)	22.4	22.4
b)	22.4/8	22.4/18
c)	8 x 22.4	9 x 22.4
d)	8 x 22.4	18 x 22.4
e)	16 x 22.4	9 x 22.4

5.9 What volume of SO_3 in liters at STP can be made from 177 grams of $SrSO_4$?

$$SrSO_4(s) \longrightarrow SO_3(g) + SrO(s)$$

a)   7.21
b)   10.8
c)   21.6
d)   23.2
e)   43.2

5.10* Oxygen gas is produced by the decomposition of potassium chlorate:

$$2 \ KClO_3(s) \longrightarrow 2 \ KCl(s) + 3 \ O_2(g)$$

What weight in grams of $KClO_3$ is required to produce 44.8 liters of O_2 at STP? The molecular weight of $KClO_3$ is 123 g/mole.

a)   164
b)   369
c)   492
d)   738
e)   1378

42

5.11 A utility company burns approximately 500 tons of coal per day.
     If the sulfur content of the coal is 1.3% by weight, how many tons
     of SO_2 are dumped into the area's atmosphere each day?
     (1 ton = 909 kilograms)

     a)   .37
     b)   3.2
     c)   6.5
     d)   13
     e)   1.3 x 10^4

5.12*How many grams of HCl are required to react completely with 1.05
     grams of Zn?

     a)   0.589
     b)   0.849
     c)   1.18
     d)   2.35
     e)   4.71

5.13 Assume the reaction responsible for acid mine drainage is

$$8H_2O + 4FeS_2 + 15O_2 \longrightarrow 2Fe_2O_3 + 8H_2SO_4$$

How many kilograms of H_2SO_4 are released to the environment
for every kilogram of iron pyrite oxidized? The molecular weights
are: H_2SO_4 - 98.0, FeS_2 - 120 g/mole.

     a)   0.61
     b)   0.82
     c)   1.0
     d)   1.2
     e)   1.6

5.14 One of the harmful effects of sulfur pollution is the corrosion
     of structural and artistic marble,

$$CaCO_3(s) + SO_2(g) + O_2(g) \longrightarrow CaSO_4(s) + CO_2(g)$$

What volume in liters of SO_2(STP) is consumed in the corrosion of
1.00 kilogram of $CaCO_3$? The molecular weight of $CaCO_3$ is
100 g/mole.

     a)   22.4 x 1000/100
     b)   22.4 x 100/1000
     c)   22.4
     d)   22.4 x 100 x 1000
     e)   none of the above

5.15*Consider the following hypothetical unbalanced reaction:

$$Ni + S + O_2 \longrightarrow NiSO_4$$

Given: 12.0 grams of Ni, 0.404 mole of S, and 14.5 liters of O_2 at STP, how many moles of $NiSO_4$ can be prepared?

a) $\dfrac{14.5}{22.4 \times 2}$

b) $\dfrac{22.4}{14.5 \times 2}$

c) .404

d) $\dfrac{12.0}{58.7}$

e) None of the above

5.16 Which of the following is <u>not</u> correct?

1.00 kg of N_2 reacts with:

a)   1.14 kg of O_2 to form 2.14 kg of NO
b)   0.57 kg of O_2 to form 1.57 kg of N_2O
c)   2.29 kg of O_2 to form 3.29 kg of NO_2
d)   0.76 kg of O_2 to form 1.76 kg of N_2O_3
e)   2.86 kg of O_2 to form 3.86 kg of N_2O_5

5.17 What volume of HCl can be made if 6.60 liters of H_2 and 16.3 liters of Cl_2 react at STP?

$$H_2(g) + Cl_2(g) \rightarrow 2\ HCl(g)$$

a)   6.60
b)   13.2
c)   16.3
d)   29.5
e)   32.6

44

5.18 How many grams of cadmium phosphide, Cd_3P_2, can be made from 375 grams
of Cd and 61.7 grams of P? The atomic and molecular weights are:
P $-$ 31.0, Cd $-$ 112, Cd_3P_2 $-$ 398 g/mole.

   a)   2.11
   b)   357
   c)   396
   d)   444
   e)   437

5.19 An industrial process for the preparation of glycol, $C_2H_6O_2$, utilizes
the following reaction

$$CO(g) \ + \ H_2(g) \ \xrightarrow{\text{catalyst}} \ C_2H_6O_2(\ell)$$

What is the percentage yield for a particular run if 1000 liters (STP)
of CO and 2000 liters (STP) of H_2 resulted in 1.1 kilograms of $C_2H_6O_2$?

   a)   26
   b)   60
   c)   79
   d)   98
   e) 100

5.20 Sodium cyanide is made from sodium carbonate, calcium carbide and
nitrogen,

$$Na_2CO_3(s) + CaC_2(s) + N_2(g) \rightarrow NaCN(s) + CaCO_3(s)$$

Calculate the percentage yield of the reaction if 1.3 tons of
NaCN resulted from the reaction of 1.0 ton of CaC_2 with excess amounts
of both Na_2CO_3 and N_2. The molecular weights of CaC_2 and NaCN are
64.1 and 49.0 g/mole respectively.

   a)   43
   b)   53
   c)   85
   d)   92
   e) 170

5.21 What mass (in grams) of Cu should be added to an excess of concentrated H_2SO_4 to generate as much $SO_2(g)$ as 8.43 grams of sodium
sulfite, Na_2SO_3? The reactions are:

$$Cu \ + \ 2 \ H_2SO_4 \ \rightarrow \ CuSO_4 \ + \ SO_2 \ + \ 2H_2O$$

$$Na_2SO_3 \ + \ H_2SO_4 \ \rightarrow \ Na_2SO_4 \ + \ SO_2 \ + \ H_2O$$

   a)   .471
   b)   2.12
   c)   4.25
   d)   8.36
   e) 33.4

5.22 The reaction between sour milk and baking soda is

$$C_3H_6O_3 + NaHCO_3(s) \rightarrow NaC_3H_5O_3 + H_2O(\ell) + CO_2(g)$$

$C_3H_6O_3$ is lactic acid, which makes sour milk taste sour. These reagents are used in baking bread. The product, $CO_2(g)$, makes the bread rise. What volume of CO_2 at STP in liters results from adding 2.25 grams of lactic acid to 1.75 grams of $NaHCO_3$? The molecular weights are: $C_3H_6O_3$ - 90.0 and $NaHCO_3$ - 84.0 g/mole.

a)   0.435
b)   0.467
c)   0.560
d)   0.600
e)   0.656

5.23 Assume equal weights of Cr and F_2 react under conditions that CrF_2 forms quantitatively. What fraction by weight of the excess reactant remains after the reaction is over? The atomic and molecular weights of Cr and F_2 are 52.0 and 38.0 g/mole respectively.

a)   0.135
b)   0.269
c)   0.365
d)   0.368
e)   0.731

5.24 Iron(III) oxide is used to catalyze the conversion of H_2S to SO_2 according to the following equations

$$H_2S + Fe_2O_3 \longrightarrow Fe_2S_3 + H_2O$$

$$Fe_2S_3 + O_2 \longrightarrow SO_2 + Fe_2O_3$$

If the efficiency of the conversion is 85%, how many kilograms of SO_2 can be obtained per kilogram of H_2S? The molecular weights of H_2S and SO_2 are 34.1 and 64.1 g/mole respectively

a)   64.1/34.1
b)   0.85 x 64.1/34.1
c)   0.85 x 3 x 64.1/34.1
d) $\dfrac{64.1}{0.85 \times 34.1}$
e)   none of the above

5.25 Ammonia can be made from the reaction of nitrides with water, for example,

$$Mg_3N_2(s) + H_2O(\ell) \longrightarrow Mg(OH)_2(s) + NH_3(g)$$

Calculate the percentage yield of this reaction if 10.5 liters of NH_3 at STP were liberated when 25.0 grams of Mg_3N_2 reacted with excess water. The molecular weight of Mg_3N_2 is 101 g/mole. Ignore the solubility of NH_3 in water.

a) 46.9
b) 47.3
c) 52.8
d) 63.1
e) 94.7

5.26 When 8.05 grams of Ca react with excess halogen, X_2, and the product of this reaction (CaX_2) reacts with excess $AgNO_3$, 75.5 grams of AgX form. What is the halogen? The atomic weights of Ca and Ag are 40.1 and 108 g/mole respectively.

a) F_2

b) Cl_2

c) Br_2

d) I_2

e) At_2

5.27 Consider the following decomposition reaction

$$H_2O_2(aq) \longrightarrow H_2O(\ell) + O_2(g)$$

If the peroxide solution is initially 30.0% by weight calculate the weight loss per kg of solution when decomposition is complete (neglect the solubility of O_2).

a) 70.6 g/kg soln
b) 141   "   "
c) 150   "   "
d) 282   "   "
e) 300   "   "

5.28* A mixture of RbBr and CsBr weighed 17.7 grams. The total bromide was determined gravimetrically by precipitation with $AgNO_3$ yielding 18.9 grams of AgBr. The composition of the mixture is:

a)   33.1% RbBr and 66.9% CsBr
b)   46.7% RbBr and 53.3% CsBr
c)   27.8% RbBr and 72.2% CsBr
d)   53.3% RbBr and 46.7% CsBr
e)   73.8% RbBr and 26.2% CsBr

5.29 A mixture consisting of $SO_2(g)$ and $SO_3(g)$ weighs 20.0 grams. After complete oxidation to $SO_3(g)$ the mixture weighs 22.8 grams. What is the mole fraction SO_2 in the mixture?

$$\left(X_{SO_2} = \frac{n_{SO_2}}{n_{SO_2} + n_{SO_3}} \right., \text{ where } X_{SO_2} \text{ is the mole fraction of } SO_2,$$

n_{SO_2} is the number of moles of SO_2 and n_{SO_3} is the number of moles of SO_3.)

a)   0.385
b)   0.614
c)   0.140
d)   0.860
e)   0.333

5.30 A mixture of methane (CH_4) and ethane (C_2H_6) was analyzed by complete combustion with O_2 to give H_2O and CO_2. A sample of the hydrocarbon mixture gave 20.0 grams of H_2O and 30.0 grams of CO_2. What is the approximate composition by weight of the mixture?

	%CH_4	%C_2H_6
a)	10	90
b)	25	75
c)	27	73
d)	35	65
e)	42	58

CHAPTER 6.   REDOX REACTIONS

Definitions

Redox - short for reduction-oxidation

Oxidation - loss of electrons, for example, $Al \longrightarrow Al^{3+} + 3e^-$

Reduction - gain of electrons, for example, $S + 2e^- \longrightarrow S^{2-}$

Disproportionation - simultaneous oxidation and reduction of an
         element, for example, $2Cu^+ \longrightarrow Cu + Cu^{2+}$

Oxidation number - An arbitrary number that indicates the charge on an
         atom

In some chemical reactions there is literally a transfer of electrons
from one atom to another, for example the reaction of sodium and chlorine.
The oxidation of the sodium atom results in the loss of an electron and
the formation of a sodium ion with a +1 charge. The sodium in sodium ion
has a +1 oxidation number. Each chlorine atom is reduced, picking up one
electron and forming a -1 ion. The oxidation number of chlorine in the
chloride ion, is -1.

In many other cases the atoms in the reactants and products share
electrons with their neighbors and so the assignment of oxidation numbers
is a somewhat arbitrary formality. It is customary in such covalent
compounds to assign oxidation numbers to the atoms as if the more
electronegative atom acquired all the shared electrons and the less electronegative atom lost them. For example, in the covalent compound H_2O, oxygen
is the more electronegative element so it is assigned oxidation number -2
as if it had taken an electron from each of the hydrogen atoms. Each
hydrogen atom is then assigned a +1 oxidation number.

Assignment of Oxidation Numbers

1.   The oxidation number of an element is zero. Examples: Al, O_2, and
     P_4 - all 0.

2.   The oxidation number of a monatomic ion is the same as the charge
     of the ion. Examples: Pb^{2+} (+2), I^- (-1), Th^{4+} (+4), and S^{2-} (-2).

3.   The metallic elements in the Li, Be, Sc, and B groups have
     oxidation numbers in their compounds of +1, +2, +3, and +3

respectively (except for $T\ell^{1+}$).

4. The oxidation number of hydrogen is +1, except for hydride (H^-) compounds, where it is -1, and H_2 where it is 0.

5. The oxidation number of oxygen is -2, with the following exceptions: O_2 (0); peroxides, O_2^{2-} (-1); and superoxides O_2^{-1} ($-1/2$).

6. All other oxidation numbers are assigned so as to make the algebraic sum of the oxidation numbers equal to the charge on the ion, or equal to zero in the case of a molecule.

__Example 6.1__   Assign oxidation numbers to the underlined elements in the following compounds and ions:

a) $\underline{Mn}O_4^{2-}$ b) $H_3\underline{P}O_2$ c) $H\underline{S}O_3^-$ d) $\underline{W}_{12}O_{39}^{6-}$

__Solution__   a) +6   b) +1   c) +4   d) +6

__Example 6.2__   Assign the oxidation number to nitrogen in the following compounds:

a) NH_3 b) N_2H_4 c) NH_2OH d) N_3^-

e) N_2 f) N_2O g) NO h) N_2O_3

i) NO_2 j) HNO_3

__Solution__   a) -3 b) -2 c) -1 d) $-1/3$

e) 0   f) +1   g) +2   h) +3

i) +4   j) +5

Balancing Redox Equations

1. The Oxidation Number Method

The coefficients are assigned using the principle that the number of electrons lost by the atoms oxidized must equal those gained by the atoms reduced. It is necessary to know all the reactants and products.

__Example 6.3__   Balance the following reaction:

$$H_2S + HNO_3 \longrightarrow S + NO + H_2O$$

<u>Solution</u>   The atoms are assigned oxidation numbers as follows:

$$\overset{+1\ -2}{H_2S} \quad + \quad \overset{+1+5-2}{HNO_3} \quad \longrightarrow \quad \overset{0}{S} + \overset{+2-2}{NO} + \overset{+1\ -2}{H_2O}$$

Each sulfur loses 2 electrons in going from -2 to 0. Each nitrogen gains 3 electrons in going from $+5$ to $+2$. Therefore the balanced reaction must have 3 S atoms for every 2 N atoms. This gives:

$$3H_2S + 2HNO_3 \longrightarrow 3S + 2NO + H_2O$$

Now the sulfur and nitrogen atoms are in balance, but hydrogen and oxygen are not. It can be seen by inspection that the coefficient of H_2O must be 4. The result is:

$$3H_2S + 2HNO_3 \longrightarrow 3S + 2NO + 4H_2O$$

---

2.   The Half-Reaction or Ion-Electron Method

In this method the oxidation and reduction reactions are considered separately. Each "half-reaction" is balanced independently and the two are added together such that the number of electrons lost in the oxidation equals the number gained in the reduction. This method is particularly useful for redox reactions that occur in aqueous solution. Water molecules and either H^+ or OH^- ions may be added as required, and the charge is balanced by adding electrons where needed. It is convenient to consider acidic and basic solutions separately.

a)   <u>Acidic Solutions</u>

Consider the reduction of dichromate, $Cr_2O_7^{2-}$, to chromium (III) in acidic solution. The following steps are recommended.

(1)   Balance the number of atoms being oxidized or reduced.

$$Cr_2O_7^{2-} \longrightarrow 2Cr^{3+}$$

(2)   In acidic solution, balance oxygen by adding H_2O to the oxygen-deficient side

$$Cr_2O_7^{2-} \longrightarrow 2\,Cr^{2+} + 7H_2O$$

(3)   In acidic solution, balance hydrogen by adding H^+ to the hydrogen deficient side

$$14H^+ + Cr_2O_7^{2-} \longrightarrow 2Cr^{3+} + 7H_2O$$

(4)   Finally, balance charges by adding electrons where needed.

$$14H^+ + Cr_2O_7^{2-} + 6e^- \longrightarrow 2Cr^{3+} + 7H_2O$$

Notice that the addition of six electrons is consistent with the reduction of two chromium (VI) atoms to chromium (III).

Example 6.4   Balance the following half-reaction in acidic solution

$$Cl_2 \longrightarrow HOCl$$

Solution $Cl_2 + 2H_2O \longrightarrow 2HOCl + 2H^+ + 2e^-$

Example 6.5   Balance the following reaction in acidic solution.

$$MnO_4^- + Pb^{2+} \longrightarrow Mn^{2+} + PbO_2$$

Solution   The balanced half reactions are:

(1) $MnO_4^- + 8H^+ + 5e^- \longrightarrow Mn^{2+} + 4H_2O$ (reduction)

(2) $Pb^{2+} + 2H_2O \longrightarrow PbO_2 + 2e^- + 4H^+$ (oxidation)

Reaction (1) is now multiplied by 2 and reaction (2) is multiplied by 5

(1') $2 MnO_4^- + 16H^+ + 10e^- \longrightarrow 2Mn^{2+} + 8H_2O$

(2') $5Pb^{2+} + 10H_2O \longrightarrow 5 PbO_2 + 10e^- + 20H^+$

Now, when (1') and (2') are added, the electrons cancel out:

$$2MnO_4^- + 5Pb^{2+} + 2H_2O \longrightarrow 2Mn^{2+} + 5PbO_2 + 4H^+$$

---

b)   Basic Solutions

Consider the oxidation of thiosulfate to sulfite in basic solution. The following steps are recommended.

(1)   Balance the number of atoms being oxidized or reduced.

$$S_2O_3^{2-} \longrightarrow 2SO_3^{2-}$$

(2)   In basic solution, balance hydrogen by adding one H_2O molecule to the hydrogen-deficient side, and one OH^- ion to the other side for each hydrogen atom.

(3)   In basic solution, balance oxygen by adding two OH^- ions to the oxygen-deficient side, and one H_2O molecule to the other side for each oxygen atom.

$$6 OH^- + S_2O_3^{2-} \longrightarrow 2SO_3^{2-} + 3 H_2O$$

(4)   Balance charges by adding electrons.

$$6 OH^- + S_2O_3^{2-} \longrightarrow 2SO_3^{2-} + 3 H_2O + 4e^-$$

Example 6.6   Balance the following half-reaction in basic solution.

$$NH_2OH \longrightarrow N_2H_4$$

Solution $2NH_2OH + 2e^- \longrightarrow N_2H_4 + 2\ OH^-$

Example 6.7   Balance the following reaction in basic solution.

$$SnO_2^{2-} + CrO_4^{2-} \longrightarrow SnO_3^{2-} + Cr(OH)_3$$

Solution   The balanced half-reactions are:

(1) $2\ OH^- + SnO_2^{2-} \longrightarrow SnO_3^{2-} + H_2O + 2e^-$

(2) $CrO_4^{2-} + 4H_2O + 3e^- \longrightarrow Cr(OH)_3 + 5\ OH^-$

Reaction (1) is multiplied by 3, and reaction (2) by 2 to give:

(1') $6\ OH^- + 3SnO_2^{2-} \longrightarrow 3SnO_3^{2-} + 3H_2O + 6e^-$

(2') $2CrO_4^{2-} + 8H_2O + 6e^- \longrightarrow 2Cr(OH)_3 + 10\ OH^-$

Now, when (1') and (2') are added, the electrons cancel out:

$$3SnO_2^{2-} + 2CrO_4^{2-} + 5H_2O \longrightarrow 3SnO_3^{2-} + 2Cr(OH)_3 + 4OH^-$$

Questions[*]

6.1 For which of the following molecules is the oxidation number of the underlined atom correctly indicated?

1. $\underline{Pb}O_2$ (+4)   2. $H_2\underline{S}O_3$ (+6)   3. $H_3\underline{P}O_2$ (+1)

    a)   1 and 2
    b)   1 and 3
    c)   2 and 3
    d)   All of them
    e)   None of them

6.2 For which of the following ions is the oxidation number of the underlined atom correctly indicated?

1. $H\underline{S}O_4^-$ (+6)   2. $\underline{Mn}O_4^{2-}$ (+6)   3. $\underline{S}_2O_3^{2-}$ (+2)

    a)   1 and 2
    b)   2 and 3
    c)   1 and 3
    d)   All of them
    e)   None of them

*Use the minimum integral coefficients in balancing these redox reactions.

6.3   Complete and balance the following half-reaction which occurs in acidic solution.

$$H_2O_2 \longrightarrow O_2(g)$$

The correct result will show:

a)   2 H^+ on the right side.
b)   2 e^- on the left side.
c)   1 H_2O on the right side.
d)   1 OH^- on the left side
e)   2 positive charges on each side.

6.4   Complete and balance the following half-reaction which occurs in a basic solution.

$$Cl^- \longrightarrow ClO_3^-$$

The correct result will show:

a)   5 H^+ on right side.
b)   5 OH^- on right side.
c)   6 electrons on left side.
d)   6 H_2O on left side.
e)   7 negative charges on each side.

6.5*   Complete and balance the following redox reaction which occurs in acidic solution. What is the sum of the coefficients of all the species in the balanced equation?

$$Cr_2O_7^{2-} + H_2SO_3 \longrightarrow Cr^{3+} + SO_4^{2-}$$

a)   4
b)   15
c)   20
d)   30
e)   24

6.6*   Complete and balance the following redox reaction which occurs in basic solution. What is the sum of the coefficients of all the species in the balanced equation?

$$MnO_4^- + I^- \longrightarrow MnO_2 + IO_3^-$$

a)   4
b)   8
c)   9
d)   20
e)   24

54

6.7   Complete and balance the following half-reaction which occurs in
basic solution. What is the sum of the coefficients of all the
species, including electrons, in the balanced half-reaction?

$$SCN^- \longrightarrow SO_4^{2-} + CO_3^{2-} + NO_3^-$$

a)   25
b)   30
c)   45
d)   48
e)   50

6.8   Complete and balance the following reaction. What is the sum of
the coefficients?

$$NH_3(g) + O_2(g) \longrightarrow NO(g) + H_2O(g)$$

a)   4
b)   8
c)   12
d)   19
e)   22

6.9   Complete and balance the following reaction which occurs in acidic
solution. What is the sum of the coefficients of all the species
in the balanced reaction?

$$FeS_2 + O_2 \longrightarrow Fe^{3+} + HSO_4^-$$

a)   25
b)   32
c)   37
d)   39
e)   43

6.10*Bromine disproportionates in alkaline solution to give bromide and
bromate ion. What is the sum of the coefficients of all species
in the balanced reaction?

a)   18
b)   24
c)   30
d)   36
e)   42

CHAPTER 7.   FARADAY'S LAWS

Definitions

coulomb – unit of electric charge; $1 \, e^- = 1.60 \times 10^{-19}$ coulomb

faraday – Avogadro's number of electrons; 96,500 coulombs
$$(6.02 \times 10^{23} \times 1.60 \times 10^{-19} = 96,500)$$

ampere – unit of electric current; 1 amp = 1 coulomb/sec

electrode – the interface of the external circuit with the electrolyte.

cathode – the electrode at which reduction occurs.

anode – electrode at which oxidation occurs.

Faraday's Laws

Faraday's laws are concerned with the interconversion of electrical and chemical energy. An electrolysis cell is one in which electrical energy is used to drive a chemical reaction which would not occur spontaneously, for example,

$$2 \, NaCl(\ell) \longrightarrow 2 \, Na(\ell) + Cl_2(g)$$

A galvanic cell, or a battery, is an apparatus in which electrical energy is derived from a chemical reaction which does occur spontaneously, for example,

$$2 \, H_2(g) + O_2(g) \longrightarrow 2 \, H_2O(\ell)$$

Faraday's laws are:

1. The mass of a particular substance involved in the chemical reaction is directly proportional to the number of coulombs of electricity.

2. The mass of a particular substance involved in the chemical reaction is directly proportional to the atomic or molecular weight of the substance, and inversely proportional to the number of electrons in the appropriate half-reaction.

For example, consider the reduction of Cu^{2+},

$$Cu^{2+} + 2e^- \longrightarrow Cu$$

On the atomic scale this equation states that two electrons are required to reduce one Cu^{2+} ion to one Cu atom. On the macroscopic scale it states that 2 faradays of electricity (2 x 96,500 coulombs) are required to reduce one mole of Cu^{2+} to one mole of Cu (63.5 grams). Faraday's first law states that the amount of Cu deposited depends on the amount of electricity used. His second law states that the mass of Cu depends on both the atomic weight and the coefficient of the electron in the redox reaction. Here, one faraday of electricity will deposit only one-half mole. (63.5/2 grams). In the reduction of silver ion,

$$Ag^+ + e^- \longrightarrow Ag$$

one faraday of electricity will deposit one mole of Ag (108 grams). In the reduction of aluminum ion,

$$Al^{3+} + 3e^- \longrightarrow Al$$

one faraday of electricity will deposit one-third mole of Al (27.0/3 = 9.0 grams).

<u>Example 7.1</u> – How many grams of Cr can be plated out in 10.0 minutes from an acidic dichromate solution using a current of 43.7 amperes?

<u>Solution</u>   The balanced half-reaction is

$$Cr_2O_7^{2-} + 14H^+ + 12e^- \longrightarrow 2Cr + 7H_2O$$

The solution can be obtained in step-wise fashion:

43.7 amps = 43.7 coulombs/sec

$$\# \text{ coulombs} = \frac{43.7 \text{ coulombs}}{\text{sec}} \times \frac{60 \text{ sec}}{\text{min}} \times 10.0 \text{ min} = 26{,}220 \text{ coulombs}$$

$$\# \text{ faradays} = \frac{1 \text{ faraday}}{96{,}500 \text{ coulombs}} \times 26{,}220 \text{ coulombs} = 0.272 \text{ faraday}$$

$$\# \text{ moles Cr} = \frac{2 \text{ moles Cr}}{12 \text{ faradays}} \times 0.272 \text{ faraday} = 4.53 \times 10^{-2} \text{ moles Cr}$$

$$\# \text{ g Cr} = \frac{52.0 \text{ g Cr}}{\text{mole}} \times 4.53 \times 10^{-2} \text{ mole} = 2.35 \text{ grams}$$

Alternatively, the unit-factor method can be used in one step:

$$\# \text{ g Cr} = \frac{52 \text{ g}}{\text{mole}} \times \frac{2 \text{ mole}}{12 \text{ far}} \times \frac{1 \text{ far}}{96{,}500 \text{ coul}} \times \frac{43.7 \text{ coul}}{\text{sec}} \times \frac{60 \text{ sec}}{\text{min}} \times 10.0 \text{ min}$$

$$= 2.35 \text{ grams}$$

<u>Example 7.2</u> – Faraday's laws are applicable as well to galvanic cells. When a lead storage battery discharges, the reduction reaction is

$$PbO_2(s) + 4H^+ + SO_4^{2-} + 2e^- \longrightarrow PbSO_4(s) + 2H_2O$$

What is the minimum mass of PbO_2 required in a battery that is guaranteed to deliver a current of 30.0 amperes for 12.0 minutes? (The molecular weight of PbO_2 is 239 g/mole).

Solution

$$\# \text{ g } PbO_2 = \frac{239 \text{ g } PbO_2}{\text{mole}} \times \frac{1 \text{ mole}}{2 \text{ far}} \times \frac{1 \text{ far}}{96,500 \text{ coul}} \times \frac{30.0 \text{ coul}}{\text{sec}} \times \frac{60 \text{ sec}}{\text{min}} \times 12 \text{ min}$$

$$= 26.7 \text{ grams}$$

Example 7.3 – How long (in minutes) will it take to produce 45.7 g of $KClO_3$ by the electrolysis of alkaline KCl solution using a current of 150 amperes? The oxidation reaction is

$$Cl^- + 6OH^- \longrightarrow ClO_3^- + 3H_2O + 6e^-$$

The molecular weight of $KClO_3$ is 123 g/mole.

Solution

$$\# \text{ min} = \frac{1 \text{ min}}{60 \text{ sec}} \times \frac{1 \text{ sec}}{150 \text{ coul}} \times \frac{96,500 \text{ coul}}{\text{far}} \times \frac{6 \text{ far}}{\text{mole}} \times \frac{1 \text{ mole}}{123 \text{ g}} \times 45.7 \text{ g} = 23.9 \text{ min}$$

Questions

7.1 What weight of zinc (in grams) will be deposited at an inert electrode on passage of a current of 33.8 amps through a solution of zinc chloride for 9.09 hours? The atomic weight of zinc is 65.4 g/mole.

The reduction equation is:

$$Zn^{2+} + 2 \; e^- \longrightarrow Zn$$

a) $\dfrac{(65.4)(33.8)(9.09)(3600)}{(96500)(2)}$

b) $\dfrac{(65.4)(33.8)(9.09)}{(96500)(2)}$

c) $\dfrac{(65.4)(33.8)(9.09)(3600)}{96500}$

d) $\dfrac{(65.4)(9.09)(3600)}{(96500)(33.8)}$

e) None of the above.

7.2 How long will it take to deposit exactly 100 grams of platinum
at an inert electrode using a constant electrolytic current of
50.0 amps. The reduction reaction is

$$PtCl_4^{2-} + 2\ e^- \longrightarrow Pt + 4Cl^-$$

The atomic weight of Pt is 195 g/mole.

a) 16.5 min
b) 33.0 min
c) 64.3 min
d) 66.0 min
e) 125 min

7.3 Aluminum is prepared commercially by electrolysis of molten bauxite,
$Al_2O_3 \cdot n(H_2O)$. The cathode reaction is:

$$Al^{+3} + 3\ e^- \longrightarrow Al$$

Given a constant current of 50,000 amps, how many kilograms of Al can
be prepared per 24 hour day?

a) 0.112
b) .884
c) 403
d) 1210
e) 10,800

7.4 The powerful oxidizing agent peroxydisulfate, $S_2O_8^{2-}$, is prepared
electrolytically by oxidation of a sulfuric acid solution. The
anion is precipitated as the potassium salt, $K_2S_2O_8$ at the
anode. What quantity of electricity, in coulombs, is required
to prepare 1.00 kilogram of $K_2S_2O_8$. The molecular weight of
$K_2S_2O_8$ is 270 g/mole.

a) 9.65×10^5

b) 1.93×10^5

c) 3.57×10^5

d) 7.15×10^5

e) 1.43×10^6

7.5 What current in amps is required to deposit gold at the rate of
1.00 kg per hour? The reduction reaction is

$$AuCl_4^- + 3\ e^- \longrightarrow Au + 4Cl^-$$

The atomic weight of Au is 197 g/mole.

a) 136
b) 408
c) 804
d) 1220
e) 2440

7.6 Samples containing Pb^{2+} can be analyzed by deposition of PbO_2. What quantity of electricity in coulombs is required to analyze a 50.0 ml sample that contains 0.250 mole of Pb^{2+} per liter of solution?

a) 965
b) 1210
c) 2410
d) 4820
e) 7720

7.7 A particular Edison battery is rated at 100 amp-hr. Assume the cell reaction is

$$Fe + NiO_2 \longrightarrow FeO + NiO$$

What weight of iron is oxidized if the battery delivers 100 amp-hr. of current?

a) 48.0 g
b) 96.0 g
c) 104 g
d) 173 g
e) 208 g

7.8* A galvanic cell is constructed which utilizes the reaction:

$$Na \longrightarrow Na^+ + e^-$$

$$2Na^+ + Na_2S_2 + 2e^- \longrightarrow 2\ Na_2S$$

For how long will this cell deliver a current of 0.750 amperes, starting with 14.5 g of Na at one electrode and 38.2 g of Na_2S_2 at the other?

a) 4.46×10^4 sec
b) 8.10×10^4 sec
c) 4.96×10^4 sec
d) 11.50×10^4 sec
e) 8.92×10^4 sec

7.9* Chlorine gas is prepared commercially by the electolysis of brine,

$$NaCl(aq) \longrightarrow NaOH(aq) + Cl_2(g) + H_2(g)$$

If a commercial electrolysis cell operates at a constant current of 80,000 amps, at 90% efficiency, what weights in kilograms of NaOH, Cl_2 and H_2 can be produced per hour?

	NaOH	Cl_2	H_2
a)	55	190	5.4
b)	60	210	6.0
c)	95	150	4.7
d)	107	95	2.7
e)	120	106	3.0

7.10* The same quantity of electricity that caused the deposition of 100 grams of silver from a $AgNO_3$ solution was passed through a solution of a mercury salt containing Hg cations of unknown charge. 186 grams of Hg were deposited. What is the apparent charge of the Hg cation? (Atomic weights: Ag – 108, Hg – 201)

a) 2
b) 1
c) 4
d) 3
e) None of the above.

CHAPTER 8. SOLUTION STOICHIOMETRY

Concentration Units

weight percent – mass of the solute divided by the mass of the solute plus the mass of the solvent multiplied by 100

molarity (M) – number of moles of solute divided by the volume of the solution in liters; moles/liter

molality (m) – number of moles of solute divided by the number of kilograms of solvent; moles/kilogram

normality (N) – number of equivalents of solute divided by the volume of the solution in liters; (equivalents/liter). The relationship between moles of solute and equivalents of solute depends on the reaction. For acid–base reactions, the equivalent is the amount of reagent that donates or accepts one mole of H^+. For example, one mole of H_2SO_4 is two equivalents. For redox reactions the equivalent is the amount of reagent that gains or loses one faraday, that is, one mole of electrons. For example, when MnO_4^- is reduced to Mn^{2+}, one mole of MnO_4^- is five equivalents.

Briefly the normality is the molarity times the functionality of the solute, where the functionality is the number of equivalents per mole.

mole fraction (X) – the ratio of the number of moles of one species to the total number of moles present. For example, for a binary solution consisting of a solute and a solvent,

$$X_1 = \frac{n_1}{n_1 + n_2} \qquad X_2 = \frac{n_2}{n_1 + n_2} \qquad X_1 + X_2 = 1$$

Here n_1 and n_2 are the number of moles of solute and solvent.

Example 8.1. Calculate the molarity of a solution made by dissolving 6.47 g of $K_2Cr_2O_7$ in water in a volumetric flask and diluting to the 500 mℓ mark. What is the normality of this solution as an oxidizing agent in reactions producing Cr^{3+} as the product? The molecular weight of $K_2Cr_2O_7$ is 294 g/mole.

<u>Solution</u>

$$\frac{\text{\# moles}}{\ell} = \frac{1 \text{ mole}}{294 \text{ g}} \times \frac{6.47 \text{ g}}{0.500 \text{ } \ell} = 0.0440 \underline{\text{M}}$$

The reduction reaction is

$$Cr_2O_7^{2-} + 14H^+ + 6e^- \longrightarrow 2Cr^{3+} + 7H_2O$$

So one mole of $Cr_2O_7^{2-}$ gains 6 Faradays.

$$\frac{\text{\# equivalents}}{\text{liters}} = \frac{6 \text{ equivalents}}{\text{mole}} \times \frac{0.0440 \text{ mole}}{\ell} = 0.264 \underline{\text{N}}$$

<u>Example 8.2.</u> Concentrated commercial HCl solution contains 37.0% HCl by weight and has a density of 1.186 g/ml. Calculate the molality, molarity, mole fraction and normality of concentrated HCl.

<u>Solution</u> Consider 1000 grams of concentrated HCl. There are 370 grams of HCl and 630 grams of H_2O.

$$\frac{\text{\# moles}}{\text{kilogram}} = \frac{1 \text{ mole}}{36.5 \text{ g}} \times \frac{370 \text{ g}}{0.630 \text{ kg}} = 16.1 \underline{\text{m}}$$

$$X_{HCl} = \frac{n_{HCl}}{n_{HCl} + n_{HCl}} = \frac{370/36.5}{370/36.5 + 630/18.0} = 0.225$$

The volume of the 1000 grams is obtained from the density

$$\text{\# liters} = \frac{1 \text{ liter}}{1000 \text{ ml}} \times \frac{1 \text{ ml}}{1.186 \text{ g}} \times 1000 \text{ g} = 0.843 \text{ } \ell$$

$$\frac{\text{\# moles}}{\text{liter}} = \frac{1 \text{ mole}}{36.5 \text{ g}} \times \frac{370 \text{ g}}{0.843 \text{ } \ell} = 12.0 \underline{\text{M}}$$

Since HCl donates one H^+ per molecule, the normality is the same as the molarity.

Solution Stoichiometry

Chapter 5 is concerned with stoichiometric problems involving substances measured in the solid, liquid or vapor state. The mole concept applies equally well to substances measured in terms of volumes of standard solution.

<u>Example 8.3.</u> How many grams of calcium carbonate are required to neutralize 40.0 ml of 0.850 M nitric acid? The molecular weight of $CaCO_3$ is 100 g/mole. Assume the reaction is quantitative.

<u>Solution</u> The balanced reaction is

$$CaCO_3(s) + 2HNO_3(aq) \longrightarrow Ca(NO_3)_2(aq) + H_2O + CO_2(g)$$

$$\text{\# g } CaCO_3 = \frac{100 \text{ g } CaCO_3}{\text{mole } CaCO_3} \times \frac{1 \text{ mole } CaCO_3}{2 \text{ moles } HNO_3} \times \frac{0.850 \text{ mole } HNO_3}{\ell}$$

$$\times \frac{1 \ \ell}{1000 \text{ ml}} \times 40.0 \text{ ml} = 1.70 \text{ grams } CaCO_3$$

<u>Example 8.4.</u> Calculate the molarity of the Na_2SO_4 solution formed by mixing 37.0 ml of 0.453 M H_2SO_4 with 2.06 g of NaOH and diluting to 250.0 ml. The molecular weight of NaOH is 40.0 g/mole. Assume the reaction is quantitative.

<u>Solution</u> The balanced reaction is

$$H_2SO_4 + 2 \text{ NaOH} \longrightarrow Na_2SO_4 + 2 H_2O$$

First, find which reagent is limiting.

$$\text{\# moles } H_2SO_4 = \frac{0.453 \text{ moles } H_2SO_4}{\ell} + \frac{1 \ \ell}{1000 \text{ ml}} \times 37.0 \text{ ml} = 0.0168 \text{ mole}$$

$$\text{\# moles NaOH} = \frac{1 \text{ moles NaOH}}{40 \text{ g NaOH}} \times 2.06 \text{ g NaOH} = 0.0515 \text{ mole}$$

The number of moles of NaOH needed is only 2 x 0.0168 or 0.0336, so NaOH is in excess and H_2SO_4 is the limiting reagent. Then

$$\frac{\text{\# moles } Na_2SO_4}{\ell} = \frac{1 \text{ mole } Na_2SO_4}{1 \text{ mole } H_2SO_4} \times \frac{0.0168 \text{ mole } H_2SO_4}{250 \text{ ml}} \times \frac{1000 \text{ ml}}{\ell} = 0.0672 \underline{M}$$

Titrations

The quantity of substance present in a solution can be determined by reacting it with a standardized solution of a reagent until it is consumed. This is feasible if the reaction is fast and there is an indicator to show when either reactant is in excess.

<u>Example 8.5.</u> Calculate the mass of $Mg(OH)_2$ present in an antacid tablet which required 13.42 ml of 0.2166 M HCl to titrate it. The molecular weight of $Mg(OH)_2$ is 58.3 g/mole.

<u>Solution</u>   The balanced reaction is

$$Mg(OH)_2 + 2\ HCl \longrightarrow MgCl_2 + 2\ H_2O$$

$$\#\ g\ Mg(OH)_2 = \frac{58.3\ g\ Mg(OH)_2}{mole\ Mg(OH)_2} \times \frac{1\ mole\ Mg(OH)_2}{2\ moles\ HCl} \times \frac{0.2166\ mole\ HCl}{\ell}$$

$$\times\ 0.01342\ \ell = 0.0847\ g\ Mg(OH)_2.$$

Standardization is the process or determining the exact molarity of a solution to be used as a titrant by reacting it with a precisely weighted amount of a very pure substance.

<u>Example 8.6.</u>   Calculate the molarity of a silver nitrate solution of which 27.18 ml was required to titrate 0.2119 g of pure KCl.   The molecular weight of KCl is 74.55 g/mole.

<u>Solution</u>   The reaction is

$$AgNO_3 + KCl \longrightarrow AgCl(s) + KNO_3$$

$$\frac{\#\ moles\ AgNO_3}{\ell} = \frac{1\ mole\ AgNO_3}{1\ mole\ KCl} \times \frac{1\ mole\ KCl}{74.55\ g} \times \frac{0.2119\ g}{0.02718\ \ell} = 0.1046\ \underline{M}$$

<u>Example 8.7.</u>   Calculate the molarity of a sodium hydroxide solution which was standardized against 0.04419 M H_2SO_4 if a 50.00 ml aliquot of the acid required 29.16 ml of the base to titrate it.

<u>Solution</u>   The reaction is

$$H_2SO_4 + 2\ NaOH \longrightarrow 2\ H_2O + Na_2SO_4$$

$$\frac{\#\ moles\ NaOH}{\ell\ NaOH\ sol'n} = \frac{2\ moles\ NaOH}{1\ mole\ H_2SO_4} \times \frac{0.04419\ mole\ H_2SO_4}{\ell\ H_2SO_4\ sol'n} \times \frac{0.050\ \ell\ H_2SO_4\ sol'n}{0.02916\ \ell\ NaOH\ sol'n}$$

$$= 0.1515\ \underline{M}$$

It is often necessary to prepare dilute solutions from more concentrated ones.   Since the number of moles of solute does not change in dilution,

$$M_1V_1 = M_2V_2$$

where M and V are molarity and volume and 1 and 2 represent concentrated and dilute solutions.

<u>Example 8.8.</u> How many ml of 15.3 M HNO_3 should be diluted to the mark in a 2.50 ℓ flask to prepare a 0.295 M solution?

<u>Solution</u>

$$V_1 = \frac{M_2 V_2}{M_1}$$

$$= \frac{(0.295 \text{ mole}/\ell)(2.50 \ \ell)}{(15.3 \text{ mole}/\ell)}$$

$$= 0.0482 \ \ell \text{ or } 48.2 \text{ ml}$$

<u>Questions</u>

8.1 Which of the following procedures will result in an aqueous solution that is 15.0% NaCl by weight? (The molecular weight of NaCl is 58.5 g/mole).

   a) Add 15.0 grams of NaCl to 100 grams of H_2O.

   b) Add 15.0 grams of NaCl to 85.0 grams of H_2O.

   c) Add 15.0 grams of NaCl to 115 grams of H_2O.

   d) Add 15.0 x 58.5 grams of NaCl to 100 grams of H_2O.

   e) None of the above

8.2 What is the molarity of a solution containing 33.3 grams of HCl in 2.43 liters of solution? (The molecular weight of HCl is 36.5)

   a) $\dfrac{(1000)(33.3)}{(36.5)(2.43)}$

   b) $(33.3)(\dfrac{36.5}{2.43})$

   c) $\dfrac{33.3}{2.43}$

   d) $\dfrac{33.3}{(36.5)(2.43)}$

   e) None of the above.

8.3 What is the molality of a solution of 69.8 grams of glycol, $C_2H_4(OH)_2$, in 927 mls of water? (Density of water = 1.00 gram per cubic centimeter; the molecular weight of $C_2H_4(OH)_2$ is 62.0)

   a) 75.4
   b) 1.21
   c) 1.22×10^{-3}
   d) 4.67×10^3
   e) 0.823

66

8.4 How many grams of HNO_3 are contained in 353 milliliters of a .465
     molar aqueous solution of HNO_3?   (Molecular weight of HNO_3 is 63.0)

   a) $\dfrac{(.465)(353)}{1000}$

   b) $\dfrac{(63.0)(.465)(353)}{1000}$

   c) $(.465)(353)$

   d) $(63.0)(.465)(353)$

   e) None of the above.

8.5 Calculate the molality of an aqueous solution for which the mole
     fraction of NaOH is 0.250.

   a) 1.06
   b) 16.4
   c) 18.5
   d) 32.3
   e) 54.0

8.6 Calculate the percentage H_2SO_4 by weight in a 2.50 molal solution.
     The molecular weight of H_2SO_4 is 98.0 g/mole.

   a) 8.90
   b) 9.80
   c) 19.7
   d) 22.3
   e) 24.5

8.7 Determine the molarity of a solution prepared by diluting 4.70 liters
     of .716 molar solution to 8.04 liters.

   a) .452
   b) 2.39
   c) 1.22
   d) .419
   e) 52.8

8.8* In the determination of chromium in an ore, all the chromium is converted to $Cr_2O_7^{2-}$, dichromate ion. This is titrated with a standard Fe^{2+} solution:

$$6\ Fe^{2+} + Cr_2O_7^{2-} + 14\ H^+ \longrightarrow 6\ Fe^{3+} + 2\ Cr^{3+} + 7\ H_2O$$

How many grams of chromium are present in an ore sample which required 37.3 ml of 0.149 $\underline{M}$ Fe^{2+} solution to titrate it? The atomic weights of Cr and Fe are 52.0 and 55.8 g/mole, respectively.

a) 0.310
b) 0.200
c) 0.0963
d) 0.0481
e) 0.289

8.9 A standard solution of .165 molar HCl is being used to determine the concentration of an unknown NaOH solution. If 25.5 milliliters of the acid are required to neutralize 15.0 milliliters of the base, what is the molarity of the NaOH?

a) $\dfrac{.165}{25.5 + 15.0}$

b) $\dfrac{15.0}{(0.165)(25.5)}$

c) $(.165)\left(\dfrac{15.0}{25.5}\right)$

d) $(.165)\left(\dfrac{25.5}{15.0}\right)$

e) None of the above

8.10 If 12.8 ml of 1.34 molar KOH is mixed with 47.8 ml of 1.67 molar LiOH the final solution will be:

a) 1.60 molar in OH^-

b) 1.60 molar in H^+

c) 1.04 molar in H^+

d) .973 molar in OH^-

e) 1.04 molar in OH^-

8.11* If 39.8 ml of 1.31 molar NaOH is mixed with 17.7 ml of 1.90 molar HI the final solution will be:

a) 0.835 molar in OH^-

b) 0.320 molar in OH^-

c) 1.49 molar in H^+

d) 1.49 molar in OH^-

e) 0.320 molar in H^+

8.12 Determine the equivalent weight (g/equiv) of a recently synthesized
acid if a solution of 0.644 gram of the acid dissolved in 50 mls
H_2O required 24.67 mls of 0.148 $\underline{M}$ NaOH for neutralization.

a) 88.2
b) 107
c) 118
d) 176
e) 353

8.13 Determine the equivalent weight (g/equiv) of a recently synthesized
base if a solution of 0.872 gram of the base dissolved in 50 mls of
H_2O required 32.47 mls of 0.168 $\underline{M}$ HCl for neutralization.

a) 79.9
b) 160
c) 169
d) 320
e) 337

8.14* A monoprotic weak acid, HA, was titrated with standard base in a
molecular weight determination. Data: # grams of HA = 2.80,
molarity of NaOH = 0.500 and volume of NaOH = 29.2 mls. What is
the molecular weight of the acid?

a) 164
b) 384
c) 192
d) 5.21
e) 96.0

8.15 A weak base, B, was titrated with standard HCl in a molecular weight
determination. The following data were obtained:

grams of B = 3.75
molarity of HCl = .625
volume of HCl (ml) = 41.1

What is the molecular weight of the base? (Assume B reacts with
only one H^+ forming the product BH^+.)

a) 292
b) 146
c) 247
d) 68.5
e) 73.0

8.16 How many grams of Na_2CO_3 (molecular weight is 106) are required to neutralize 3.94 liters of 0.156 molar HCl? (Assume the reaction

$$CO_3^{2-} + 2\ H^+ \longrightarrow H_2O + CO_2(g) \text{ is complete.})$$

   a) (2)(3.94)(0.156)(106)

   b) (3.94)(0.156)(106)

   c) $\dfrac{(3.94)(0.156)(106)}{(2)}$

   d) $\dfrac{(0.156)(106)}{(2)(3.94)}$

   e) None of the above

8.17* Concentrated $HClO_4$ is 70.0 percent $HClO_4$. Its density is 1.668 grams per milliliter. What is the molality of $HClO_4$?

   a) 23.2
   b) .295
   c) 38.7
   d) 6.97
   e) 11.6

8.18* What volume of a concentrated NaOH solution must be diluted to prepare 3.50 liters of a solution that is 0.100 molar in NaOH? The concentrated NaOH solution if 50.0 percent by weight and has density 1.525 grams per milliliter.

   a) 18.4 ml
   b) 230 ml
   c) .0184 ml
   d) 109 ml
   e) 7.00 ml

8.19 If 23.5 ml of 0.100 molar Sn^{2+} solution is added to 56.0 ml of 0.200 molar Fe^{3+} solution, the solution will contain (after reaction) which one of the choices below? Assume no shrinkage in volume upon mixing. The reaction is:

$$Sn^{2+} + 2\ Fe^{3+} \longrightarrow Sn^{4+} + 2\ Fe^{2+}$$

   a) .0409 molar Sn^{4+}

   b) .0818 molar Fe^{3+}

   c) .252 molar Sn^{2+}

   d) .252 molar Fe^{2+}

   e) .111 molar Sn^{2+}

70

8.20 A 2.09 gram sample of iron ore is dissolved in acid and the iron
     quantitatively converted to the Fe(II) state. The solution is then
     titrated against standard Ce^{4+} in an acidic medium. The titration
     requires 39.0 milliliters of 0.280 molar Ce^{4+}.
     The titration reaction is:

$$Fe^{2+} + Ce^{4+} \longrightarrow Fe^{3+} + Ce^{3+}$$

What is the percentage iron in the ore?

   a) 87.9
   b) 43.9
   c) 38.1
   d) 29.2
   e) 35.1

8.21*A 1.06 gram sample of limestone is dissolved in an acidic solution, and
     then the calcium is precipitated as calcium oxalate, CaC_2O_4. After
     filtering and washing the precipitate, it is dissolved in acid and
     35.7 ml of .123 molar MnO_4^- is required to titrate it. What is the
     percentage Ca in the sample? The titration reaction (not necessarily
     balanced) is:

$$H_2C_2O_4 + MnO_4^- \longrightarrow CO_2 + Mn^{2+}$$

   a) 19.9
   b) 58.5
   c) 41.5
   d) 18.3
   e) 81.7

8.22*A 5.41 gram sample containing arsenic (atomic weight 74.9 g/mole) is
     dissolved in an acidic solution and the arsenic is quantitatively
     oxidized to H_3AsO_4 (molecular weight 142 g/mole). An excess of KI is
     then added, and the H_3AsO_4 is quantitatively reduced to H_3AsO_3. The
     resulting I_3^- is then titrated with 35.97 milliliters of 0.865
     molar $S_2O_3^{2-}$ (thiosulfate) solution. What is the percentage arsenic
     by weight in the sample? The relevent reactions are

$$2\ H^+ + H_3AsO_4 + 2\ I^- \longrightarrow H_3AsO_3 + I_2 + H_2O$$

$$I_2 + I^- \longrightarrow I_3^-$$

$$I_3^- + 2\ S_2O_3^{2-} \longrightarrow 3\ I^- + S_4O_6^{2-}$$

   a) 43.1
   b) 21.6
   c) 16.2
   d) 10.8
   e) 32.3

8.23 One method for neutralizing rivers and streams polluted by acid mine drainage is to add limestone, $CaCO_3$:

$$2 \ H^+ + CO_3^{2-} \longrightarrow CO_2(g) + H_2O$$

If the average hydrogen ion concentration in a polluted stream is 8.15×10^{-4} moles per liter, how many liters can be purified per ton of limestone? (1 ton = 908 kg; molecular weight of $CaCO_3$ = 100 g/mole; assume the limestone is pure $CaCO_3$ and the reaction goes to completion.)

a) 14.8 liters
b) 2.23×10^4 liters
c) 100 liters
d) 1.11×10^7 liters
e) 2.23×10^7 liters

8.24*Suppose 0.40 moles of Zn metal is added to 400 ml of a solution which is 0.10 molar in Cu^{+2} and 6.0 molar in acid. The first reaction is:

$$Zn(s) + Cu^{+2}(aq) \longrightarrow Zn^{+2}(aq) + Cu(s)$$

Then the excess Zn reacts with the acid:

$$Zn(s) + 2 \ H^+(aq) \longrightarrow Zn^{+2}(aq) + H_2(g)$$

(Assume both reactions go to completion.) After the zinc reacts which one computation below is <u>incorrect</u>?

a) 0.36 moles of H_2 gas is produced.
b) The concentration of Zn^{+2} is 1.0 $\underline{M}$.
c) The final concentration of acid is 5.0 $\underline{M}$.
d) 0.040 mole of Cu metal is produced.
e) The solution contains 26 g of Zn^{2+}.

8.25*In the determination of iron ore, all the iron is brought into the Fe^{2+} state, whereupon it can be titrated with a standard oxidizing agent to the Fe^{3+} state. How many milligrams of iron are present in an ore sample which required 22.1 ml of .0844 N oxidizing agent to titrate it? (The atomic weight of Fe is 55.8 grams/mole).

a) .0335 milligrams
b) 2.99×10^4 milligrams
c) 104 milligrams
d) 1.87 milligrams
e) Cannot be determined from the information provided.

CHAPTER 9.   HEATS OF REACTION

Definitions

Joule (J)   – S.I. unit for energy; $1 \text{ J} = 1 \text{ kg m}^2/\text{sec}^2$; 1 J = `
                 1×10^7 ergs; 1 J = 0.239 cal; 1 J = 1 watt-sec

Calorie (cal)   – amount of heat required to raise the temperature of 1
                 gram of H_2O 1°C; 1 cal = 4.184 J.

Specific heat   – the quantity of heat energy that must be absorbed by
                 1 gram of substance to raise its temperature by 1°C.

Molar heat of fusion – the quantity of heat energy that must be absorbed
                 by a substance to convert 1 mole from solid to liquid
                 at its normal melting point.

Molar heat of vaporization – the quantity of heat energy that must be
                 absorbed by a substance to convert 1 mole from liquid
                 to vapor at its normal boiling point.

Heat of reaction (ΔH) – the quantity of heat energy absorbed or released
                 when a reaction occurs at constant temperature and pres-
                 sure. More precisely ΔH is termed the enthalpy change,
                 or the enthalpy of reaction.

Exothermic reaction – a reaction in which heat energy is released;
                 ΔH is negative.

Endothermic reaction – a reaction in which heat energy is absorbed;
                 ΔH is positive.

Standard state – the pure substance at 1.00 atm and 25°C.

Standard enthalpy of reaction ($\Delta H°$) – the quantity of heat energy
                 absorbed or released when the reactants and products
                 are in their standard states.

Standard enthalpy of formation (ΔH_f^o) – the quantity of heat energy absorbed
                 or released when a compound is formed in its standard
                 state by reaction of its constituent elements, all
                 in their standard state; ΔH_f^o for the elements in their
                 standard state is zero.

Bond energy –   the energy required to break a chemical bond.

72

Calorimetry

A calorimeter is a device designed to measure accurately specific heats and heats of reaction. The relationship between heat, mass, and temperature is

$$\text{Heat} = (\text{specific heat})(\text{mass})(\text{temperature change})$$

$$J = \left(\frac{J}{g\text{-deg C}}\right)(g)(\text{deg C})$$

__Example 9.1__   Calculate the specific heat of aluminum if it took 398 joules of heat to raise the temperature of 138 grams of Al from 24.6°C to 37.9°C.

__Solution__ $\text{Specific heat} = \dfrac{\text{Heat}}{(\text{Mass})(\text{Temperature change})}$

$$= \frac{398 \text{ J}}{(138g)(37.9-24.6°C)} = 0.217 \text{ J/g-deg C}$$

__Example 9.2__   A slug of metal weighing 39.7 g was warmed to 85.5°C and dropped into an insulated container with 45.3 g of water initially at 22.1°C. The final temperature of the metal slug and the water was 29.0°C. Calculate the specific heat of the metal, assuming no heat losses. The specific heat of water is 4.18 J/g-deg C.

__Solution__   Let X be the specific heat of the metal.

Heat lost by metal = heat gained by water

$$\frac{X \text{ J}}{g\text{-deg C}} \times (39.7 \text{ g}) \times (85.5 - 29.0°C) = \frac{4.18 J}{g\text{-degC}} \times (45.3 \text{ g}) \times (29.0-22.1°C)$$

$$X = \frac{4.18 \times 45.3 \times (29.0-22.1)}{39.7 \times (85.5-29.0)} \quad \frac{J}{g\text{-deg C}}$$

$$= 0.58 \text{ J/g-deg C}$$

Hess' Law

Hess's law, also called the law of constant heat summation, states that the heats of chemical reactions are additive. For example,

Reaction	$\Delta H°$ (kJ/mole)
(1) $S(s) + O_2(g) \longrightarrow SO_2(g)$	-297
(2) $SO_2(g) + \frac{1}{2} O_2(g) \longrightarrow SO_3(g)$	$\underline{-98}$
(3) $S(g) + \frac{3}{2} O_2(g) \longrightarrow SO_3(g)$	-395

Reaction (3) is the sum of reactions (1) and (2). Hess's law states that the heat of reaction 3 is the sum of the heats of reaction 1 and 2.

<u>Example 9.3</u>   Calculate the $\Delta H_f^°$ of ethyl alcohol, C_2H_5OH, given:

Reaction	$\Delta H°$ (kJ)
(1) $C(s) + O_2(g) \longrightarrow CO_2(g)$	-393
(2) $H_2(g) + \frac{1}{2} O_2(g) \longrightarrow H_2O(g)$	-286
(3) $C_2H_5OH(\ell) + 3\,O_2(g) \longrightarrow 2CO_2(g) + 3H_2O(g)$	-1370

<u>Solution</u>   The reaction of interest is

$$2C(s) + 3H_2(g) + \frac{1}{2}O_2(g) \longrightarrow C_2H_5OH(\ell)$$

This reaction and the corresponding heat of reaction is obtained by the following summation:

Reaction	$\Delta H°$ (kJ)
2 x (1): $2C(s) + 2\,O_2(g) \longrightarrow 2CO_2(g)$	$2(-393)$
3 x (2): $3H_2(g) + \frac{3}{2}\,O_2(g) \longrightarrow 3H_2O(g)$	$3(-286)$
-1 x (3): $2CO_2(g) + 3H_2O(g) \longrightarrow C_2H_5OH(\ell) + 3\,O_2(g)$	$-1(-1370)$
$2C(s) + 3H_2(g) + \frac{1}{2}O_2(g) \longrightarrow C_2H_5OH(\ell)$	-274 kJ

The heat of formation of ethanol is -274 kJ/mole

The most important appplication of Hess' law is the calculation of heats of reactions from tables of $\Delta H_f^°$. The standard heat of reaction is the sum of the heats of formation of the products minus the sum of the heats of formation of the reactants.

Example 9.4   Calculate the heat of the following reaction from the tabulated values of ΔH_f^o.

$$4NH_3(g) + 5\ O_2(g) \longrightarrow 4NO(g) + 6H_2O(g)$$

Compound	$NH_3(g)$	$NO(g)$	$H_2O(g)$
ΔH_f^o (kJ/mole)	−46.2	90.4	−242

Solution

$$\Delta H^\circ = 4\Delta H_{f_{NO}}^o + 6\Delta H_{f_{H_2O}}^o - [4\Delta H_{f_{NH_3}}^o + 5\Delta H_{f_{O_2}}^o]$$

$$= (4)(90.4) + 6(-242) - [4(-46.2) + 5(0)]$$

$$= -906\ kJ$$

That is, 906 kJ of heat energy are released when 4 moles of $NH_3(g)$ react with 5 moles of $O_2(g)$ to form 4 moles of $NO(g)$ and 6 moles of H_2O (g) at 25°C and 1.0 atm.

Example 9.5   Calculate the heat released when 1.00 kg of iron(III) oxide is reduced with excess aluminum in the "thermite" reaction,

$$Fe_2O_3(s) + 2Al(s) \longrightarrow 2\ Fe(s) + Al_2O_3(s)$$

Compound	$Fe_2O_3(s)$	$Al_2O_3(s)$
ΔH_f^o (kJ/mole)	−822	−1670

Solution   The standard heat of reaction is

$$\Delta H^\circ = \Delta H_{f_{Al_2O_3}}^o - \Delta H_{f_{Fe_2O_3}}^o = -848\ kJ$$

$$\#kJ = \frac{-848\ kJ}{mole\ Fe_2O_3} \times \frac{1\ mole\ Fe_2O_3}{159.7\ g} \times 1000\ g = -5310\ kJ$$

Bond Energies

Heats of gas-phase reaction can be estimated using tables of bond energies. Bond energies are always endothermic since energy must be provided to break a chemical bond. The procedure is to consider all the bonds formed and broken in the chemical reaction and calculate the sum of the corresponding bond energies. The structural formulas of the reactant and product molecules must be known.

<u>Example 9.6</u>   Estimate ΔH for the following reaction using the bond energies listed below.

$$C_2H_4 + HCl \longrightarrow CH_3CH_2Cl$$

Bond	C–C	C=C	C–H	C–Cl	H–Cl
kJ mole^{-1}	347	619	414	326	431

The structural formulas are: C_2H_4 – $\begin{smallmatrix}H\\H\end{smallmatrix}{>}C{=}C{<}\begin{smallmatrix}H\\H\end{smallmatrix}$, and CH_3CH_2Cl – H–C–C–Cl (with H H above and H H below)

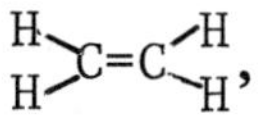

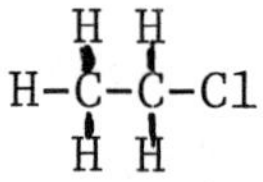

<u>Solution</u>   In this reaction the HCl bond is broken, C–H and C–Cl bonds are formed. A C=C (double) bond is converted to a C–C (single) bond; consider this the breaking of the C=C bond and the making of a C–C bond. Bond making is an exothermic process; bond breaking is an endothermic process.

$$\Delta H° = -[\Delta H_{C-H} + \Delta H_{C-Cl} + \Delta H_{C-C}] + [\Delta H_{H-Cl} + \Delta H_{C=C}]$$

$$= -[414 + 326 + 347] + [431 + 619]$$

$$= -37 \text{ kJ}$$

The reaction is exothermic.

Questions

9.1 The standard heat of formation of Fe_2O_3 is –824 kJ/mole. How much heat (in kJ) is evolved when 332 grams of Fe_2O_3 are formed? Assume the reactants are the constituent elements in their standard states; the molecular weight of Fe_2O_3 is 160 g/mole.

a)   396
b)   824
c) 1.71×10^3
d) 2.74×10^5
e) 1.72×10^6

9.2 Given that the standard heats of formation of NO and N_2O_4 are 90.4 and 9.16 kJ/mole respectively, find the standard heat of reaction for the following reaction:

$$2\ NO + O_2 \longrightarrow N_2O_4$$

a)   9.16 + 90.4
b)   9.16 –(2 x 90.4)
c)   –9.16 + (2 x 90.4)
d)   9.16
e)   None of the above

9.3* Given the following information:

Compound	Standard heat of formation (kJ/mole)
$CO_2(g)$	-394
$H_2O(\ell)$	-286
$C_4H_8(g)$	$+16.0$

Calculate the heat of combustion of C_4H_8 in kJ. The combustion reaction is:

$$C_4H_8(g) \ + \ 6 \ O_2(g) \longrightarrow \ 4 \ CO_2(g) + 4 \ H_2O(\ell)$$

a) -2740
b) 696
c) 16.0
d) -2700
e) -696

9.4 Given the reaction

$$C_3H_8(g) + 5 \ O_2(g) \longrightarrow \ 3CO_2(g) + 4H_2O(g) \quad \Delta H = -2030 \text{ kJ}$$

Calculate the number of kilojoules of heat evolved when 16.4 grams of CO_2 are produced. The molecular weight of CO_2 is 44.0 g/mole.

a) $\dfrac{(16.4)(22.4)}{(3)(2030)}$

b) $\dfrac{(16.4)(2030)}{(3)(44.0)}$

c) $\dfrac{(16.4)(2030)}{44.0}$

d) $(16.4)(2030)$

e) None of the above

9.5 What quantity of heat (in kJ) is required to convert 1.00 kilogram of ice at $-10°C$ to steam at $150°C$?

Data:
Specific heat of ice	= 2.09 J/gram-°C
Specific heat of liquid water	= 4.18 J/gram-°C
Specific heat of water vapor	= 2.09 J/gram-°C
Specific heat of fusion	= 335 J/gram
Specific heat of vaporization	= 2260 J/gram

a) 879
b) 3120
c) 3140
d) 3760
e) 5.65×10^4

9.6* What is the final temperature if 75 grams of ice at −25°C are mixed with 600 grams of water at 45°C.

Data: Specific heat of ice   =   2.09 J/gram-°C
             Specific heat of liquid water =   4.18 J/gram-°C
             Specific heat of fusion   =   335 J/gram

a)   10.0°C
b)   31.1°C
c)   29.7°C
d)   33.4°C
e)   −10.3°C

9.7 A scientist heated a piece of Fe to a temperature of 95.8°C. The metal was then dropped into an insulated calorimeter containing a thermometer, a stirrer, and 602 grams of water as its calorimeter fluid. The original temperature of the water was 22.3°C and the final temperature was measured to be 23.5°C. Neglecting all calorimeter corrections, what must have been the weight of Fe in grams? (The specific heat of Fe is 0.460 J/g-degree).

a)   161 grams
b)   90.8 grams
c)   73.9 grams
d)   141 grams
e)   114 grams

9.8 A scientist heated 95.5 grams of Zn to an unknown temperature and dropped the metal into an insulated calorimeter which contained a thermometer, a stirrer, and 287 grams of water as its calorimeter fluid. The original temperature of the water was 23.5°C and the final temperature was measured to be 24.8°C. Neglecting all calorimeter corrections, what must have been the temperature in °C to which the metal was heated? (The specific heat of Zn is 0.387 J/g-degree).

a)   67.1
b)   75.5
c)   83.0
d)   18.3
e)   96.4

9.9* A scientist heated 119 grams of an unknown metal to a temperature of 86.6°C. The metal was then dropped into an insulated calorimeter containing a thermometer, a stirrer, and 390 grams of water as its calorimeter fluid. The original tempeature of the water was 23.2°C and the final temperature was measured to be 24.3°C. Neglecting all calorimeter corrections, what is the specific heat of the metal in J/gram-degree?

(see next)

a) 0.198
b) 0.242
c) 0.291
d) 0.385
e) 0.452

9.10*Calculate the heat evolved or absorbed when 91.0 grams of SO_2Cl_2 are formed in the reaction:

$$SO_2 + Cl_2 \longrightarrow SO_2Cl_2$$

The heats of formation in kJ per mole are:

SO_2 −297

Cl_2 0

SO_2Cl_2 −389

The molecular weight of SO_2Cl_2 is 135 g/mole

a) 690 kJ absorbed
b) 62 kJ evolved
c) 62 kJ absorbed
d) 90 kJ absorbed
e) 460 kJ evolved

9.11 The heat of combustion of methane, the principal component of natural gas is 891 kJ/mole. Estimate the number of liters of CH_4 at STP required to heat 1.00 kg of H_2O from 25°C to its boiling point. The specific heat of liquid water is 4.184 J/g-deg. Assume the natural gas is 100% CH_4 and the efficiency of combustion is 100%.

a) 1.89
b) 6.26
c) 7.89
d) 31.4
e) 352

9.12 One estimate of the energy consumption in the United States for the year 2000 is 160 quads. One quad is 10^{15} BTU or 1.06×10^{15} kJ. Estimate the number of tons of coal required to satisfy this demand. Assume coal is 100% carbon, and that the efficiency of combustion is 100%. The heat of combustion of carbon is 394 kJ/mole. (One ton is 2000 x 454 grams)

a) 3×10^6
b) 5×10^8
c) 6×10^9
d) 1×10^{10}
e) 3×10^{12}

80

9.13*Calculate $\Delta H°$ for the reaction: $FeO(s) + CO(g) \longrightarrow Fe(s) + CO_2(g)$
given:

$$\Delta H° \text{ (kJ)}$$

(1) $Fe_2O_3(s) + 3\ CO(g) \rightarrow 2\ Fe(s) + 3CO_2(g)$ -27.6

(2) $3Fe_2O_3(s) + CO(g) \rightarrow 2Fe_3O_4(s) + CO_2(g)$ -58.6

(3) $Fe_3O_4(s) + CO(g) \rightarrow 3FeO(s) + CO_2(g)$ $+38.1$

a) $38.1 - (58.6 + 27.6)$ b) $-38.1/3 + 58.6/6 - 27.6/2$
c) $38.1/3 - 58.6/6 + 27.6/2$ d) $-38.1/3 - 58.6/6 + 27.6/2$
       e) None of the above

9.14*Estimate ΔH in kJ for the following reaction:

$$CH_2CH_2 + Cl_2 \rightarrow CH_2ClCH_2Cl$$

Given the following bond energies:

Bond	Cl–Cl	C–Cl	C–C	C–O	C–H	H–Cl	H–H	O–H	O=O	C=C	C=O
kJ/mole	243	326	347	351	414	431	435	464	494	619	707

The structural formulas are

$$\begin{array}{ccc} H\!\!>\!\!C\!\!=\!\!C\!\!<\!\!H & & H\ H \\ H & & \\ \end{array}$$

   $\overset{H}{\underset{H}{>}}C=C\overset{H}{\underset{H}{<}}$ $Cl-Cl$ $Cl-\overset{H}{\underset{H}{C}}-\overset{H}{\underset{H}{C}}-Cl$

a) 62.8   b) -137 c) -381 d) -62.8
     e) -757

9.15*Given the following heat of reaction and the bond energies listed
below, calculate the energy of the C–O bond. All numerical values
are in kilojoules per mole and all substances are in the gas
phase.

$$CH_4 + HOOH \rightarrow CH_3OH + H_2O \qquad \Delta H = -276\ kJ$$

Bond	O–O	O–H	C–H
kJ/mole	138	464	414

The structural formulas are $H-\overset{H}{\underset{H}{C}}-H$ $H-O-O-H$ $H-\overset{H}{\underset{H}{C}}-O-H$ $H-O-H$

a) 87.9   b) 364   c) 322   d) 740   e) 1020

$$\text{CHAPTER 10. COLLIGATIVE PROPERTIES}$$

Definitions

<u>Colligative properties</u> - those properties of a solution which depend
on the concentration of particles present and not upon the chemical
nature or mass of those particles. The colligative properties of
a solution containing one mole of chloroform in one kilogram of
benzene are identical to those of a solution containing one mole
of cyclohexane in one kilogram of benzene.

<u>Vapor pressure</u> - the equilibrium pressure in an enclosed vessel containing
a liquid with free space above it for its vapor. For a given liquid
the vapor pressure is a function of temperature only. If more than
one volatile component is present the total vapor pressure is the
sume of the partial pressures of the various volatile components.

<u>Freezing point</u> - temperature at which crystals of pure solvent are in
equilibrium with the liquid phase (pure solvent, or solution);
called the normal freezing point if the pressure on the system is
one atmosphere. Since the freezing point is only slightly dependent
on pressure, no distinction will be made between freezing point and
normal freezing point.

<u>Boiling point</u> - the normal boiling point is the temperature at which the
equilibrium vapor pressure of the liquid phase is one atmosphere.
(Unless otherwise stated "boiling point" will mean "normal boiling
point").

Vapor Pressure of a Solution

An ideal solution is one in where there are no solute-solute, solute-
solvent, or solvent-solvent interactions. Consider a two-component ideal
solution. Let one component be denoted A, the other B. The mole fractions
are X_A and X_B, and the equilibrium vapor pressures of pure A and B are
denoted P_A° and P_B°, respectively. For an ideal solution the vapor pressure
of a component is the product of the mole fraction of that component times
the equilibrium vapor pressure of the pure component. For the binary system
A and B,

$$P_A = X_A P_A^\circ \qquad \text{and} \qquad P_B = X_B P_B^\circ$$

the total vapor pressure over the solution is

$$P_{TOT} = P_A + P_B$$

<u>Example 10.1</u>   Calculate the total vapor pressure above a solution containing 125 grams of benzene (C_6H_6) and 175 grams of carbon tetrachloride (CCl_4) at 25°C.   The vapor pressures of the pure liquids at this temperature are: C_6H_6 – 92.5 torr, CCl_4 – 109 torr. (1 atmosphere is 760 torr).   The molecular weights are C_6H_6 – 78.0, CCl_4 – 154 g/mole.

<u>Solution</u>   The mole fractions are:

$$X_{C_6H_6} = \frac{125/78}{125/78 + 175/154} = 0.585 \qquad X_{CCl_4} = 1 - X_{C_6H_6} = 0.415$$

The partial pressures are

$$P_{C_6H_6} = 0.585 \times 92.5 = 54.1 \text{ torr} \qquad P_{CCl_4} = 0.415 \times 109 = 45.2 \text{ torr}$$

The total pressure is

$$P_{TOT} = P_{C_6H_6} + P_{CCl_4} = 54.1 + 45.2 = 99.3 \text{ torr}$$

Vapor Pressure Lowering

If an ideal solution contains just two components, a solvent A and a nonvolatile solute B, its vapor pressure will be less than that of the pure solvent by an amount ΔP.   It can be shown that:

$$\Delta P = P_A^\circ X_B$$

<u>Example 10.2</u>   The vapor pressure of pure water at 21°C is 18.7 torr. Estimate the vapor pressure at 21°C of a solution containing 65.0 g of sucrose ($C_{12}H_{22}O_{11}$) dissolved in 275 g of water.   The molecular weight of $C_{12}H_{22}O_{11}$ is 342 g/moles.

<u>Solution</u>   The mole fraction sucrose is

$$X_{sucrose} = \frac{65.0/342}{65.0/342 + 275/18} = 0.0123$$

The vapor pressure lowering is

$$\Delta P_{H_2O} = P_{H_2O}^\circ X_{sucrose}$$

$$= 18.7 \times 0.0123 = 0.230 \text{ torr}$$

The vapor pressure of the solution is 18.7–0.230 or 18.5 torr.

Boiling Point Elevation

It is observed that the boiling point of a solution is greater than that of the pure solvent. This is consistent with the vapor pressure lowering discussed above. The elevation of the boiling point is a much easier physical property to measure than the lowering of the vapor pressure. For dilute solutions:

$$\Delta T_b = K_b m_{TOT}$$

Here ΔT_b is the increase in the boiling point (°C), K_b is the boiling point elevation constant (°C/molal), and m_{TOT} is the total molality of the solution. If the solute is a nonelectrolyte, that is, the solute molecules do not dissociate into ions in solution, then m_{TOT} is the same as the molality. If the solute is a strong electrolyte, then m_{TOT} is an integer times the molality. For example, consider 1 molal solutions of NaCl and $CaCl_2$,

$$NaCl \xrightarrow{\text{100\%}} Na^+ + Cl^- \qquad m_{TOT} = 2 \text{ molal}$$

$$CaCl_2 \xrightarrow{\text{100\%}} Ca^{2+} + 2Cl^- \qquad m_{TOT} = 3 \text{ molal}$$

The constant K_b is measured for each solvent. For example, for H_2O, K_b is 0.512°C/molal.

Example 10.3 What is the boiling point of the solution in example 10.2? K_b for water is 0.512°C per mole kg^{-1}, (or 0.512°C/m).

<u>Solution</u>   The molality of the solution is

$$\frac{\text{\# moles}}{\text{kg solvent}} = \frac{1 \text{ mole}}{342 \text{ g}} \times \frac{65.0 \text{ g}}{275 \text{ g}} \times \frac{1000 \text{ g}}{1 \text{ kg}} = 0.691 \text{ m}$$

Since sucrose is a nonelectrolyte, $m_{TOT} = 0.691$ m

$$\Delta T = K_b \, m_{TOT}$$

$$= 0.512 \text{ °C/molal} \times 0.691 \text{ molal} = 0.354 °C$$

The boiling point of the solution is 100° + 0.354 or 100.354°C

Freezing Point Depression

It is also observed that the freezing point of a solution is less than that of the pure solvent. For dilute solutions,

$$\Delta T_f = K_f \, m_{TOT}$$

Here K_f is a property of the solvent, and is always negative. For example, K_f for H_2O is $-1.86°C/molal$. The freezing point depression and the boiling point elevation are convenient measurements for obtaining molecular weights of new compounds.

Example 10.4   A new compound, a nonelectrolyte, is soluble in carbon tetrachloride. A solution consisting of 7.08 grams of the new compound dissolved in 125 grams of CCl_4 has a freezing point of $-32.7°C$. Calculate the molecular weight of the new compound. The normal freezing point of CCl_4 is $-22.3°C$ and K_f for CCl_4 is $-29.8°C/molal$.

Solution   Let M be the molecular weight. Since the compound is a nonelectrolyte the total molality is the molality of the compound.

$$m = \frac{\Delta T}{K_f} = \frac{-32.7-(-22.3)°C}{-29.8°C/molal} = 0.349 \ m$$

$$0.349 = \frac{7.08/M}{125/1000}$$

$$M = 162 \ g/mole$$

The freezing point depression is usually more useful for obtaining molecular weights for the following reasons:

1) K_f is usually much greater than K_b so that ΔT_f is a more sensitive measurement;
2) solutes that are thermally unstable will not be decomposed during the measurements; and
3) solutes that are volatile will interfere with ΔT_b but not ΔT_f.

Osmotic Pressure

Consider a solution separated from pure solvent by a membrane which is permeable to solvent but not to solute. There will be a net flow of pure solvent into the solution unless compensated by an extra pressure on the solution side, which is called the osmotic pressure. An osmometer for measuring this pressure, π, is picture below. At equilibrium, the extra pressure π corresponds to the extra height of the solution above the solvent.

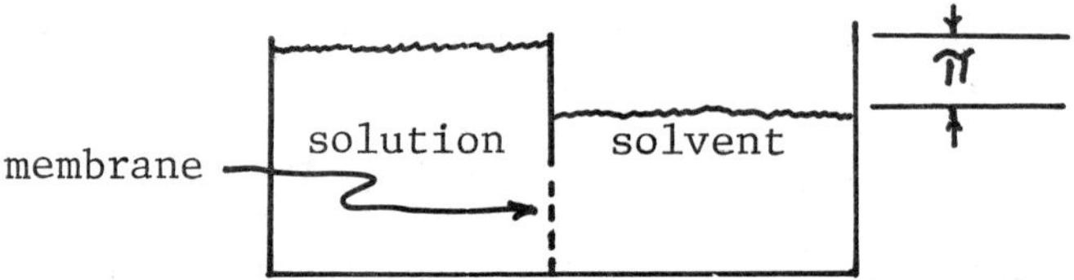

For dilute solutions the osmotic pressure is related to the molarity, M, of the solution by the equation:

$$\pi = MRT$$

where π is in atm and R is the gas constant 0.0821 ℓ-atm/mole-K, and T is the absolute temperature. The measurement of osmotic pressure is particularly useful for measuring molecular weights of high molecular weight compounds such as proteins and synthetic polymers.

Example 10.5   A biochemist prepared a solution containing 0.376 g of a protein fragment in 200. ml of solution. The solution exerted an osmotic pressure of 1.52 torr at 22°C. Calculate the molecular weight of the protein.

Solution

$$M = \frac{\pi}{RT} = \frac{1.52 \text{ torr x 1 atm/760 torr}}{(0.0821\ell\text{-atm/mole-K})(273 + 22 \text{ K})}$$

$$= 8.26 \times 10^{-5} \text{ mole/}\ell$$

$$\text{moles solute} = 8.26 \times 10^{-5} \text{ mole/}\ell \times 0.200\ell = 1.65 \times 10^{-5} \text{ mole}$$

$$\text{molecular weight} = \frac{\text{grams solute}}{\text{moles solute}} = \frac{.376 \text{ g}}{1.65 \times 10^{-5} \text{ mole}} = 22,800 \text{ g/mole}$$

Questions

10.1 Calculate the boiling point of a .128 molal solution of a non-electrolyte in benzene. The boiling point and boiling point elevation constant for benzene are 80.1°C and 2.53°C per molal respectively.

    a)   80.1 − 2.53
    b)   80.1
    c)   80.1 −(2.53)(0.128)
    d)   80.1 +(2.53)(0.128)
    e)   None of the above

10.2 A solution of 14.7 grams of nonelectrolyte in 0.180 kilogram of H_2O boils at 100.920°C at standard pressure. What is the molecular weight of the compound? (The boiling point elevation constant for H_2O at standard pressure is 0.512°C per molal)

    a)   147 grams/mole
    b)   15.0 grams/mole
    c)   8.20 grams/mole
    d)   45.4 grams/mole
    e)   174 grams/mole

10.3*A solution of 11.3 grams of nonelectrolyte in .105 kilogram of H_2O
freezes at $-2.36°C$ at standard pressure. What is the molecular weight
of the compound? (The freezing point depression constant for H_2O at
standard pressure is $-1.86°C$ per molal.)

    a)   1.51 grams/mole
    b)   24.5 grams/mole
    c)   93.5 grams/mole
    d)   84.8 grams/mole
    e)   200 grams/mole

10.4 What weight of ethylene glycol ($HOCH_2CH_2OH$, molecular weight is
62.1 g/mole) must be added to 8000 grams of water to produce an
antifreeze solution that would protect a car radiator down to
$-17.8°F$ ($= -27.7°C$)? Ethylene glycol is a nonelectrolyte.

a) $\dfrac{(1.86)(62.1)}{(17.8)(8)}$

b) $\dfrac{(27.7)(62.1)(8)}{1.86}$

c) $\dfrac{(17.8)(62.1)}{(1.86)(8)}$

d) $\dfrac{(1.86)(62.1)(8)}{17.8}$

e) None of the above

10.5 A solution containing 2.52 grams of a nonelectrolyte with empirical
formula CH_2 was dissolved in 250 grams of benzene. The freezing
point of the solution was $5.2°C$. What is the molecular formula
of the compound. The normal freezing point of benzene is $5.5°C$ and
the freezing point depression constant for benzene is $4.9°C$ per
molal.

    a) C_4H_8
    b) C_6H_{12}
    c) $C_{10}H_{20}$
    d) $C_{12}H_{24}$
    e) $C_{15}H_{30}$

10.6 A new compound, A, a nonelectrolyte, has been prepared. A is
slightly soluble in solvent S. The boiling point of a 0.1 m
solution of another nonelectrolyte in S boils at $134.0°C$. The
normal boiling point of S is $133.5°C$. A solution containing
7.5 grams of A dissolved in 250 grams of S boils at $134.5°C$.
Estimate the molecular weight of A.

    a)   75
    b)   100
    c)   125
    d)   150
    e)   300

10.7 If one mole of ethyl alcohol (a nonelectrolyte) is dissovled in
     one kilogram of water, the freezing point is lowered 1.86°C.
     If one mole of Na_3PO_4 is dissolved in a liter of water, the freezing
     point is lowered approximately:

     a)   4 x 1.86°C
     b)   3 x 1.86°C
     c)   2 x 1.86°C
     d)   5 x 1.86°C
     e)   None of the above

10.8* Determine the vapor pressure (in torr) of water in equilibrium
     with a solution containing 449 grams of a nonvolatile nonelectrolyte
     dissolved in 850 grams of H_2O at 45.0°C.

        The equilibrium vapor pressure of pure H_2O is 71.0 torr at
        45.0°C.

        The molecular weight of the solute is 54.0 grams/mole

     a)   71.0
     b)   60.4
     c)   83.5
     d)   10.6
     e)   131

10.9 A solution containing 507 grams of a solute dissolved in 305 grams
     of a solvent has an equilibrium vapor pressure of 109 torr at 50.0°C.
     Estimate the molecular weight of the solute given the following
     information:

        The equilibrium vapor pressure of the pure solvent is
        181 torr at 50.0°C.

        The molecular weight of the solvent is 25.4 grams/mole.

        The solute is a nonvolatile nonelectrolyte.

     a)   183
     b)   7.90
     c)   105
     d)   42.3
     e)   64.0

10.10* An aqueous solution containing 2.00 grams of a protein in 150
     milliliters of solution exerts an osmotic pressure of 10.5 torr at
     27.0°C. Estimate the molecular weight of the protein.

     a)   2,140 g/mole
     b)   576,000 g/mole
     c)   26,800 g/mole
     d)   23,800 g/mole
     e)   313 g/mole

CHAPTER 11.   MIXED STOICHIOMETRY

The quantities of substances involved in chemical reactions can be specified in various ways. For pure solids and liquids the mass is usually given, but for gases it is more likely the volume, along with the temperature and pressure at which the volume was measured. When substances are in solution the volume and concentration of the solution is usually specified.

The common link is the number of moles. All quantities should be converted to moles so they can be compared and the appropriate stoichiometric relationships applied. The final result is then converted from moles to the units desired.

Example 11.1   A 16.0 gram samples of CaO was mixed with water and treated with 5.00 liters of HCl(g) at STP. How much additional 0.855 H_2SO_4 solution must now be added to complete the neutralization of the CaO? The neutralization reactions may be written:

$$CaO \; + \; 2HCl(g) \; \longrightarrow \; CaCl_2 \; + \; H_2O$$

$$CaO \; + \; H_2SO_4 \; \longrightarrow \; CaSO_4 \; + \; H_2O$$

The molecular weight of CaO is 56.0 g/mole.

Solution

$$\# \text{ moles CaO} = \frac{1 \text{ mole}}{56.0 \text{ g}} \; \times \; 16.0 \text{ g} = 0.286 \text{ mole}$$

$$\# \text{ moles HCl} = \frac{1 \text{ mole}}{22.4 \ell} \; \times \; 5.00 \; \ell = 0.223 \text{ mole}$$

$$\# \text{ moles CaO neutralized by HCl} = \frac{1 \text{ mole CaO}}{2 \text{ moles HCl}} \; \times \; 0.223 \text{ mole HCl} = 0.112 \text{ mole CaO}$$

$$\# \text{ moles CaO left} = 0.286 - 0.112 = 0.174 \text{ mole CaO}$$

$$\# \text{ moles } H_2SO_4 \text{ required} = \frac{1 \text{ mole } H_2SO_4}{1 \text{ mole CaO}} \; \times \; 0.174 \text{ moles CaO} = 0.174 \text{ mole } H_2SO_4$$

$$\# \text{ liters } H_2SO_4 = \frac{1 \; \ell}{0.855 \text{ mole}} \; \times \; 0.174 \text{ moles} = 0.204 \; \ell$$

$$\text{or } 204 \text{ ml}$$

<u>Example 11.2</u>   Given the following neutralization reaction,

$$CaCO_3 \; + \; H_3PO_4 \; \longrightarrow \; H_2O \; + \; CO_2(g) \; + \; Ca_3(PO_4)_2$$

If 1.00 kg of $CaCO_3$ are added to 1.00 ℓ of "concentrated" H_3PO_4 solution, what volume of $CO_2(g)$ at STP is released, and what are the masses of the products and excess reagent after reaction. Assume the reaction is quantitative. "Concentrated" H_3PO_4 solution is 85.% H_3PO_4 by weight and has a density of 1.689 g/ml.

<u>Solution</u> First calculate the number of moles of $CaCO_3$ and H_3PO_4

$$\text{\# moles } CaCO_3 \; = \; \frac{1 \text{ mole}}{100 \text{ g}} \times 1000 \text{ g} = 10.0 \text{ moles}$$

$$\text{\# moles } H_3PO_4 \; = \; \frac{1 \text{ mole}}{98.0g} \times \frac{1.689 \text{ g}}{ml} \times 1000 \text{ ml} \times \frac{85g \; H_3PO_4}{100g \text{ sol'n}} = 14.6 \text{ moles}$$

The balanced equation is

$$3CaCO_3 \; + \; 2H_3PO_4 \; \longrightarrow \; 3H_2O + 3CO_{2(g)} + Ca_3(PO_4)_2$$

$CaCO_3$ is the limiting reagent because 14.6 moles of H_3PO_4 would require

3 x 14.6/2 or 21.9 moles of $CaCO_3$.

$$\text{\# ℓ } CO_2 \; = \; \frac{22.4ℓ}{\text{mole } CO_2} \times \frac{3 \text{ moles } CO_2}{3 \text{ moles } CaCO_3} \times 10.0 \text{ moles } CaCO_3 = 224 \text{ liters } CO_2$$

$$\text{\# g } H_2O \; = \; \frac{18.0 \text{ g}}{\text{mole}} \times \frac{3 \text{ moles } H_2O}{3 \text{ moles } CaCO_3} \times 10.0 \text{ moles } CaCO_3 = 180 \text{ g } H_2O$$

$$\text{\# g } Ca_3(PO_4)_2 = \frac{310 \text{ g}}{\text{mole}} \times \frac{1 \text{ mole } Ca_3(PO_4)_2}{3 \text{ moles } CaCO_3} \times 10.0 \text{ moles } CaCO_3 = 1030 \text{ g } Ca_3(PO_4)_2$$

$$\text{\#moles } H_3PO_4 \; = \; \frac{2 \text{ moles } H_3PO_4}{3 \text{ moles } CaCO_3} \times 10.0 \text{ moles } CaCO_3 = 6.67 \text{ moles } H_3PO_4 \text{ reacted}$$

$$\text{\# g } H_3PO_4 \text{ left } = \; \frac{98 \text{ g}}{\text{mole}} \times (14.6 - 6.67) \text{ mole} = 780 \text{ g } H_3PO_4 \text{ left}$$

Questions

11.1*Which of the following samples contains the greatest amount of mercury? (atomic weight of Hg = 201 per mole)

    a)   40.5 grams of Hg metal.
    b)   0.460 moles of Hg metal
    c) 1.83×10^{23} atoms of Hg metal
    d)   6.0 liters of $Hg(CH_3)_2(g)$ at 1.00 atmosphere and 273°C
    e)   5.0 liters of 0.020 M $Hg(NO_3)_2$ solution

90

11.2*Suppose 0.10 faraday of charge are passed through 500 ml of 1.00 M NaCl solution. The electrode reactions are:

$$\text{Anode} \qquad 2\ Cl^- \longrightarrow Cl_2 + 2\ e^-$$

$$\text{Cathode} \qquad 2H_2O + 2e^- \longrightarrow 2\ OH^- + H_2$$

Only one statement below is true for this reaction. Which is the correct statement?

a) The solution afterwards contains 0.40 mole of Cl^-.
b) 7.1 g of Cl_2 are produced.
c) 0.20 moles of OH^- are produced.
d) The concentration of OH^- in the solution is 0.10 molar
e) 2.24 ℓ of Cl_2 are produced (STP).

11.3*A piece of sodium (Na) weighing 7.48 grams was dropped into a 2.96 liter flask containing chlorine gas (Cl_2) at STP. The following reaction took place to completion:

$$2\ Na(s) + Cl_2(g) \longrightarrow 2\ NaCl(s)$$

What was the weight of solid in the flask after reaction?

a)   1.41 grams
b) 12.16 grams
c) 15.45 grams
d) 19.00 grams
e) 16.84 grams

11.4*10.6 grams of $Ca(HCO_3)_2$ (molecular weight 162) are added to 131 ml. of 0.117 molar HCl and the following reaction proceeds to completion:

$$Ca(HCO_3)_2 + 2HCl \longrightarrow CaCl_2 + 2\ H_2O + 2\ CO_2(g)$$

How many liters of $CO_2(g)$ at STP will be liberated?

a)   .171 liters
b)   .686 liters
c)   .343 liters
d) 10.6   liters
e)   1.46 liters

11.5*A solution of NaOH can be used to absorb Cl_2 gas by the reaction:

$$2\ OH^- + Cl_2 \longrightarrow H_2O + OCl^- + Cl^-$$

If 11.2 liters of Cl_2 gas (STP) are bubbled into 1.00 liter of 6.00 molar NaOH, which of the following is incorrect?

(see next)

a)   The concentration of Na^+ will be 6.00 M.
b)   The solution will contain 6.00 moles of NaOCl.
c)   All the chlorine gas will be absorbed.
d)   The concentration of Cl^- will be .500 M.
e)   The concentration of OCl^- will be .500 M.

11.6 One of the processes used in sulfur recovery is the following absorption-regeneration reaction sequence:

$$H_2S(g) + Na_4As_2S_5O_2 \longrightarrow Na_4As_2S_6O + H_2O$$

$$O_2(g) + 2Na_4As_2S_6O \longrightarrow 2S + 2Na_4As_2S_5O_2$$

Calculate the ratio of grams sulfur recovered per liter of H_2S at STP assuming this process is 100% efficient.

a)   32.1/22.4
b)   2 x 32.1/22.4
c)   32.1/(2 x 22.4)
d)   4 x 32.1/22.4
e)   None of the above

11.7*How many grams of zinc are required to react with H_2SO_4 to produce the hydrogen gas necessary to reduce 314 grams of Fe_3O_4 to iron?

a)   .0832
b)   89.0
c)   1.90 x 10^7
d) 356
e)   22.2

11.8 The solubility of $Ca(OH)_2$ at room temperature is **0.18 g** $Ca(OH)_2$**/liter** of solution. What volume of $CO_2(g)$ in liters at STP can be collected in a **25 liter** $Ca(OH)_2$ trap if the solution is initially saturated and the following reaction is quantitative?

$$Ca(OH)_2(aq) + CO_2(g) \longrightarrow CaCO_3(s) + H_2O$$

The molecular weight of $Ca(OH)_2$ is **74.1 g/mole.**

a)   15
b)   12
c)   2.7
d)   1.4
e)   0.061

11.9 A common laboratory procedure for the preparation of chlorine gas is the redox reaction

$$MnO_2 + HCl \longrightarrow MnCl_2 + Cl_2 + H_2O$$

What volume of Cl_2 (in liters at STP) can be made by reacting excess manganese(IV) oxide with 1.0 ℓ of 6.0 $\underline{M}$ HCl?

a)   22
b)   34
c)   45
d)   67
e)   130

11.10 SO_2 is quantitatively collected by bubbling the gas through a solution of sodium tetrachloromercurate (II),

$$SO_{2(g)} + HgCl_4^{2-} + H_2O \longrightarrow HgCl_2SO_3^{2-} + 2H^+ + 2Cl^-$$

What volume of SO_2 in liters at STP can be absorbed in a 1.0 x 10^3 liter vessel filled with 0.10 $\underline{M}$ Na_2HgCl_4?

a)   22
b)   44
c)   220
d) 2200
e) 4400

11.11   Rainwater is becoming increasingly acidic due to such reactions as

$$SO_2(g) + H_2O \longrightarrow H^+ + HSO_3^-$$

What volume of SO_2 (STP) would account for an average $[H^+]$ of 1.0 x 10^{-4} $\underline{M}$ in a rainstorm that dropped 1.0 x 10^9 liters, assuming this is the only reaction that contributes H^+?

a)   22.4
b)   22.4 x 10^3
c)   22.4 x 10^5
d)   22.4 x 10^7
e)   22.4 x 10^9

11.12   What mass of Cu (in grams) should be added to an excess of concentrated H_2SO_4 to generate as much SO_2 as is formed when 8.43 grams of Na_2SO_3 react with excess concentrated H_2SO_4? The reactions are:

$$Cu + H_2SO_4 \longrightarrow CuSO_4 + SO_2 + H_2O$$

$$Na_2SO_3 + H_2SO_4 \longrightarrow Na_2SO_4 + SO_2 + H_2O$$

a)   0.471
b)   2.12
c)   4.25
d)   8.36
e)   33.4

11.13* A metallurgist is given two alloys containing Co. Alloy #1 is 49.0%
Co (by weight) and alloy #2 is 72.0% Co. What quantities of the two
alloys must be mixed to prepare 6.25 kg. of a third alloy which is
59.0% Co?

a)   5.58 kg of alloy #1 and .067 kg. of alloy #2.
b)   3.53 kg of alloy #1 and 2.72 kg of alloy #2.
c)   4.50 kg of alloy #1 and 1.75 kg of alloy #2.
d)   0.67 kg of alloy #1 and 5.58 kg of alloy #2.
e)   2.72 kg of alloy #1 and 3.53 kg of alloy #2.

11.14* An alloy contained an unknown mixture of Be and Sc. A sample of
the alloy, weighing 4.60 grams, was treated with an excess of hydrochloric acid. The resulting reaction released 7.95 liters of hydrogen
gas at STP. What is the weight-percentage of Be in the sample?
(Hint: Assume the oxidation products are Be(II) and Sc(III).)

a)   62.9%
b)   43.6%
c)   37.1%
d)   56.4%
e)   69.2%

11.15* A solution labelled "0.100 $\underline{M}$ Hg_2^{2+}" has been partially oxidized
by oxygen in air to Hg^{2+}. Calculate the concentration of botn
Hg_2^{2+} and Hg^{2+} if 40.0 ml of the solution required 45.8 ml of
0.100 $\underline{M}$ $SnCl_2$ for complete reduction to Hg. (Sn^{2+} is oxidized to
Sn^{4+}.)

	$[Hg_2^{2+}]$	$[Hg^{2+}]$
a)	0.0855 $\underline{M}$	0.0290 $\underline{M}$
b)	0.0710 $\underline{M}$	0.0290 $\underline{M}$
c)	0.0855 $\underline{M}$	0.0145 $\underline{M}$
d)	0.0667 $\underline{M}$	0.0333 $\underline{M}$
e)	0.0500 $\underline{M}$	0.0500 $\underline{M}$

CHAPTER 12. LABORATORY STOICHIOMETRY

Beginning students frequently obtain an appreciation of stoichiometry by carrying out reactions in the laboratory and obtaining accurate weights of reactants and products. Such operations are never free from error. The objectives of this section are to calculate the percent error between observed results and the theoretical results, and to estimate qualitatively or quantitatively the effect of a particular manipulative error upon the final result.

<u>Example 12.1</u> A student weighed 0.9496 g of $CaCO_3$, dissolved it in HCl, and added excess $(NH_4)_2C_2O_4$ and NH_3 to precipitate calcium oxalate. The dried product, assumed to be $CaC_2O_4 \cdot H_2O$, weighed 1.4008 g. Calculate the percent yield, and consider some explanations for this result. The molecular weights of $CaCO_3$ and $CaC_2O_4 \cdot H_2O$ are 100.1 and 146.1 g/mole, respectively.

<u>Solution</u> The net reaction is

$$CaCO_3 + H_2O + C_2O_4^{2-} \rightarrow CaC_2O_4 \cdot H_2O + CO_3^{2-}$$

The theoretical yield is

$$\#g \; CaC_2O_4 \cdot H_2O =$$

$$\frac{146.1 \; g \; CaC_2O_4 \cdot H_2O}{mole \; CaC_2O_4 \cdot H_2O} \; x \; \frac{1 \; mole \; CaC_2O_4 \cdot H_2O}{1 \; mole \; CaCO_3} \; x \; \frac{1 \; mole \; CaCO_3}{100.1g \; CaCO_3} \; x \; .9496g \; CaCO_3$$

$$= 1.3860 \; g \; CaC_2O_4 \cdot H_2O$$

The percentage yield is

$$\frac{1.4008 \; x \; 100}{1.3860} = 101.07\%$$

The student's product weighs 1.4008 − 1.3860 or 0.0148 gram (1.07%) more than theoretical yield. This error cannot be blamed on the analytical balance which is accurate to about 0.0002 gram. If the starting material contained moisture, a very common impurity, the percent yield would have been less than 100%. However, if the starting material contained some CaO, this would result in an excess mass of $CaC_2O_4 \cdot H_2O$.

It is also possible that the product has not been dried sufficiently. On the other hand, if the oven temperature is too high, the product will dehydrate, thereby decreasing the yield,

$$CaC_2O_4 \cdot H_2O(s) \underset{\Delta}{\rightarrow} CaC_2O_4(s) + H_2O(g)$$

Every experimental result has some error associated with it. A good chemist plans his experiments carefully and conducts them in such a way as to minimize the total experimental error in the result.

Questions

12.1* It is possible to estimate Avogadro's number by adding a few drops of an alcoholic solution of oleic acid to a water surface dusted by lycopodium powder. To calculate the result, some numerical values are necessary and some assumptions have to be made, such as those listed below. Select the one item which does <u>not</u> belong because it is either incorrect, unnecessary, or irrelevant.

    a)   the molecular weight of oleic acid.
    b)   assume oleic acid is insoluble in water.
    c)   the density of oleic acid.
    d)   the volume of a drop of solution.
    e)   the mass of the electron.

12.2* A chemist analyzed a sample of $CaCO_3 \cdot 6H_2O$ by the standard gravimetric procedure and determined the percentage of water to be 52.3%. The theoretical value is 51.9%. A possible source of error is:

    a)   After weighing the sample, a piece of rust fell from the tongs into the crucible.
    b)   Moisture driven from the sample condensed on the inside of the crucible cover.
    c)   All the weighings were made on a balance which is high by 0.4%.
    d)   The sample was not heated enough to drive off all the water.
    e)   The original sample was wet.

12.3* A team of students determined x in the formula $PbCl_x$ using Job's method and weighing the solid $PbCl_x$ produced. (In this method, various molar ratios of Pb^{2+} and Cl^- are mixed, keeping the sum of the moles constant. A maximum yield is obtained at the correct stoichiometric ratio.) They found x = 2.09 rather than 2.00, which most other chemists have found. A possible source of error is:

    a)   The NaCl solution from the stockroom was actually less concentrated than labelled.

(over)

b) None of the students dried their precipitates thoroughly, leaving about 4.5% moisture, on the average.

c) None of the students collected their precipitates thoroughly leaving about 4.5% behind, on the average.

d) All of the burets in the lab were 4.5% high.

e) All of the burets in the lab were 4.5% low.

12.4*A team of students prepared iron from Fe_2O_3 by reduction with aluminum. The product was recovered from the by-products and unreacted reagents by separating all the metallic iron particles with a magnet. The yield was 103% based on the weight of Fe_3O_3. Which one of the following can <u>not</u> explain this excessive yield?

a) The aluminum was in excess and some alloyed with the iron.

b) The Fe_2O_3 was in excess, so not all the oxygen was removed by the aluminum.

c) Particles of Al_2O_3 were trapped within the particles of iron metal.

d) Some of the Fe_2O_3 was reduced to magnetite, Fe_3O_4, which is ferromagnetic and was attracted to the magnet and weighed with the iron.

e) The Fe_2O_3 was not pure but contained considerable Fe_3O_4 to begin with.

12.5*Which one of the cations from Groups I, II, or III is present in an unknown if the following observations are made?

No precipitate forms when HCl is added.
A precipitate forms when H_2S is added (acidic solution).
The precipitate does not dissolve when excess NaOH is added.
The precipitate dissolves when excess HNO_3 is added.
A precipitate forms when excess NH_3 is added.

a) Cd^{2+}

b) Fe^{3+}

c) Hg_2^{2+}

d) Bi^{3+}

e) Ag^+

CHAPTER 13. THE GAS LAWS

Definitions and Units

Volume   A gas will uniformly fill its container whatever the volume.
     If gas is placed in a flexible container, for example, a ballon or
     a cylinder with a movable piston, it will expand or contract until
     the pressure of the gas inside the container is equal to the pressure
     outside. The volume is usually measured in liters (ℓ) or milliliter
     (ml). One liter is a cubic decimeter, $(0.1 \text{ m})^3$.

Pressure – the force exerted per unit area by the molecules of the gas on
     the walls of the container. The common units of pressure are the
     torr, the mm of of Hg, and the atmosphere (atm).

$$1 \text{ atm} = 760 \text{ torr} = 760 \text{ mm of Hg} = 101,325 \text{ pascals}$$

The pressure of one atmosphere supports a column of liquid mercury
760 mm (29.9 inches) high. The S.I. unit is the pascal, equal to
one newton per square meter.

Temperature   If the pressure is constant, the volume of an ideal gas is
     proportional to the absolute temperature. Both Celsius (centigrade)
     and Kelvin (absolute) temperature scales are commonly used.

Ideal Gas – a hypothetical gas in which there are no forces of interaction
     between molecules and the volume of the gas molecules is a negligible
     fraction of the volume of the container. Real gases approach ideal
     behavior at low pressures and high temperatures. The ideal gas
     approximation works reasonably well for most gases at 1.0 atm and
     $0°C$ (STP).

Molar Volume at STP   22.4 liters/mole for an ideal gas.

Boyle's Law

     Boyle's law states that the pressure (P) of an ideal gas is inversely
proportional to the volume (V) if the temperature (T) it constant.

$$P \propto \frac{1}{V} \qquad (T \text{ constant})$$

Boyle's law can also be written

$$PV = C \qquad (T \text{ constant})$$

or

$$P_1 V_1 = P_2 V_2 \qquad (T \text{ constant})$$

Here C is a constant that depends on the amount of gas and the absolute
temperature. The subscripts 1 and 2 refer to two conditions and the
corresponding pressure-volume values.

Example 13.1 Helium is received from a natural gas well at a pressure of
2.17 atmospheres. To what pressure should it be pumped in order to store
12,500 ft^3 in a storage tank of 92.7 ft^3 capacity? Assume the temperature
is constant.

Solution Let condition 1 be the well and condition 2 the storage
tank. Then

$$P_1 = 2.17 \text{ atm} \qquad P_2 = ?$$

$$V_1 = 12,500 \text{ ft}^3 \qquad V_2 = 92.7 \text{ ft}^3$$

$$P_2 = P_1 V_1 / V_2 = (2.17 \text{ atm})(12,500 \text{ ft}^3)/92.7 \text{ ft}^3 = 293 \text{ atm}$$

Charles' Law

Charles' law states that the volume of an ideal gas is directly
proportional to the absolute temperature if the pressure is constant.

$$V \propto T \qquad (P \text{ constant})$$

Charles' law can also be written

$$V = C'T \qquad (P \text{ constant})$$

or

$$\frac{V_1}{T_1} = \frac{V_2}{T_2} \qquad (P \text{ constant})$$

Here C' is a constant that depends on the amount of gas and the pressure.

Example 13.2 A toy balloon filled with air occupies a volume of 3.75 liters
inside a house at 27°C. If brought outdoors on a cold winter day, when the
temperature is -17°C, what will its volume be? You may assume constant
pressure since a gas inside a rubber balloon has essentially the same
pressure as the external pressure.

Solution Let condition 1 be inside the house and condition 2 the outdoors.

$$T_1 = 273 + 27 = 300 \text{K}$$

$$T_2 = 273 - 17 = 256 \text{K}$$

$$\frac{V_1}{T_1} = \frac{V_2}{T_2}$$

$$V_2 = \frac{V_1 T_2}{T_1} = \frac{(3.75\ell)(256K)}{300K} = 3.20\ell$$

Avogadro's Law

Avogadro's law states that the volume of an ideal gas is directly proportional to the number of moles (n) of the gas if the temperature and pressure are constant.

$$V \propto n \qquad (T, P \text{ constant})$$

The Ideal Gas Law

Combining Boyle's, Charles', and Avogadro's laws,

$$V \propto \frac{nT}{P}$$

or

$$PV \propto nT$$

or

$$PV = nRT$$

Here R is the proportionality constant, called the ideal gas constant, and has the value 0.0821 ℓ-atm/mole-K. This equation, called the equation of state, can be used to calculate P, V, n or T for a gas, given the other variables. The result will be correct to the extent that the real gas behaves ideally under these conditions of pressure and temperature.

If a fixed quantity of gas is tranformed from condition 1 to 2, then

$$\frac{PV}{T} = nR = \text{constant}$$

then

$$\frac{P_1 V_1}{T_1} = \frac{P_2 V_2}{T_2} \qquad (n \text{ constant})$$

Example 13.3 A tank of hydrogen gas has a volume of 17.9 liters and is filled to a pressure of 152 atmospheres at 21°C. It is to be emptied into an evacuated furnace to provide a reducing atmosphere. If the furnace has a volume of 6,230 liters and operates at 500°C what will be the hydrogen pressure in the furnace?

<u>Solution</u> Let condition 1 be the tank and condition 2 the furnace.

$$T_1 = 21 + 273 = 294 \text{ K} \qquad T_2 = 500 + 273 = 773 \text{ K}$$

$$P_1 = 152 \text{ atm} \qquad P_2 = ?$$

$$V_1 = 17.9 \text{ } \ell \qquad V_2 = 6230 \text{ } \ell$$

$$P_2 = P_1 \text{ x } \frac{T_2}{T_1} \text{ x } \frac{V_1}{V_2} = 152 \text{ atm x } \frac{773 \text{ K}}{294 \text{ K}} \text{ x } \frac{17.9 \text{ } \ell}{6230 \text{ } \ell} = 1.15 \text{ atm}$$

<u>Example 13.4</u> A sample of an unknown organic liquid weighing 1.2176 g
is inserted into an evacuated flask, 586 ml in volume. At a temperature
of 172°C the liquid is completely vaporized and exerts a pressure of
673 torr. Calculate the number of moles of gas in the flask, and from
that the molecular weight of this unknown organic compound.

<u>Solution</u> First convert to the appropriate units.

$$T = 172 + 273 = 445 \text{ K}$$

$$V = 586 \text{ ml x } 1 \text{ } \ell/1000 \text{ ml} = 0.586 \text{ } \ell$$

$$P = 673 \text{ torr x } 1 \text{ atm}/760 \text{ torr} = 0.886 \text{ atm}$$

$$n = \frac{PV}{RT} = \frac{(0.886 \text{ atm})(0.586 \text{ } \ell)}{0.0821 \frac{(\ell\text{-atm})}{(\text{mole-K})}(445 \text{ K})} = 0.0142 \text{ mole}$$

$$\text{The molecular weight} = \frac{1.2176 \text{ g}}{0.0142 \text{ mole}} = 85.7 \text{ g/mole}$$

Dalton's Law

Dalton's law states that in a mixture of ideal gases the total
pressure is the sum of the partial pressures (P) of the constituent gases,

$$P_{tot} = P_1 + P_2 + P_3 + \ldots\ldots$$

The partial pressure of a component of the mixture is related to the
total pressure by its mole fraction,

$$P_i = X_i P_{tot} \qquad i = 1, 2, 3, \ldots\ldots$$

Consider for example a mixture consisting of 0.25 mole of H_2, 0.25 mole of
N_2 and 0.50 mole of CO_2 in a 22.4 ℓ container at 0°C. The total pressure
would be 1.0 atmosphere, assuming ideal behavior. The partial pressures
of H_2, N_2 and CO_2 would be 0.25, 0.25 and 0.50 atm, respectively.

<u>Example 13.5</u> 0.368 g of methylene chloride (CH_2Cl_2) and 0.516 g of dichloroethane ($C_2H_4Cl_2$) are added to an evacuated 4.50 ℓ vessel. The vessel is then heated to 210°C causing both liquids to vaporize completely. Calculate the total pressure of the mixture and the partial pressures of the component gases. The molecular weights of CH_2Cl_2 and $C_2H_4Cl_2$ are 84.9 and 98.9, respectively.

<u>Solution</u> First calculate the number of moles of both compounds, then use the ideal gas law to calculate the partial pressure of each, and then add them to get the total pressure.

$$n_{CH_2Cl_2} = \frac{1 \text{ mole}}{84.9 \text{ g}} \times 0.368 \text{ g} = 4.33 \times 10^{-3} \text{ mole } CH_2Cl_2$$

$$n_{C_2H_4Cl_2} = \frac{1 \text{ mole}}{98.9 \text{ g}} \times 0.516 \text{ g} = 5.22 \times 10^{-3} \text{ mole } C_2H_4Cl_2$$

$$T = 210 + 273 = 483 \text{ K}$$

$$P = \frac{nRT}{V}$$

$$P_{CH_2Cl_2} = \frac{(4.33 \times 10^{-3}\text{mole})(0.0821 \text{ } \ell\text{-atm/mole-K})(483 \text{ K})}{4.50 \text{ } \ell} = 0.0382 \text{ atm}$$

$$P_{C_2H_4Cl_2} = \frac{(5.22 \times 10^{-3})(0.0821)(483)}{4.50} = 0.0460 \text{ atm}$$

$$P_{tot} = P_{CH_2Cl_2} + P_{C_2H_4Cl_2} = .0382 + .0460 = 0.0842 \text{ atm}$$

Dalton's law must be applied to calculate the amount of a gas that has been collected by the displacement of water and hence is mixed with water vapor. Liquid water and water vapor are in equilibrium,

$$H_2O(\ell) \rightleftharpoons H_2O(g)$$

The pressure of the water vapor depends on the temperature. If a gas is collected by water displacement,

$$P_{tot} = P_{gas} + P_{H_2O}$$

<u>Example 13.6</u> A chemist collected 167 mls of nitrogen by displacement of water. The temperature was 18°C and the atmospheric pressure was 741 torr. The equilibrium vapor pressure of water at 18°C is 16 torr. Calculate the volume of dry N_2 at STP.

102

<u>Solution</u> First calculate the partial pressure of N_2 in the N_2-H_2O gas mixture. Then use the ideal gas law to convert the volume to STP conditions.

$$P_{N_2} = P_{tot} - P_{H_2O} = 741 - 16 = 725 \text{ torr}$$

$$V_{STP} = 167 \text{ ml} \times \frac{725 \text{ torr}}{760 \text{ torr}} \times \frac{273 \text{ K}}{(273 + 18)\text{K}} = 149 \text{ ml}$$

<u>Volume-Volume Stoichiometry</u>

If gases are maintained at constant temperature and pressure, the volumes which react are proportional to the moles that react. Since reactions involve simple ratios of moles, gas reactions will involve simple ratios of volumes. Gas stoichiometry problems can be worked using volumes instead of moles as long as all reactants and products are measured under the same conditions of temperature and pressure.

<u>Example 13.7</u> What volume of $CO_2(g)$ can be formed by complete combustion of 650 ml of acetylene, $C_2H_2(g)$? Assume the reactants and products are measured at the same temperature and pressure.

<u>Solution</u> The balanced equation is

$$C_2H_2(g) + 2\tfrac{1}{2}\, O_2(g) \longrightarrow 2\,CO_2(g) + H_2O(g)$$

$$\# \text{ ml } CO_2 = \frac{2 \text{ ml } CO_2}{1 \text{ ml } C_2H_2} \times 650 \text{ ml } C_2H_2 = 1300 \text{ ml } CO_2$$

<u>Example 13.8</u> 75.0 ml of $C_3H_8(g)$ are mixed with 400 ml of $O_2(g)$ in a flexible container and ignited. Assuming the combustion reaction is quantitative, calculate the final volume of the container when the temperature and pressure have returned to their initial values. The combustion reaction is

$$C_3H_8(g) + 5\,O_2(g) \longrightarrow 3\,CO_2(g) + 4\,H_2O(g)$$

<u>Solution</u> This is a limiting reagent problem. C_3H_8 is the limiting reagent because there is more than the required volume of oxygen.

$$V_{O_2} \text{ required} = \frac{5 \text{ ml } O_2}{1 \text{ ml } C_3H_8} \times 75 \text{ ml } C_3H_8 = 375 \text{ ml } O_2$$

$$V_{O_2} \text{ excess} = 400 \text{ ml} - 375 \text{ ml} = 25 \text{ ml}$$

$$V_{CO_2} = \frac{3 \text{ ml } CO_2}{1 \text{ ml } C_3H_8} \times 75 \text{ ml } C_3H_8 = 225 \text{ ml } CO_2$$

$$V_{H_2O} = \frac{4 \text{ ml } H_2O}{1 \text{ ml } C_3H_8} \times 75 \text{ ml } C_3H_8 = 300 \text{ ml } H_2O$$

$$V_{tot} = V_{O_2} + V_{CO_2} + V_{H_2O} = 25 + 225 + 300 = 550 \text{ ml}$$

Questions

13.1 The pressure of a gas in a 25.0 liter tank is 745 torr. If all the gas is transferred to a 38.7 liter container at the same temperature, what is the final pressure of the gas (in torr)?

a) 745 torr
b) .634 torr
c) 482 torr
d) 8.30×10^{-3} torr
e) More information is required.

13.2 To what centigrade temperature must 10.0 liters of ideal gas at 25°C be heated to expand to 50.0 liters? Assume the pressure is constant.

a) 25 x 50/10
b) 25 x 10/50
c) (25 + 273) x 50/10
d) (25 + 273) x 50/10 - 273
e) None of the above

13.3 A 1.00 liter gas container will explode if the pressure exceeds 12.5 atm. At 25°C, the helium-filled container is at a pressure of 2.07 atm. To what centigrade temperature can the container be heated before it will explode?

a) 151
b) 347
c) 424
d) 1530
e) 1800

13.4* A 37.8 liter vessel of hydrogen gas at a fixed temperature and pressure and a 17.3 liter vessel of oxygen gas at the same temperature and pressure are mixed in a reaction chamber and sparked. After the temperature and pressure of the resulting water vapor are returned to the initial temperature and pressure, what is the volume (in liters) of the resultant gas(es)? (Assume all the water is present as water vapor.)

    a) 37.8 liters
    b) 17.3 liters
    c) 34.6 liters
    d) 3.2 liters
    e) More information is required.

13.5 What is the volume occupied by .154 mole of an ideal gas at 745 torr and 30.0°C?

    a) 2.58 liters
    b) .387 liters
    c) 5.15×10^{-3} liters
    d) .256 liters
    e) 3.91 liters

13.6 A fixed amount of an ideal gas undergoes the following changes:

Initial Conditions:	Final Conditions:
P_1 = 844 torr	P_2 = 721 torr
T_1 = 49.9°C	T_2 = 35.1°C
V_1 = 39.3 liters	V_2 = ? liters

What is the value of V_2?

    a) $(39.3)\left(\dfrac{844}{721}\right)\left(\dfrac{35.1 + 273}{49.9 + 273}\right)$

    b) $(39.3)\left(\dfrac{844}{721}\right)\left(\dfrac{35.1}{49.9}\right)$

    c) $(39.3)\left(\dfrac{721}{844}\right)\left(\dfrac{35.1 + 273}{49.9 + 273}\right)$

    d) $(39.3)\left(\dfrac{721}{844}\right)\left(\dfrac{49.9 + 273}{35.1 + 273}\right)$

    e) $(39.3)\left(\dfrac{844}{721}\right)\left(\dfrac{49.9 + 273}{35.1 + 273}\right)$

13.7 At what temperature will 20.6 grams of an ideal gas occupying 11.9 liters exert a pressure of 406 torr? (The molecular weight of the hypothetical gas is 79.6 grams per mole.)

    a) 300°C
    b) 27°C
    c) 9.52×10^{-4} °C
    d) 0.367°C
    e) 2.28×10^{5} °C

13.8 Using modern vacuum techniques, a 4.69 liter flask was evacuated to 2.37×10^{-10} torr at 62.9°C. How many molecules of an ideal gas would be present under these conditions?

a) 1.13×10^{37} molecules
b) 3.20×10^{10} molecules
c) 1.30×10^{14} molecules
d) 1.71×10^{11} molecules
e) 2.43×10^{13} molecules

13.9*Determine the molecular weight of a compound given the following gas phase data: mass – 11.8 grams; temperature = 39.1°C; pressure = 630 torr; and volume = 1.31 liters.

a) .046
b) 34.9
c) 36.7
d) 279
e) 11.9

13.10 Determine the density of a gas in grams per liter given the following data: molecular weight = 80.9 grams; pressure = 758 torr; temperature = 30.5°C.

a) 2.46×10^{3}
b) 32.2
c) 3.24
d) 0.309
e) 3.61

13.11 What is the molecular formula of a gas with empirical formula S_2O if its density at STP is 10.7 g/liter?

a) S_4O_2

b) S_6O_5

c) S_2O

d) S_3O_6

e) S_6O_3

13.12 A chemist received an unknown pure liquid which analyzed 85.6% C and 14.4% H. At a temperature of 545 K and a pressure of 1.01 atmosphere the density of its vapor was 0.631 g/1. The molecular formula of the liquid is:

a) C_2H_4

b) CH_4

c) C_3H_6

d) C_2H_6

e) C_3H_8

13.13*A 1.87 gram sample of a gaseous compound containing 49.0% carbon, 2.00% hydrogen, and 49.0% oxygen occupies 670.3 milliliters at 153°C and 740 torr pressure. What is the molecular formula of the compound?

The atomic weights of carbon, hydrogen, and oxygen are 12.0, 1.0, and 16.0 grams/mole respectively.

    a) $C_7H_6O_2$

    b) $C_6H_{10}O_2$

    c) $C_4H_{10}O$

    d) C_2H_6O

    e) $C_4H_2O_3$

13.14 The quantity, 2.91 grams, of a gaseous compound of boron and hydrogen occupies 1.22 liters at 1.09 atmospheres and 25°C. The compound is:

    a) B_6H_{12}

    b) B_5H_9

    c) B_2H_6

    d) B_4H_{10}

    e) B_6H_{10}

13.15*The density of a gas is 4.38 grams per liter at 10.6°C and 542 torr. What is the density of the gas at STP?

    a) .122
    b) .239
    c) 3.00
    d) 5.91
    e) 6.38

13.16 Calculate the volume (in liters) of a gas bulb given the following data. The bulb, filled with an ideal gas, registered a pressure of 805 torr. A portion of the gas was transferred to a 425 ml bulb at 375 torr. The pressure of the gas remaining in the original bulb was 698 torr. Assume the temperature is constant.

    a) 0.121
    b) 0.699
    c) 0.710
    d) 1.06
    e) 1.49

13.17 The Haber process for preparing ammonia involves the direct
conversion of hydrogen and nitrogen gases at high temperature and
pressure using a catalyst:

$$N_2(g) + 3 H_2(g) \rightleftharpoons 2 NH_3(g)$$

How many liters of ammonia can be prepared from 14.3 liters of N_2 and
39.9 liters of H_2, assuming complete conversion at a constant
temperature of 1000 K and constant external pressure of 500 atm?

a) (.333)(39.9) liters
b) 39.9 liters
c) 14.3 liters
d) (.667)(39.9) liters
e) None of the above

13.18 The total pressure of a mixture of gases at 332 K is 1670 torr. The
mixture is analyzed and found to contain 2.79 moles of H_2S, 3.02 moles
of H_2, and 2.30 moles of Ne. What is the partial pressure of Ne
in torr?

a) 474
b) 725
c) 3.84×10^3
d) 1.67×10^3
e) 237

13.19* Suppose 7.70 grams of H_2O and 6.70 grams of CH_2Cl_2 are injected into
a large evacuated container maintained at 25°C. After all the H_2O
and CH_2Cl_2 have vaporized, the total pressure is found to be 124
torr. What is the partial pressure of CH_2Cl_2? Molecular weights of
H_2O and CH_2Cl_2 are 18.0 and 84.9 g/mole respectively.

a) 31.4 torr
b) 672 torr
c) 19.3 torr
d) .0698 torr
e) 22.9 torr

13.20 A sample of O_2 was collected over water at 45.2°C when the barometric
pressure was 827 torr. The volume of the gas as it was collected was
3.03 liters. What would the volume of dry O_2 be at STP? (The vapor
pressure of H_2O at 45.2° is 73.6 torr).

a) 2.58 liters
b) 3.51 liters
c) 2.83 liters
d) 18.2 liters
e) 3.25 liters

13.21* A 1.91 liter tank containing O_2 at 320 torr is connected to a 2.11 liter tank containing O_2 at 199 torr and the contents allowed to mix at constant temperature. The final pressure in torr is:

a) 488
b) 103
c) 257
d) 520
e) 2.74

13.22* A 3.09 liter flask contained He at 795 torr. A 3.98 liter flask at the same temperature contained Ar at 856 torr. The contents of the two flasks were mixed by opening a stopcock connecting them. What is the mole fraction of Ar in the mixture? (Assume ideal gas behavior, and neglect any gas in the connecting stopcock.)

a) .581
b) 2.06
c) .519
d) .481
e) .419

13.23* A 4.20 gram sample of pure NH_4Cl was placed in an evacuated container and heated until it completely decomposed as follows:

$$NH_4Cl(s) \longrightarrow NH_3(g) + HCl(g)$$

If the total pressure in the container is now 883 torr, what is the partial pressure of HCl?

a) 59.7 torr
b) 1.16 torr
c) 294 torr
d) 69.3 torr
e) 442 torr

13.24 A 5.00 liter reaction vessel contains 25.0 grams of I_2 vapor at 200°C. How many grams of $I_2(s)$ will form if the temperature is lowered to 100°C? The vapor pressure of I_2 at 100°C is 45.5 torr. The molecular weight of I_2 is 254 g/mole.

a) 15.0
b) 17.5
c) 20.0
d) 22.5
e) 25.0

13.25*A 7.75 ℓ reaction vessel contains a mixture of propane C_3H_8, and
excess O_2 at 127°C and at a pressure of 5.36 atm. The mixture is
ignited and the propane is quantitatively oxidized. The pressure in
the reaction vessel after it has been cooled to 127°C is 6.06 atm.
Calculate the number of grams of propane in the original mixture.
The molecular weight of propane is 44.0 g/mole.

a)   7.3
b)   14
c)   22
d)   24
e)   48

CHAPTER 14. KINETIC MOLECULAR THEORY

Many observed properties of gases can be explained by the kinetic molecular theory of gases. There are four assumptions:

1. Gas molecules can be considered point masses, that is, the volume occupied by the molecules of the gas is insignificant compared to the volume of the container.

2. Gas molecules are in a state of continual randon motion. The trajectory of a gas molecule is determined only by the collisions with other molecules of the gas and with the walls of the container.

3. There are neither attractive nor repulsive forces between gas molecules.

4. All collisions are perfectly elastic.

Graham's Law

Graham's law is the empirical observation that the rate of effusion of gas molecules from a small aperture is inversely proportional to the square root of the density of the gas.

$$v \propto \frac{1}{\sqrt{\rho}}$$

Here v is the velocity of the gas molecules in volume of gas per unit time, for example, liters/hr. For an ideal gas the density (in g/ℓ) is related to the molecular weight (M) by the following equation

$$\rho = \frac{MP}{RT}$$

So, if the pressure and temperature are constant, for 2 gases, 1, and 2,

$$\frac{v_1}{v_2} = \sqrt{\frac{M_2}{M_1}}$$

This empirical result is consistent with the expression for the root mean square velocity that can be derived using the kinetic molecular theory.

$$v_{rms} = \sqrt{\frac{3RT}{M}}$$

so

$$\frac{(v_{rms})_1}{(v_{rms})_2} = \sqrt{\frac{M_2}{M_1}}$$

<u>Example 14.1</u> Compare the rate at which helium gas will be lost through a leaky valve compared to the rate of loss of oxygen gas.

<u>Solution</u> The molecular weights of He and O_2 are 4.0 and 32.0 g/mole, respectively.

$$\frac{v_{He}}{v_{O_2}} = \sqrt{\frac{M_{O_2}}{M_{He}}} = \sqrt{\frac{32}{4}} = \sqrt{8} = 2.83$$

The helium gas will be lost 2.83 times as rapidly as oxygen gas.

<u>Example 14.2</u> A vessel contains an equal number of H_2 and He molecules. Compare the contributions of He and H_2 with respect to the following:

    a) mass,
    b) pressure,
    c) average molecular speed,
    d) average rate of impact on the walls,
    e) average force of impact on the walls.

<u>Solution</u> a) The molecular weight of He is twice that of H_2. Therefore He contributes twice as much to the total mass.

  b) The mole fractions are $X_{He} = X_{H_2} = 0.5$. Therefore according to Dalton's law, $P_{He} = P_{H_2} = P_{tot}/2$.

  c) From Graham's law,

$$\frac{v_{H_2}}{v_{He}} = \sqrt{\frac{M_{He}}{M_{H_2}}} = \sqrt{\frac{4}{2}} = \sqrt{2} = 1.41$$

  The hydrogen molecules are moving 1.41 times as rapidly as the helium atoms, on the average.

  d) The average rate of impact of H_2 molecules on the walls will be 1.41 times that of He because they traverse the container that much faster.

e) The force of the impact depends on the momentum of the atom or molecule,

$$p = mv$$

Here p is the momentum of the atom or molecule and m is the corresponding mass.

$$\frac{p_{H_2}}{p_{He}} = \frac{m_{H_2}}{m_{He}} \times \frac{v_{H_2}}{v_{He}} = \frac{2}{4} \times 1.41 = 0.705 = \frac{1}{1.41}$$

The average force of the impact of H_2 is less than that of He. However, the faster rate of impact of H_2 exactly compensates, and the net result is that both contribute equally to the pressure as predicted by Dalton's law.

Questions

14.1* According to Graham's Law of gas effusion, helium gas will effuse x times as fast as HCl. What is the value of x?

a) .331
b) 146
c) 3.02
d) 9.13
e) .110

14.2 Given:

Gas	Molecular weight (grams/mole)
Ar	39.9
O_2	32.0

What is the rate of effusion of Ar compared to O_2?

a) .555
b) 1.25
c) .896
d) 1.12
e) .802

14.3 If 75.0 liters of CO_2 effuse through a porous partition in 25.0 min., what volume of CO will effuse through the same partition at the same temperature and pressure in 30.0 min?
The molecular weights of CO and CO_2 are 28.0 and 44.0, respectively.

a) 57.3
b) 7.98
c) 141
d) 113
e) 71.8

14.4*Two hypothetical gases effuse into a glass tube 120 cm long. Gas A, with molecular weight 64.0 g/mole, enters at the end marked 0 cm. At the same time, gas B, with molecular weight 36.0 g/mole, enters the end marked 120. The point at which the gases meet is marked by a white deposit. How far from the end marked 0 cm is the deposit located?

   a) 51.4 cm
   b) 76.8 cm
   c) 1.20 cm
   d) 68.6 cm
   e) 43.2 cm

14.5 The speed at which molecules can escape the earth's gravitational field, the escape velocity, is approximately 1.1×10^6 cm/sec. At what temperature will N_2 molecules have v_{rms} equal to the escape velocity?

   (Hint: $v_{rms} = \sqrt{\dfrac{3RT}{M}}$, $R = 8.314 \times 10^7$ erg/mole-K, 1 erg = 1 $g\text{-}cm^2/sec^2$).

   a) 14,600 K
   b) 45,000 K
   c) 68,000 K
   d) 136,000 K
   e) 408,000 K

14.6 The root mean square velocity of CO molecules at 25°C is 1110 miles per hour. What is the rms velocity in miles per hour of He molecules at 25°C?

   a) 7770 mph
   b) 1110 mph
   c) 420 mph
   d) 2940 mph
   e) 13200 mph

14.7*A gas mixture contains an equal number of CO molecules and CO_2 molecules, and no others. Assuming ideal gas behavior, which of the following statements is true?

   a) The partial pressures of CO and CO_2 are the same.
   b) The total mass of CO_2 is the same as the CO.
   c) If the temperature is increased at constant volume, the pressure due to CO_2 will rise more rapidly than that of CO.
   d) Since CO molecules have less mass, they have higher velocity, make more frequent impacts, and hence contribute more pressure.
   e) None of the above.

114

14.8*Two rigid vessels of equal volume are at the same temperature and
pressure. One contains H_2(gas) and the other contains He(gas). Which
of the following statements is correct?

   a) A rise in temperature will decrease the density in both cases to the
      same extent, but the increased molecular velocity will compensate
      and lead to the pressure remaining constant.
   b) The two vessels contain the same mass of gas.
   c) The helium vessel is reduced to half its volume by pushing in the
      walls. As a result, the helium pressure is doubled and the helium
      makes more impacts per unit wall area than the hydrogen.
   d) If both vessels are cooled, the helium, because it is denser, will
      condense at a higher temperature.
   e) The average momentum transferred per molecule per impact is the
      same for both gases.

14.9*Which one of the following statements is a required postulate for
the theoretical derivation of the ideal gas law?

   a) When a molecule penetrates the wall and sticks there, all its
      kinetic energy gets transferred to the wall.
   b) The mass of the molecules is so small that it can be neglected.
   c) Molecules are spherical in shape and incompressible.
   d) Because of van der Waal's forces all molecules are attracted to
      each other, expecially at short distances.
   e) The average kinetic energy of the molecules is proportional to the
      absolute temperature.

14.10*A sealed container partly filled with water is allowed to come to
equilibrium. The partial pressure of the water vapor in the space
above the liquid water depends on:

   a) the temperature
   b) the nature of the interior wall of the vapor space
   c) the fraction of the total volume occupied by the liquid
   d) the nature of the other gases in the vapor space
   e) none of the above

CHAPTER 15. CRYSTALLINE SOLIDS

Crystal Lattice and Unit Cell

In the solid state each atom is confined to a region of space approximately equal to the atomic volume. An atom leaves its position very rarely, and diffusion of atoms through the solid is very slow. In an amorphous (glassy) solid the arrangement of atoms is random, but in a crystalline solid atoms are arranged in a geometric pattern called a lattice. The smallest portion of space in which this pattern is displayed is called the unit cell.

If there is only one kind of atom there are only a few simple lattices possible. Almost all monatomic crystals fall into one of the three following lattice types. Note that an atom is placed in a lattice site so that its center (nucleus) is at a lattice point.

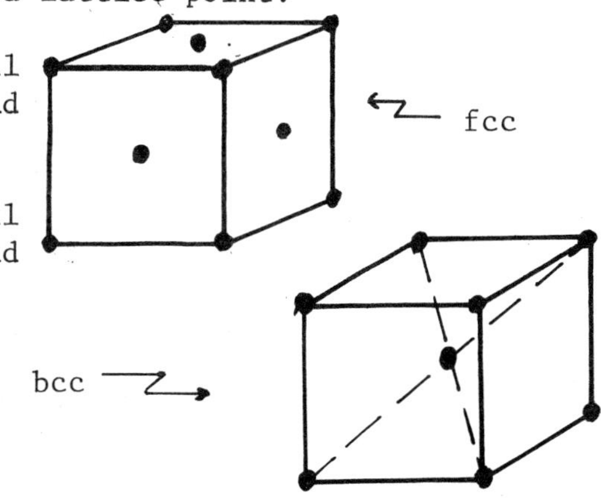

Face-centered cubic (fcc) - the unit cell is a cube with an atom on each corner and one in the center of each face.

Body-centered cubic (bcc) - the unit cell is a cube with an atom on each corner and one in the center of the cube.

Hexagonal closest packed (hcp) - the unit cell is not a cube. The lattice is most easily described as a stack of closest-packed layers with atoms in alternate layers directly above one another.

If there are two or more different atoms, such as in a salt, a variety of lattice types is possible. Each lattice is named according to some prototype salt, or mineral, which exemplifies it. Three examples are considered here.

Cesium chloride lattice - the unit cell is a cube with an anion in each corner and a cation at the center (or vice versa).

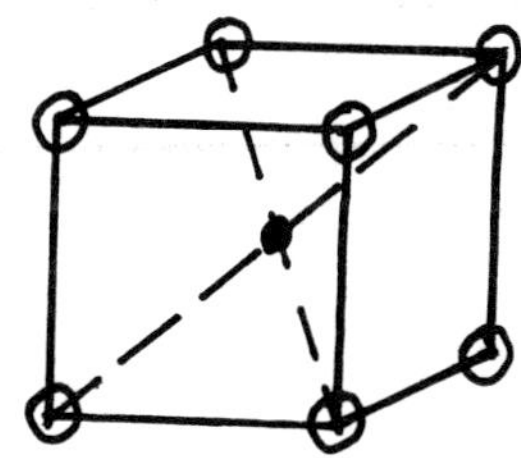

<u>Sodium chloride lattice</u> (or rock salt lattice) - the unit cell is a cube with anions at all corners and in the center of each face (as in the fcc), and cations in the center of each edge and in the center of the cube (or vice versa).

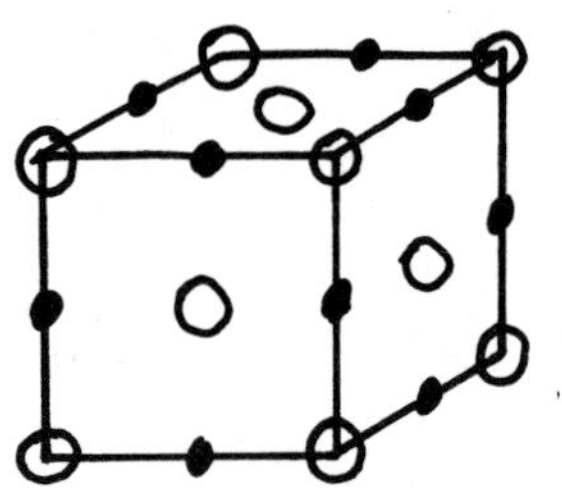

<u>Counting Atoms per Cell</u>

An atom whose center is at the corner of a cube will only have 1/8 of its volume inside the unit cell. The other seven eighths are in the seven other unit cells which share this lattice point. Similarly, edge atoms are only 1/4 inside the unit cell, and face atoms 1/2 inside the cell.

<u>Example 15.1</u> (a) Copper crystallizes in a face-centered cubic lattice. How many Cu atoms are in a unit cell? (b) How many calcium and oxide ions are in a unit cell of CaO, which crystallizes in the rock salt lattice?

Solution: (a) Corner atoms: 8 x 1/8 = 1

Face atoms: 6 x 1/2 = 3

Total = 1 + 3 = 4

(b) Oxide ions: corners: 8 x 1/8 = 1

face: 6 x 1/2 = 3

total oxide ions = 4

Calcium ions: edges: 12 x 1/4 = 3

center: 1 x 1 = 1

total calcium ions = 4

<u>Nearest Neighbors and Coordination Number</u>

Nearest neighbors are those atoms that touch a given atom.

<u>Example 15.2</u>   How many nearest neighbors does a copper atom have in the
fcc lattice?

<u>Solution</u>: Refer to the sketch of the unit cell. Consider the center
atom in the top face. It touches the four atoms in the corners of the
top face. It also touches the four atoms in the centers of the four
vertical faces, and the four in the centers of the four vertical faces
in the unit cell just above it (now shown). The total is 12.

The coordination number is the number of closest equidistant ions
of opposite charge surrounding a given ion.

<u>Example 15.3</u> What is the coordination number of cesium in the cesium
chloride lattice?

<u>Solution</u>: The cesium ion is at the center of the unit cell. It is sur-
rounded by the 8 chloride ions on the corners. The coordination number
of Cs^+ is eight.

<u>Atomic, Ionic, and Cell Dimensions</u>

Although atoms and ions do not have definable boundaries it is
convenient to consider them as hard spheres and to assign a radius to
each. These assignments are compromise values so that optimum agreement
is obtained between internuclear distances measured in most crystals and
those calculated from tabulated values of atomic and ionic radii.
Values listed in different text books may differ slightly from one
another or from the values used in this book.

<u>Example 15.4</u> Derive the numerical relationship between the radius r, of
a metal atom in a fcc lattice and the edge length, a, of the unit cell.
What is the atomic radius of copper, given that the edge length of the
unit cell is 3.62 Å. One Angstrom (Å) is 1.0×10^{-8} cm.

<u>Solution</u>: A "space-filling" sketch of the
face of a unit cell is shown at the right
with a diagonal drawn across the face.
The length of the diagonal is 4r, and
the edge of the face is a. By plane
geometry

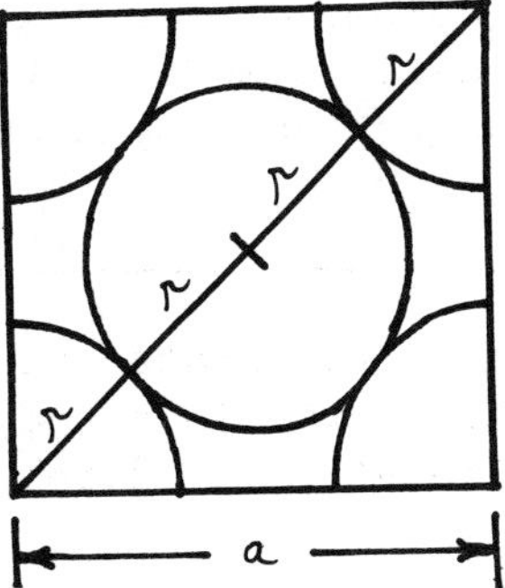

$$(4r)^2 = 2a^2$$

$$4r = a\sqrt{2}$$

$$r = a\frac{\sqrt{2}}{4}$$

For copper $r = \dfrac{3.62 \times \sqrt{2}}{4} = 1.28$ Å.

Radius Ratio

In the case of salts, the type of lattice is determined largely by the radius ratio, r_c/r_a, the ratio of cation radius to anion radius. In most salts the anion is larger. When the radius ratio is close to unity, that is, when the cation is almost as large as the anion, then the coordination number must be large, because a large number of anions is required to surround the cation. However, if the anion is very much larger than the cation very few anions are required to surround it.

<u>Example 15.5</u> Calculate the minimum radius ratio for the NaCl lattice.

<u>Solution</u>: The minimum radius ratio is that at which the anions contact the cations as well as each other as shown in the figure. (If the cations are somewhat larger the anions lose contact with each other but the lattice is still favored. If the cations are smaller they lose all contact, and the ZnS lattice is favored).

$$\text{Face diagonal} = 4r_a = a\sqrt{2}$$

$$\text{But } a = 2r_a + 2r_c$$

$$4r_a = 2\sqrt{2}\,r_a + 2\sqrt{2}\,r_c$$

$$(2-\sqrt{2})r_a = \sqrt{2}\,r_c$$

$$r_c/r_a = \sqrt{2} - 1 = 0.414$$

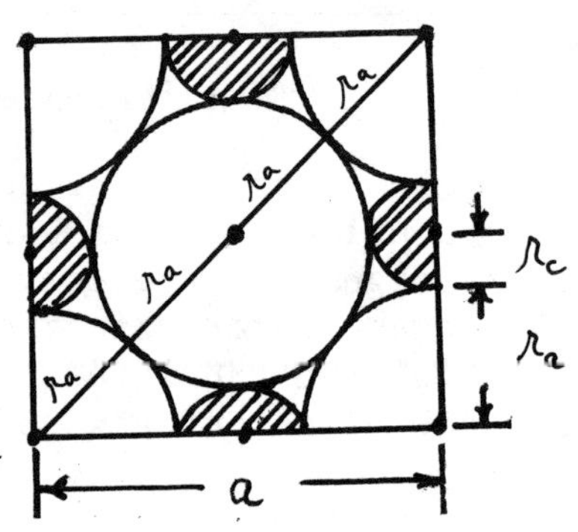

Density

If the lattice type and the dimensions of the unit cell are known it is possible to calculate the density of the crystal. The basis of the calculation is one unit cell. For the cubic examples considered here, the volume of the unit cell is a^3. The number of atoms contained in a unit cell depends on the lattice, and the mass of the unit cell is obtained using the atomic or molecular weight and Avogadro's number.

<u>Example 15.6</u> Calculate the density in g/cm^3 of pure iron metal (α iron), which crystallizes in a body centered cubic lattice with a unit cell edge length, a = 2.861 Å.

<u>Solution:</u>

The volume of the unit cell = $(2.861 \times 10^{-8} \text{ cm})^3 = 23.42 \times 10^{-24} \text{ cm}^3$

The unit cell contains: 8 corner atoms x 1/8 = 1

plus 1 internal atom x 1 = 1

Total of 1 + 1 = 2 atoms

Mass of the unit cell $= \dfrac{2 \text{ atoms} \times 55.85 \text{ g/mole}}{6.02 \times 10^{23} \text{ atoms/mole}}$

$= 18.55 \times 10^{-23} \text{ g}$

The density $= \dfrac{\text{mass}}{\text{volume}} = \dfrac{18.55 \times 10^{-23} \text{ g}}{23.42 \times 10^{-24} \text{ cm}^3} = 7.92 \text{ g/cm}^3$

The observed density is 7.86 g/cm^3. Usually observed densities are less than the theoretical or "crystal" density because in the density measurement, the sample is made up of many small imperfect crystals. The larger and more perfect the crystals, the closer the observed density approaches the theoretical density.

Questions

15.1 Cs metal crystallizes in a body centered cubic lattice with the unit cell edge length equal to 6.17 angstroms. Assuming the atoms are identical hard spheres, calculate the atomic radius of the metal. (Hint: The square roots of 2 and 3 are 1.414 and 1.731 respectively.)

a) $\dfrac{(1.731)(6.17)}{2}$

b) $\dfrac{(1.731)(6.17)}{4}$

c) $\dfrac{6.17}{2}$

d) $\dfrac{(1.414)(6.17)}{4}$

e) None of the above.

120

15.2 NaBr crystallizes in the NaCl lattice with the unit cell edge length
equal to 5.80 angstroms. If the ionic radius of Br is 1.95 angstroms,
what is the ionic radius of Na? (Hint: Assume the cation to anion
ratio is such that the cations and anions touch.)

a)   3.07 Angstroms
b)   2.90 Angstroms
c)   .950 Angstroms
d)   1.90 Angstroms
e)   3.85 Angstroms

15.3*Pd metal crystallizes in a face centered cubic lattice with the
unit cell edge length equal to 3.87 Angstroms. What is the density
of Pd metal? The atomic weight of Pd is 106 g/mole.

a)   12.1 g/cm^3
b)   27.2 g/cm^3
c)   90.9 g/cm^3
d)   .165 g/cm^3
e)   6.06 g/cm^3

15.4 Cu metal crystallizes in a face centered cubic lattice. The density
of Cu is 8.89 $grams/cm^3$. What is the length of an edge of the unit
cell of Cu metal? The atomic weight of Cu is 63.5 g/mole

a)   3.62 Angstroms
b)   47.4 Angstroms
c) 3.06×10^8 Angstroms
d)   .907 Angstroms
e)   4.56 Angstroms

15.5 The edge length of a unit cell of Au metal is 4.13 Angstroms and its
density is 18.6 $grams/cm^3$. How many atoms of Au are present in the
unit cell? The atomic weight of Au is 197 g/mole.

a)   2
b)   14
c)   4
d)   9
e)   5

15.6*Most metals crystallize in the body centered cubic (bcc), the face
centered cubic (fcc), or the hexagonal closest packed (hpc) lattice.
Select the one right statement below; where

a = edge length of unit cell, and
r = atomic radius

a)   In fcc there are 6 atoms per unit cell.
b)   In fcc, a $\sqrt{3}$ = 2r.
c)   In bcc there are 6 second nearest neighbors at a distance a.
d)   In fcc there are 10 nearest neighbors at distance 2r.
e)   In bcc there are 6 nearest neighbors at distance 2r.

15.7 One may classify crystalline solids into five categories: ionic, polar molecular, nonpolar molecular, covalent network, and metallic. Which one of the following statements concerning these solids is correct?

    a)   All ionic solids are diamagnetic.

    b)   All polar molecular solids are semiconductors with electrical conductivity values between that of insulators and that of metals.

    c)   Nonpolar molecular solids have high melting points because the molecules are packed in the crystal lattice very efficiently.

    d)   The transition metals are characterized by high melting points.

    e)   The atoms in a covalent network solid are joined by forces due to a sea of delocalized valence electrons.

15.8 CsI crystallizes in the CsCl lattice. The length of an edge of the unit cell is 5.44 Angstroms. What is the density of CsI? The molecular weight of CsI is 260 g/mole.

    a)   10.7 g/cm^3

    b)   317 g/cm^3

    c)   79.3 g/cm^3

    d)   .397 g/cm^3

    e)   2.68 g/cm^3

15.9* NaI crystallizes in the NaCl lattice. The density is equal to 4.14 grams/cm^3. What is the length of an edge of the unit cell of NaI? The molecular weight of NaI is 150 g/mole.

    a)   6.22 Angstroms

    b)   1.17 Angstroms

    c)   3.92 Angstroms

    d)   .525 x 10^9 Angstroms

    e)   241 Angstroms

15.10 AgCl crystallizes in the NaCl lattice. The length of an edge of the unit cell is 5.547 Å. Which of the following is _not_ consistent with these facts?

    a)   The shortest Ag^+-Cl^- distance is 5.547/2 Å.

    b)   The shortest Ag^+-Ag^+ distance is 5.547 Å.

    c)   The theoretical density of AgCl is 5.57 g/cm^3.

    d)   The theoretical molar volume of AgCl is 25.7 cm^3/mole.

    e)   The shortest Cl^--Cl^- distance is 3.92 Å.

CHAPTER 16. ATOMIC STRUCTURE

Introduction

An atom consists of a nucleus and one or more electrons surrounding
the nucleus. Most chemical properties can be explained in terms of changes
in the electronic structure of atoms and molecules. It is not possible
to observe electrons directly. However, it is possible, through
spectroscopy, to observe the energy required to remove or rearrange
electrons in atoms and molecules. A theory of atomic structure is
accepted by scientists if the theory can be used to make quantitative
predictions of electron energies that agree with experimental results.

The Wave Nature of Light

Radio waves, infrared, visible light, ultraviolet radiation, x rays
and γ rays are all forms of electromagnetic radiation. All can be
characterized in terms of a travelling wave, analogous to waves on the
surface of water. These forms of electromagnetic radiations all travel
at the speed, c, which is 186,000 miles/sec or 3.00×10^{10} cm/sec in a
vacuum. The various types of electromagnetic radiation are distinguished
by their values for wavelength (λ), and frequency (ν),

$$\nu = c/\lambda$$

The frequency is the number of waves passing a given point in space per
unit time. The unit for ν is $\sec^{-1}$, called the hertz. The wavelength is
the peak-to-peak distance and is given in dimensions of length
(cm, m, nm, Å, etc).

Example 16.1 Television is transmitted via electromagnetic radiation. A
typical wavelength is 2.00 meters. Calculate the frequency of a TV station
that transmits at 2.00 meters. Assume the speed of light in air is the
same as in a vacuum.

Solution $\nu = c/\lambda$

$$= \frac{3.00 \times 10^{10} \text{cm sec}^{-1}}{2.00 \text{ m} \times 100 \text{ cm/m}} = 1.50 \times 10^{8} \text{ sec}^{-1}, \text{ or } 1.50 \times 10^{8} \text{ hertz}$$

Planck theorized in 1900 that radiant energy was quantized, that is, could be emitted or absorbed only in certain discrete quantities, called quanta. The energy (E) of a quantum of radiation is proportional to the frequency,

$$E = h\nu$$

Here h is Planck's constant, 6.63×10^{-27} erg-sec. The unit of energy, the erg, is 1 $g\text{-}cm^2/sec^2$. The joule is equivalent to 10^7 ergs. Light radiation can also be conceived of as a stream of particles called photons, each of which has an energy of one quantum.

Example 16.2 A carbon-carbon single bond has a bond energy of 347 kJ/mole (83 kcal/mole). Calculate the frequency and wavelength (cm) of a photon with sufficient energy to break the C-C bond.

Solution The bond energy in ergs/bond is

$$E(\text{erg/bond}) = \frac{10^7 \text{ ergs}}{J} \times \frac{1000 \text{ J}}{kJ} \times \frac{347 \text{ kJ}}{\text{mole}} \times \frac{1 \text{ mole}}{6.02 \times 10^{23} \text{ bonds}}$$

$$= 5.76 \times 10^{-12} \text{ erg/bond}$$

$$\nu = E/h = \frac{5.76 \times 10^{-12} \text{ erg}}{6.63 \times 10^{-27} \text{ erg-sec}} = 8.69 \times 10^{14} \text{ sec}^{-1}$$

$$\lambda = c/\nu = \frac{3.00 \times 10^{10} \text{ cm/sec}}{8.69 \times 10^{14} \text{ sec}} = 3.45 \times 10^{-5} \text{ cm}$$

A unit often used in spectroscopy is the wavenumber $(\bar{\nu})$,

$$\bar{\nu} = 1/\lambda$$

The wave number has units of cm^{-1}. The wavenumber is proportional to frequency and energy,

$$E = h\nu = \frac{hc}{\lambda} = hc\bar{\nu}$$

Example 16.3 Express each of the following in terms of wave numbers.

(a) $\nu = 1.98 \times 10^{14} \text{ sec}^{-1}$ (b) $E = 3.81 \times 10^{-12}$ erg   (c) $\lambda = 5575$ Å

Solution   a) $\bar{\nu} = \dfrac{\nu}{c} = \dfrac{1.98 \times 10^{14} \text{ sec}^{-1}}{3.00 \times 10^{10} \text{ cm sec}^{-1}} = 6600 \text{ cm}^{-1}$

b) $\bar{\nu} = E/hc = \dfrac{3.81 \times 10^{-12} \text{ erg}}{(6.63 \times 10^{-27} \text{ erg-sec})(3.00 \times 10^{10} \text{ cm/sec})}$

$= 19200 \text{ cm}^{-1}$

c) $\bar{\nu} = \dfrac{1}{\lambda} = \dfrac{10^{8} \text{ Å}}{\text{cm}} \times \dfrac{1}{5575 \text{ Å}} = 17940 \text{ cm}^{-1}$

The Wave Nature of Matter

DeBroglie theorized that particles of mass m, moving with velocity v, would show wave properties. The DeBroglie matter-wave relationship is

$$\lambda = \frac{h}{mv}$$

Example 16.4 What is the wavelength of a beam of neutrons moving with velocity 2.5×10^{5} cm/sec (5600 mph)? The mass of the neutron is 1.67×10^{-24} gram.

Solution $\lambda = \dfrac{h}{mv} = \dfrac{6.63 \times 10^{-27} \text{ g-cm}^{2}/\text{sec}}{(1.67 \times 10^{-24} \text{ g})(2.5 \times 10^{5} \text{ cm/sec})} = 1.6 \times 10^{-8} \text{ cm}$

Note that this wavelength, 1.6 Å, is the same order of magnitude as the unit cell dimensions in crystals. Neutron diffraction is used to determine crystal structures particularly of magnetic materials.

The Bohr Atom

In 1913 Bohr formulated a theory for the atom that was based on Rutherford's planetary model of the atom. Bohr hypothesized that electrons moved in circular orbits about the nucleus. This model contradicts classical physics according to which the electron should lose energy by radiation and spiral into the nucleus. Bohr countered by stating that classical physics simply was not applicable to phenomena on the atomic scale. According to Bohr, only orbits of certain radii were allowed

$$r_n = \frac{n^2 h^2}{4\pi^2 m_e e^2} = n^2 \times 0.529 \text{ Å} = n^2 a_o$$

Here n is a positive integer (1, 2, 3, ...), h is Planck's constant, m_e and e are the electronic mass and charge, and a_o, 0.529 Å, is the first Bohr radius. Bohr's integer n is called the principal quantum number. The energy associated with these orbiting electrons is

$$E_n = -\frac{2\pi^2 m_e e^4}{n^2 h^2} = \frac{-13.6}{n^2} \text{ eV/atom} = \frac{-1312}{n^2} \text{ kJ/mole}$$

The energy unit, eV, is the electron volt.

1 eV/atom = 1.60 x 10^{-12} erg/atom = 96.5 kJ/mole = 23.1 kcal/mole. The energy of the hydrogen atom approached zero as the principal quantum number approached infinity, that is, the zero of energy on this scale is the ionized atom, H^+ and e^-. If a photon of just the right energy ($h\nu$) is absorbed by a hydrogen atom in the ground state (n=1), its energy is used to promote the electron to a higher orbital (n>1). For example,

$$H(n=1) + h\nu \rightarrow H^*(n=2)$$

$$\Delta E = h\nu = E_2 - E_1 = 13.6 \left\{ \frac{1}{1^2} - \frac{1}{2^2} \right\} = 10.2 \text{ eV/atom}$$

In general, when light is absorbed or emitted by a hydrogen atom.

$$\Delta E = 13.6 \left\{ \frac{1}{n_1^2} - \frac{1}{n_2^2} \right\} \qquad \text{(eV/atom)}$$

Here n_1 and n_2 are the principal quantum numbers of the lower and upper energy state.

<u>Example 16.5</u> Calculate the energy, frequency, wavelength and wavenumber of the photon emitted when a hydrogen atom undergoes the electronic transition n=5 to n=2.

<u>Solution</u> $\Delta E = 13.6 \left\{ \frac{1}{2^2} - \frac{1}{5^2} \right\} = 2.86 \text{ eV/atom}$

$$\Delta E(\text{ergs}) = \frac{1.60 \times 10^{-12} \text{ erg}}{1 \text{ eV}} \times 2.86 \text{ eV} = 4.57 \times 10^{-12} \text{ erg}$$

$$\nu = \frac{\Delta E}{h} = \frac{4.57 \times 10^{-12} \text{ erg}}{6.63 \times 10^{-27} \text{ erg-sec}} = 6.89 \times 10^{14} \text{ sec}^{-1}$$

$$\lambda = \frac{c}{\nu} = \frac{3.00 \times 10^{10} \text{ cm/sec}}{6.89 \times 10^{14} \text{ /sec}} = 4.35 \times 10^{-5} \text{ cm}$$

$$\bar{\nu} = 1/\lambda = 23000 \text{ cm}^{-1}$$

Modern Atomic Theory

The Bohr model of the atom, with one quantum number, was successful
in explaining the electronic behavior in the hydrogen atom, and in
hydrogen-like atoms (He^+, Li^{2+}, Be^{3+}, etc). In modern atomic theory
four quantum numbers are used. They are:

Quantum Number	Name	Allowed Values
n	Principal	1, 2, 3, 5, ...
ℓ	Orbital	0, 1, 2, 3, ...(n-1)
m	Magnetic	$-\ell$, $-(\ell-1)$,...0...$\ell-1$, ℓ
s	Spin	$\pm 1/2$

The principal quantum number, n, may be any positive integer. The
region in space in which an electron is most likely to be found is
called an orbital. The size and energy of the orbital is determined by
n. The shape of the orbital is determined by the value of ℓ, which may
be zero or a positive integer less than n. For example, if n=4, ℓ may
be 0, 1, 2, or 3. The atomic orbitals are labelled according to the
value of ℓ as follows:

ℓ	0	1	2	3
Orbital	s	p	d	f

The orientation of the orbital with respect to a set of reference axes is
indicated by the magnetic quantum number m. The magnetic quantum number
may take on all positive and negative integral values, including zero,
between $-\ell$ and $+\ell$, inclusive. For example, if $\ell=2$, then m could be
-2, -1, 0, 1, or 2. The spin quantum number may be either +1/2 or -1/2.

Example 16.6 Which of the following sets of quantum numbers is not allowed
for describing an electron in an orbital in a hydrogen atom.

Case	n	ℓ	m	s
1	4	2	-3	1/2
2	1	1	0	-1/2
3	3	2	2	0

Solution None of them, for the following reasons:

1. If $\ell=2$, m can only be -2, -1, 0, 1, or 2.

2. If n=1, ℓ can only be 0.

3. The only allowed values for s are +1/2 and -1/2

Electron Configurations

For atoms beyond hydrogen the structure is approximated by assigning the electrons to hydrogen-like orbitals. According to the Aufbau (building up) principle electrons are placed in orbitals in order of increasing energy.

The Pauli exclusion principle states that no two electrons in an atom can have all four quantum numbers the same. As a result an orbital can be occupied by at most two electrons, and these must be of opposite spin. Thus for a given n there can be:

one s orbital	2 electrons
three p orbitals	6 electrons
five d orbitals	10 electrons
seven f orbitals	14 electrons

Hund's rule states that a set of orbitals of the same energy will be filled with one electron each until pairing is necessary.

Generally orbitals increase in energy with increasing n, and for any given n increase with increasing ℓ. However, the 4s level is lower than the 3d, and generally ns is lower than (n-1)d. Also ns is lower than (n-2)f.

The periodic table is a guide to the electron configuration of the elements as shown below.

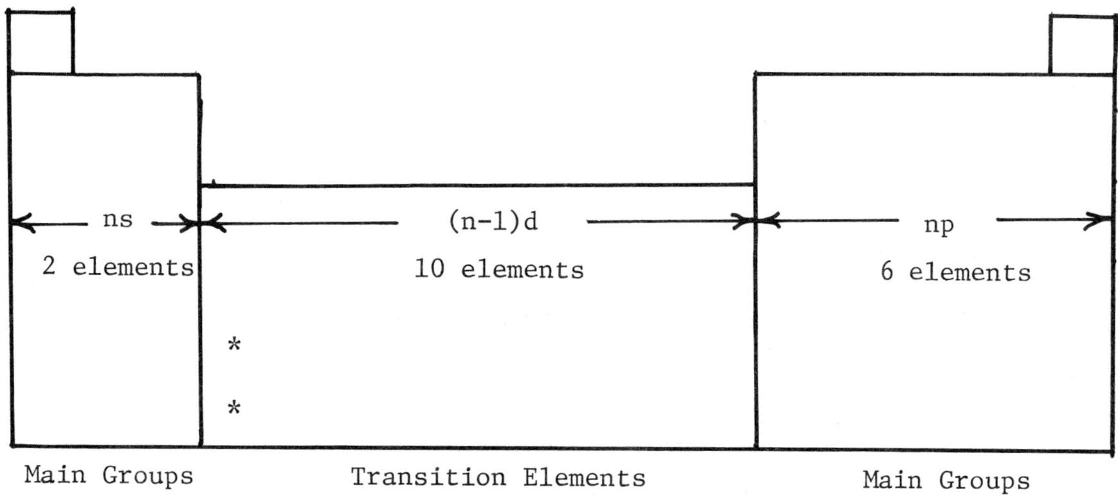

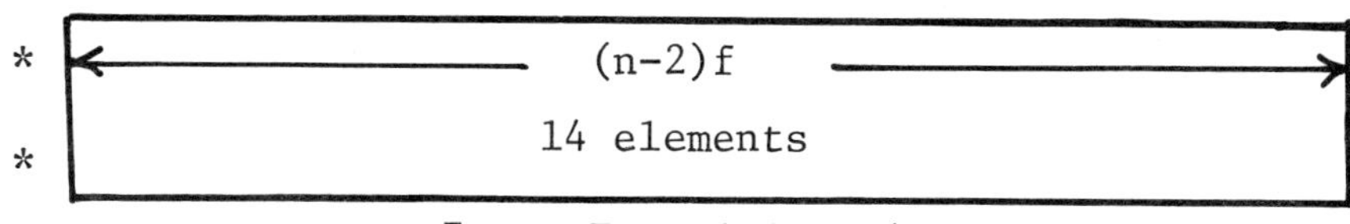

Inner Transition Elements

Orbital occupancy is indicated by a right superscript. For example $6p^4$ means 4 electrons in 6p orbitals.

For negative ions, the extra electrons are added in the usual way. For positive ions, electrons are removed from the highest n orbital first

---

Example 16.7 Write the electron configuration for the $_{82}Pb$ atom.

Solution Look in the periodic table and you will find the noble gas with the next lower atomic number to 82 is $_{54}Xe$. This means the first 54 electrons in $_{82}Pb$ are arranged exactly as in $_{54}Xe$ and this symbol can be used to depict them. Since $_{54}Xe$ completes the 5th period you next start the 6th period. Reading from the left there are 2 electrons in the 6s orbital, and (noting the asterisk) 14 electrons in the 4f orbitals, 10 electrons in the 5d orbitals, and finally 2 in 6p orbitals. The final configuration then is:

$$_{82}Pb = [_{54}Xe]6s^2 4f^{14} 5d^{10} 6p^2$$

As a check, the sum of the electrons is 54 plus the superscripts, i.e. $54 + 2 + 14 + 10 + 2 = 82$.

Note that period n always fills in the order ns, (n-2)f, (n-1)d, and np. The 6th period is the first one in which f orbitals are encountered, and the 4th period is the first one in which d orbitals are encountered.

Example 16.8 Write the electron configurations for

    a)   the $_{15}P$ atom

    b)   the $_{35}Br^-$ ion

    c)   the $_{44}Ru^{3+}$ ion

Show the detailed outer electron configuration of (a)

Solution

    a) $_{15}P:\ 1s^2 2s^2 2p^6 3s^2 3p^3$, or $[_{10}Ne]\ 3s^2 3p^3$

    b) $_{35}Br^-:\ [_{18}Ar]4s^2 3d^{10} 4p^6$, or $[_{36}Kr]$

    c) $_{44}Ru:\ [_{36}Kr]5s^2 4d^6$

           $_{44}Ru^{3+}:\ [_{36}Kr]4d^5$

The last three electrons in $_{15}$P are added one each to the three p orbitals

$$_{15}\text{P: }[_{10}\text{Ne}] \quad \underset{3s}{\uparrow\downarrow} \quad \underset{3p_x}{\uparrow} \quad \underset{3p_y}{\uparrow} \quad \underset{3p_z}{\uparrow}$$

There are a few exceptions to the rules for electron configurations which are usually attributable to the tendency to form filled or half-filled sub-shells of electrons.

Some of these are:

Cr: $[\text{Ar}]4s^13d^5$ Cu: $[\text{Ar}]4s^13d^{10}$

Mo: $[\text{Kr}]5s^13d^5$ Ag: $[\text{Kr}]5s^13d^{10}$

Pt: $[\text{Xe}]6s^14f^{14}5d^9$ Au: $[\text{Xe}]6s^14f^{14}3d^{10}$

Questions

16.1 Certain electromagnetic radiation has frequency equal to 1.68×10^{10} sec^{-1}. What is the wavelength of this radiation in angstroms?

  a) 1.11×10^{-16}
  b) 6.23×10^{-17}
  c) 5.04×10^{20}
  d) 1.79×10^{8}
  e) 0.560

16.2 Certain electromagnetic radiation has wavelength equal to 3.76×10^{6} angstroms. What is the frequency of this radiation in sec^{-1}?

  a) 1.25×10^{-12}
  b) 1.76×10^{-25}
  c) 7.98×10^{11}
  d) 1.89×10^{14}
  e) 5.29×10^{-15}

16.3*What is the wave number of electromagnetic radiation which has a frequency of 5.69×10^{17} sec^{-1}?

  a) 1.76×10^{-18}
  b) 3.77×10^{-9}
  c) 1.26×10^{-19}
  d) 1.90×10^{7}
  e) 5.27×10^{-8}

130

16.4 Certain electromagnetic radiation has frequency equal to 1.05×10^{18}
sec^{-1}. What is the energy of a quantum of this radiation in ergs?

a) 3.14×10^{28}
b) 2.86×10^{-8}
c) 6.94×10^{-9}
d) 1.90×10^{-34}
e) 2.42×10^{-1}

16.5 Which of the following combinations of quantum numbers do <u>not</u> represent
permissible solutions of the Schroedinger wave equation for the
hydrogen atom:

	n	ℓ	m	s
1.	7	-3	0	-1/2
2.	-2	3	-1	1/2
3.	4	3	-2	-1/2

a) 2
b) 3
c) 1 and 2
d) 1 and 3
e) None of them is permissible

16.6 The Rydberg equation for the hydrogen atom in frequency units is

$$\nu = 3.29 \times 10^{15} \left\{ \frac{1}{n_1^2} - \frac{1}{n_2^2} \right\} \ \sec^{-1}$$

What is the energy in ergs of a quantum of electromagnetic radiation
that corresponds to a transition from the fourth excited state ($n_1 = 5$)
to the fifth excited state ($n_2 = 6$)?

a) 1.78×10^{-9}
b) 4.02×10^{13}
c) 2.67×10^{-13}
d) 1.98×10^{-12}
e) 2.40×10^{-25}

16.7* Which of the following electron configurations is <u>incorrect</u>?

a) Alkaline earth metals: ns^2 n>1

b) Rb: (Kr) $5s^1$

c) Na$^+$: (He) $2s^2 2p^6$

d) Cr^{6+}: (Ar) $4s^1 3d^5$

e) All the above configurations are correct.

16.8*Electromagnetic radiation of sufficient energy can be used to break
chemical bonds. Calculate the wavelength in angstroms of photons with
enough energy to break H-O bonds. The H-O bond energy
is **427 kJ/mole**   (1   kJ/mole = **1.66** x 10^{-14} erg/molecule)

a) 3.56 x 10^{12} Å

b) 35600 Å

c) 2810 Å

d) 2.81 x 10^{-5} Å

e) 7.03 x 10^{-4} Å

16.9*In what region of the electromagnetic spectrum would you look for the
spectral line resulting from the electronic transition from the
energy level with n_1 = 1 to the level with n_2 = 9 in the hydrogen
atom? The Rydberg constant in wavenumbers is 110000 cm^{-1}.

a) microwave   0.10 to 10.0 cm^{-1}

b) infrared   10 to 1.4 x 10^4 cm^{-1}

c) visible   1.4 x 10^4 to 2.5 x 10^4 cm^{-1}

d) ultraviolet 2.5 x 10^4 to 1.0 x 10^6 cm^{-1}

e) x-rays   1.0 x 10^6 to 1.0 x 10^8 cm^{-1}

16.10*Each of the following statements describes the significance of a
certain key experiment or idea. Which one is <u>not</u> correct?

a) Moseley's measurements on the characteristic X-rays of the elements
   showed that certain isotopes of different elements could have the
   same mass.
b) J. J. Thompson used a primitive mass spectrometer to measure the
   charge-to-mass ratio of the electron. Subsequently R. A.
   Millikan's oil drop experiment provided a value for the
   electronic charge.
c) The phenomenon of radioactivity was discovered by Becquerel who
   observed that uranium salts emitted radiation that penetrated
   the black paper coverings of photographic plates and darkened
   them as if they had been exposed to light.
d) Planck's successful explanation of the black-body radiation spectrum
   introduced the hypothesis that radiation is emitted not
   continuously but in discrete quanta.
e) The Davisson-Germer electron diffraction experiment demonstrated
   deBroglie's hypothesis that electrons behave as waves.

16.11*Bohr proposed a theory of atomic structure for the hydrogen atom that
was influenced by Rutherford's nuclear model of the atom, the quantum
hypothesis and the experimental results of atomic spectroscopy. Which
of the below statements is <u>not</u> a correct statement of some aspect of
Bohr's theory?

(over)

a) Although the theory worked well for hydrogen it failed to explain the spectra of atoms containing two or more orbiting electrons.
b) The energy of the electron in the hydrogen atom depends on only one quantum number, the principle quantum number, n.
c) The orbiting electron radiates its energy into space and falls closer and closer to the nucleus.
d) In a given quantum state the electron is constrained to move in an orbit with a precisely fixed radius.
e) The angular momentum of the electron in the hydrogen atom is restricted to values which are integral multiples of a universal constant, $h/2\pi$.

16.12 Electrons in atoms are visualized as occupying atomic orbitals. Which of the following general statements concerning orbitals is not true?

a) The quantum number ℓ (from which we get the s, p, d, f... orbital designation) describes the shape of an orbital.
b) Orbitals are filled in the order of increasing energy even if it means occasionally jumping to a higher n and returning to the lower n later in the sequence.
c) At most two electrons may be assigned to a single orbital in accordance with the Pauli exclusion principle.
d) In atoms beyond hydrogen, all orbitals with the same n do not have the same energy but increase with increasing ℓ.
e) Although the 1s orbital has no nodes, the 2s orbital is split into two equal halves by a nodal plane, the 3s into quarters by two nodal planes, etc.

16.13 The Balmer series of emission spectral lines arises from transitions in the hydrogen atom in which the excited electron falls into the first excited state (n=2). Calculate the quantum number, n, of the initial state for the Balmer line at 2.31×10^4 wavenumbers. The Rydberg constant in wavenumbers is 110000 cm^{-1}.

a) 5
b) 8
c) 4
d) 7
e) 6

16.14*A typical electron diffraction experiment is conducted with electrons accelerated through 24600 volts, i.e. with 24600 eV energy. What is the wavelength of the electrons in Angstroms?

(1 electron volt = 1.60×10^{-12} erg; electron mass = 9.11×10^{-28} g; h = 6.63×10^{-27} erg-sec)

(see next)

a) 12.8 Angstroms

b) 7.83×10^{-10} Angstroms

c) 9.39×10^{-10} Angstroms

d) 0.0783 Angstroms

e) 0.626 Angstroms

16.15 Einstein's explanation of the photoelectric effect is expressed by the following equation, where E_k is the kinetic energy of the ejected electron.

$$E_k = \frac{1}{2} mv^2 = h\nu - W$$

The work function (W) for Na is 2.35 eV. What is the maximum wavelength of light which will eject electrons from metallic sodium? (1 eV = 1.60×10^{-12} erg)

a) 3.76×10^{-12} cm

b) 5.29×10^{-5} cm

c) 1.29×10^{-4} cm

d) 2.92×10^{-4} cm

e) 0.1128 cm

CHAPTER 17. PERIODIC PROPERTIES

Introduction

The chemical and physical properties of atoms are determined by
their electronic configurations. The periodic properties of interest here
are the atomic dimensions, ionization energy, electron affinity and
electronegativity. A knowledge of these atomic properties is essential
to understand the phenomena of chemical bonding.

The atomic size is usually obtained from the crystal structure of the
element or of a compound containing the element. The atomic, metallic
and ionic radii are obtained from the unit cell dimensions of atomic,
metallic and ionic crystals, respectively. For example the atomic radius
of Ar, the metallic radius of Pt, and the ionic radii of Na^+ and Cl^- are
related to the crystal structures of Ar, Pt, and NaCl. The covalent radius
is related to the distance between nuclei in molecules for example O_2 or
CO. The van der Waals radius measures how close an atom can come to
another atom to which it is <u>not</u> chemically bonded. The following sketch
illustrates the difference between the covalent radius, r_{cov}, and the
van der Waals radius, r_{vdw}, for Br_2 and BrCl.

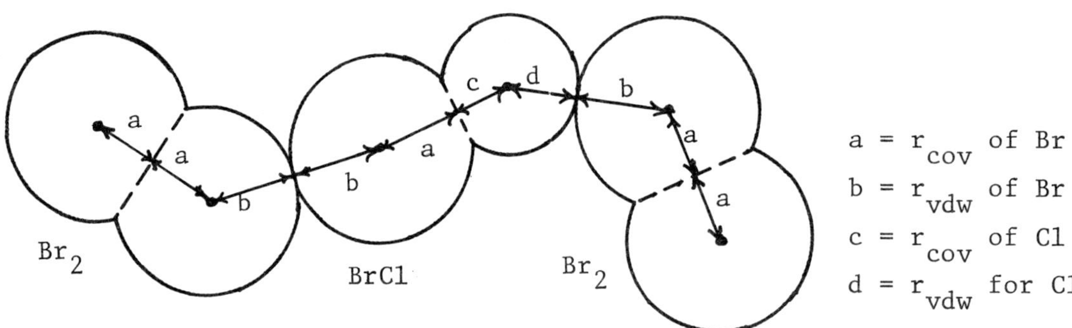

The ionization energy (IE) is the minimum energy required to remove
an electron from an atom in the gaseous state,

$$A(g) \longrightarrow A^+(g) + e^- \qquad\qquad \Delta H = IE$$

This process is always endothermic (IE>0). The energy required to
remove a second electron is called the second ionization energy, and
so on for successive electrons. The electronic affinity (EA) is the
energy absorbed or released when an electron is added to an atom in
the gaseous state.

$$A(g) + e^- \longrightarrow A^-(g) \qquad\qquad \Delta H = EA$$

This process is exothermic (EA<0) for most elements. The EA is endothermic for some metallic elements and for the formation of dianions, A^{2-}, from A^-. Some examples are given below.

Element	Cl_2	O_2	H_2	Be	Zn	O_2	S
Ion Formed	Cl^-	O^-	H^-	Be^-	Zn^-	O^{2-}	S^{2-}
EA (kJ/mole)	-350	-140	-73	59	88	780	590

The electronegativity of an atom (EN) is a dimensionless defined quantity which indicates the tendency of an atom to attract electrons to itself. The elements in the lower left corner of the periodic table have low values (they are electropositive) and those in the upper right have high values (they are electronegative). For example,

Element	Cs	Ca	Fe	C	N	O	F
EN	0.7	1.0	1.8	2.5	3.0	3.5	4.0

Periodic Trends

Horizontal rows in the periodic table are called "periods", vertical columns are called "groups". The trends in periodic properties are indicated in the following table.

Periodic Property	Period (Left to Right)	Group (Top to Bottom)
Size	Decrease	Increase
Ionization Energy	Increase	Decrease
Electronegativity	Increase	Decrease

At the start of each period, for example, from $_{18}$Ar to $_{19}$K, there is a sudden large increase in size. The trends in electronegativity parallel those for ionization energy because the electron affinity makes only a minor contribution. These trends are generalizations and there are several exceptions. There are irregularities in size in the transition metals that can be explained in terms of the orientation in space of the d orbitals and the "lanthanoid contraction". The ionization energies of Be, Mg, and Ca <u>exceed</u> those of B, Al, and Ga, respectively, because more energy is required to remove an electron from the ns^2 configuration than from the ns^2np^1 configuration. The ionization energies of N, P, and S <u>exceed</u> those of O, S, and Si, respectively, because more energy is required to remove an electron from the np^3 (half-filled) configuration than from the np^4 ($\uparrow\downarrow$ $\downarrow$ $\downarrow$) configuration. The same electron configuration argument explains why it takes more energy to ionize Zn, Cd, and Hg, than Ga, In, and Tl, respectively.

136

<u>Example 17.1</u>   Arrange the following elements in order of increasing size:
Ge, Si, Ba, S, Pb

<u>Solution</u>   S < Si   (same period)
                Si < Ge < Pb   (same group)
                Pb < Ba   (same period)

The correct sequence is:

                S < Si < Ge < Pb < Ba

            (smallest)   (largest)

<u>Example 17.2</u> Arrange the following elements in order of increasing
ionization energy: In, C, Al, Rb, O

<u>Solution</u>   Rb < In   (same period)
                In < Al   (same group)
                Al < Si   (same period)
                Si < C   (same group)
                C < O   (same period)

The correct sequence is:

                Rb < In < Al < Si < C < O
            (smallest)   (largest)

Classification of Bonds

Chemical bonds can be classified as primarily ionic or covalent.
Covalent bonds are further divided into polar and nonpolar covalent. The
type of bonding depends on the difference in electronegativity of the
bonded atoms. The greater the difference the more polar the bond. Ionic
bonds are formed in those cases where there is a large difference in
electronegativity. Some examples of compounds in which the bonds are
primarily ionic or polar covalent are given below.

Ionic: CsF, $RbCl$, $BaCl_2$, Rb_2O.

Polar Covalent: CO, N_2O, SO_2, $POCl_3$, S_2Cl_2

The homonuclear diatomic molecules have nonpolar covalent bonding, for
example, H_2, N_2, O_2 and I_2.

<u>Example 17.3</u> Classify the bonding in each of the following compounds:
SiO_2, NaI, SrS, PF_3, Br_2.

<u>Solution</u>   Ionic: NaI and SrS

                Polar Covalent: SiO_2, PF_3

                Nonpolar Covalent: Br_2

Example 17.4 Arrange the following bonds in order of increasing polarity:
C-O, P-H, S-Cl.

Solution The electronegativities are:

Element	H	C	O	P	S	Cl
Electro-negativity	2.1	2.5	3.5	2.1	2.5	3.0

The electronegativity differences are:

$$C-O: \quad |2.5 - 3.5| \quad = 1.0 \qquad \text{(most)}$$

$$P-H: \quad |2.1 - 2.1| \quad = 0 \qquad \text{(least)}$$

$$S-Cl: \quad |2.5 - 3.0| \quad = 0.5 \qquad \text{(intermed.)}$$

The order of increasing polarity is P-H < S-Cl < C-O

Oxidation Number

Most elements gain or lose the number of electrons required to achieve a noble gas configuration. Several examples follow:

[He]: H^+, Li^+, Be^{2+}, B^{3+}, C^{4+}, N^{5+}

[Ne]: N^{3-}, O^{2-}, F^-, Na^+, Mg^{2+}, Al^{3+}, Si^{4+}, P^{5+}, S^{6+}, Cl^{7+}

[Ar]: P^{3-}, S^{2-}, Cl^-, K^+, Ca^{2+}, Sc^{3+}, Ti^{4+}, V^{5+}, Cr^{6+}, Mn^{7+}

Several elements prefer the ns^2 configurations, for example

$$Sb^{3+} \qquad [Kr]5s^2 4d^{10}$$
$$Pb^{2+} \qquad [Xe]6s^2 4f^{14} 5d^{10}$$

Most of the transition metals have stable +2 and/or +3 oxidation numbers. Most of the lanthanoids and actinoids have stable +3 oxidation numbers.

Example 17.5 Predict the oxidation state of the following elements:
Se, Y, I, W

Solution Noble gas configurations are achieved by
Se^{2-}, Y^{3+}, I^-, and W^{6+}

Questions

17.1 Atom A has 2 valence electrons and atom B has 5 valence electrons. The formula expected for an ionic compound of A and B is:

a) A_2B

b) A_3B_2

c) A_2B_3

d) AB_3

e) AB_2

17.2 Which one of the following trends in atomic and ionic radii is <u>not</u> correct?

a) $I^- < Br^- < Cl^- < F^-$
b) $As < Ga < Ca < K$
c) $Cl < As < Pb < Cs$
d) $Al^{3+} < Na^+ < O^{2-} < N^{3-}$
e) $N < P < As < Sb$

17.3 Which of the following trends in first ionization energy is <u>not</u> correct?

a) $Pb < Sn < Si < C$
b) $K < Ca < Ge < As$
c) $Te < Se < S < O$
d) $Li < Be < C < O$
e) $F < P < Al < Sr$

17.4 For which of the following reactions is the enthalpy change correctly indicated?

1. $F + e^- \longrightarrow F^-$ exothermic

2. $He + e^- \longrightarrow He^-$ endothermic

3. $Se^- + e^- \longrightarrow Se^{2-}$ endothermic

a) 1
b) 1 and 2
c) 2 and 3
d) 1 and 3
e) All of them

17.5 The first ionization energy of Se is 941 kJ/mole. Which of the values below would you predict for the first ionization energy of Br?

a) -941 kJ/mole
b) 745 kJ/mole
c) 941 kJ/mole
d) 1140 kJ/mole
e) 2300 kJ/mole

17.6 *The first ionization energy of neon is 22 eV. Which of the
following values would you predict for the second ionization
energy of neon?

a) 23 eV
b) 10 eV
c) 22 eV
d) 41 eV
e) 88 eV

17.7 *The second ionization energy of oxygen is 35 eV. Which of the
following values would you predict for the second ionization
energy of fluorine?

a) 12 eV
b) 20 eV
c) 35 eV
d) 61 eV
e) 75 eV

17.8 Given the following table of Pauling electronegativity values

Atom	Be	In	Pb	Bi	B	H	Se	C	Br	N	O	F
Electro-negativity	1.5	1.7	1.8	1.9	2.0	2.1	2.4	2.5	2.8	3.0	3.5	4.0

Which one of the following sequences would you predict to have
decreasing ionic character of the bond indicated?

a) In-Se > In-C > In-O > In-F
b) In-Se > In-F > In-O > In-C
c) In-F > In-O > In-C > In-Se
d) In-Se > In-O > In-F > In-C
e) In-F > In-C > In-O > In-Se

17.9 Which one of the following oxidation states would <u>not</u> be
expected?

a) Tl(II)
b) Ar(0)
c) B(III)
d) Al(III)
e) All of these oxidation states would be expected

17.10 Which one of the following sets of oxidation numbers would you
predict for these elements:

	Pb	V	Nb	Sr	La
a)	5	6	7	4	5
b)	4	5	5	2	3
c)	5	5	7	4	3
d)	4	7	6	2	3
e)	None of the above.				

17.11* Identify element X given the following information:

Property	Ga	Bi	X
First ionization energy (kJ/mole)	577	774	832
Covalent radius (Angstroms)	1.26	1.46	1.38
Pauling electronegativity	1.6	1.9	1.9
Most stable non-zero oxidation state	3	3	3

Hint: X is a representative element with atomic number less than 82.

a) B
b) Ti
c) C
d) Tl
e) Sb

17.12 Indicate whether the bonding in the following compounds is primarily ionic (I), polar covalent (PC), or non-polar covalent (NC).

	$SrCl_2$	NO_2	Br_2	PH_3	CO
a)	I	PC	PC	PC	NC
b)	I	PC	NC	PC	PC
c)	PC	I	NC	NC	PC
d)	I	NC	NC	PC	NC
e)	PC	PC	PC	NC	PC

17.13* The observed dipole moment for FBr is 1.29 debyes, and FBr bond distance is 1.76 Angstroms. Estimate the percent ionic character in the FBr bond. (Electron charge, e = 4.80 x 10^{-10} esu; 1 debye = 1.0 x 10^{-18} esu cm.)

a) 15.3%
b) 73.3%
c) 13.6%
d) 28.4%
e) 26.9%

17.14 Given the following hypothetical ionization energy and electron affinity values, defined as follows:

$$X \rightarrow X^+ + e^- \qquad \Delta H = IE$$
$$X + e^- \rightarrow X^- \qquad \Delta H = EA$$

Atom	IE(eV)	EA(eV)
X	10.5	−3.6
Y	15.0	−2.0
Z	14.5	−0.6

(see next)

Which of the following sequences indicates the correct order
of increasing nonmetallic behavior?

a) Z < Y < X
b) Y < Z < X
c) X < Y < Z
d) X < Z < Y
e) Y < X < Z

17.15*Given the following hypothetical ionization energy and electron
affinity values, defined as follows:

$$X \rightarrow X^+ + e^- \qquad \Delta H = IE$$
$$X + e^- \rightarrow X^- \qquad \Delta H = EA$$

Atom	IE(eV)	EA(eV)
X	10.0	−2.4
Y	9.5	−2.2
Z	17.0	−2.2

Which of the following sequences indicates the correct order
of increasing covalent character?

a) XY < XZ < YZ
b) XZ < XY < YZ
c) XZ < YZ < XY
d) XY < YZ < XZ
e) YZ < XZ < XY

CHAPTER 18. CHEMICAL BONDING

Lewis Structures

Lewis structures are electron-dot representations of the electronic configurations of atoms, ions, and molecules. All valence electrons are included. Chemical bonds are shared pairs of electrons and are represented by a line. Unshared electrons are represented by dots. The total number of electrons around an atom including both electrons in a shared pair, should equal the outer electronic configuration of the noble gases. This is the octet rule. Some examples of Lewis structures are given below.

Molecule	F_2	O_2	N_2	H_2O	C_2H_4
Number of Valence Electrons	14	12	10	8	12
Lewis Structure	$:\!\ddot{F}\!-\!\ddot{F}\!:$	$:\!\ddot{O}\!=\!\ddot{O}\!:$	$:\!N\!\equiv\!N\!:$	$H\!-\!\ddot{O}\!-\!H$	

Ethylene (C_2H_4):

$$\begin{matrix} H & & & H \\ & \diagdown & & \diagup \\ & & C=C & \\ & \diagup & & \diagdown \\ H & & & H \end{matrix}$$

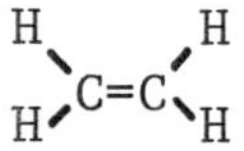

Note that hydrogen needs only two electrons to achieve the stable He configuration. Sometimes more than one Lewis structure is possible. In these cases the molecule or ion is represented as a "resonance hybrid" of two or more Lewis structures, for example,

NO_2^- (5 + 2 x 6 + 1 = 18 valence electrons)

$$:\!\ddot{O}\!-\!\ddot{N}\!=\!\ddot{O}\!: \longleftrightarrow :\!\ddot{O}\!=\!\ddot{N}\!-\!\ddot{O}\!:$$

The octet rule is frequently violated for compounds containing elements from the third or higher periods. For example,

Molecule	PF_5	SF_6	IF_7
Number of Valence Electrons	40	48	56
Lewis Structure			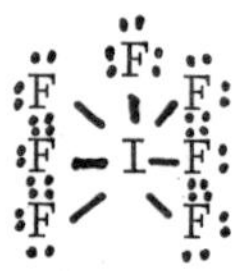

In these Lewis structures the central atoms are surrounded by 10 (P), 12 (S), and 14 (I) electrons. The electronic configurations $ns^2np^6(n-1)d^{10}$ can account for 18 valence electrons, so this "violation" of the octet rule is not a serious matter. However the octet rule must be followed for the second row elements C, N, O, and F.

The formal charge is a useful device for understanding molecules with nonuniform distributions of charge. The formal charge on an atom in a molecule or ion is the number of valence electrons minus the number of unshared electrons minus one-half the number of shared electrons. For example,

$$CO \quad (4 + 6 = 10 \text{ valence electrons}) \quad :C\equiv O:$$
$$-1 \ +1$$

The formal charge on C is $(4 - 2 - 6/2 = -1)$. The formal charge on O is $(6 - 2 - 6/2 = +1)$. The sum of the formal charges on each atom must be the same as the charge of the molecule or ion. A Lewis structure with a formal charge that exceeds 2 in absolute value is not a reasonable structure. Neither is one which has formal charges of the same sign on adjacent atoms. For example,

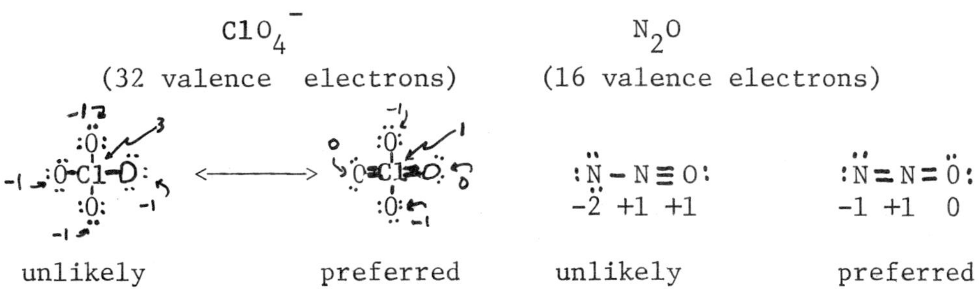

ClO_4^-		N_2O	
(32 valence electrons)		(16 valence electrons)	
unlikely	preferred	unlikely	preferred

<u>Example 18.1</u> Draw Lewis structures of the following molecules and ions and assign the formal charges if other than zero:

a) CO_2 b) H_2O_2 c) CN^- d) HCO_2^-

<u>Solution</u>   a) $:O=C=O:$ b) $H-\ddot{O}-\ddot{O}-H$

c) $:C\equiv N:$ d) $H-C$ structures

-1

Molecular Geometry – VSEPR Theory

The Valence Shell Electron Pair Repulsion (VSEPR) theory is useful for predicting the geometry of a molecule or ion. If a molecule or ion can be represented by the formula, AB_n, with no unshared pairs of electrons on the central atom A, then the following geometries are predicted:

n	Geometry	<BAB	Examples
2	Linear	180°	CO_2, BeH_2, $HgCl_2$
3	Trigonal Planar	120°	BCl_3, SO_3, NO_3^-
4	Tetrahedral	109.5°	CH_4, SO_4^{2-}, ClO_4^-
5	Trigonal Bipyramidal	120° & 90°	PF_5, $SbCl_5$, $CuCl_5^{3-}$
6	Octahedral	90°	SF_6, PF_6^-, $CrCl_6^{3-}$

If the central atom has one or more unshared pairs of electrons, it can be represented by the formula, AB_nE_m, where E stands for an unshared pair of electrons. The electron pair repulsive interactions follow the sequence

$$E\text{-}E \;>\; E\text{-}B \;>\; B\text{-}B$$

That is, pairs of unshared electrons repel more strongly than pairs of bonding electrons. Molecules and ions with formula AB_nE_m will orient their shared and unshared pairs of electrons according to the geometries given above, but the actual molecular shapes will be distorted due to the stronger repulsive E-E and E-B interactions. These rules are consistent with the following structures.

Example	NH_3	H_2O	XeF_2	XeF_4
Formula	AB_3E	AB_2E_2	AB_2E_3	AB_4E_2
Geometry	Trigonal Pyramidal	Bent	Linear	Square Planar
<BAB	107°	104.5°	180°	90°
Structure	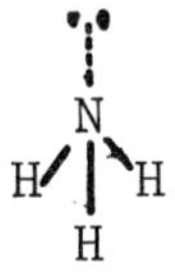	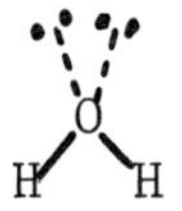	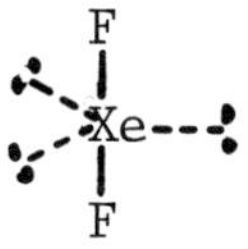	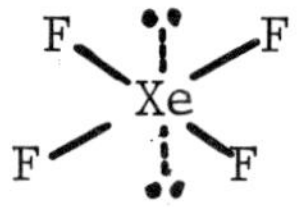

The CH_4 ($\angle HCH = 109.5°$), NH_3 ($\angle HNH = 107°$), H_2O ($\angle HOH = 104.5°$) series illustrates the electron repulsion trend, B-B < E-B < E-E.

<u>Example 18.2</u> Predict the molecular geometry of the following molecules and ions:

a) MnO_4^- b) ClF_3 c) SF_4 d) BrF_5

<u>Solution</u>

a) MnO_4^- is AB_4, $\therefore$ tetrahedral

b) ClF_3 has the Lewis structure

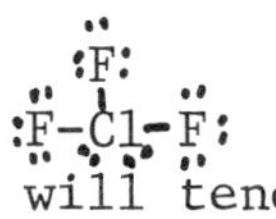

or AB_3E_2. The electron pairs will tend to assume a trigonal bipyramidal structure. The unshared pairs will prefer to be in the trigonal plane (equatorial) rather than on the axial positions. Therefore the molecules will be "T-shaped".

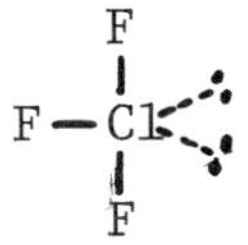

c) SF_4 has the Lewis structure

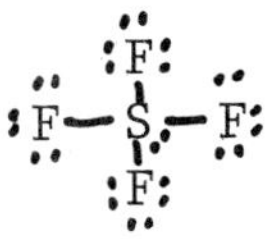

or AB_4E. The unshared pair will prefer the equatorial position. Therefore the molecule will be "see-saw" shaped,

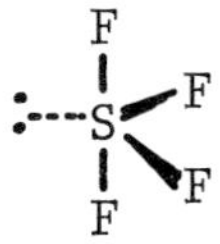

d) BrF_5 has the Lewis structure, or AB_5E

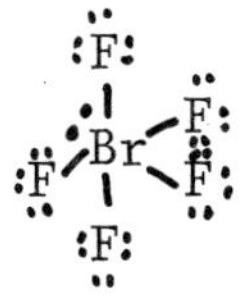

Therefore the electron pairs will point to the corners of an octahedron with one corner unoccupied; the structure is square pyramidal.

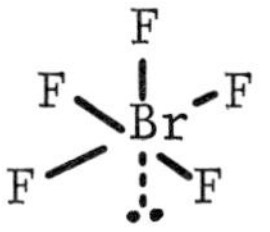

Hybridization

The geometry and bonding of many covalent compounds can be qualitatively described using a set of "hybrid" orbitals. If the molecule has formula AB_n or AB_mE_{n-m}, then the central atom forms a set of n hybrid orbitals. The sigma bonding in the molecule arises from the overlap between the hybrid orbitals of A and atomic (or hybrid) orbitals on B. Several hybridization schemes are listed below (the molecule is in the x-y plane).

n	Observed Geometry	Hybrid Orbitals	Atomic Orbitals	Example
2	Linear	sp	$s + p_x$	BeH_2, CO_2
3	Trigonal Planar	sp^2	$s + p_x + p_y$	CO_3^{2-}, BCl_3
4	Tetrahedral	sp^3	$s + p_x + p_y + p_z$	$SiCl_4$, PO_4^{3-}
4	Square Planar	dsp^2	$d_{x^2-y^2} + s + p_x + p_y$	$PtCl_4^{2-}$, $Au(CN)_4^-$
5	Trigonal Bipyramidal	dsp^3	$d_{z^2} + s + p_x + p_y + p_z$	PCl_5, $SbCl_5$
6	Octahedral	d^2sp^3	$d_{x^2-y^2} + d_{z^2} + s + p_x + p_y + p_z$	SF_6, $Co(NH_3)_6^{3+}$

Consider, for example the octahedral molecule SF_6. The Lewis structure is:

$$\ddot{\underset{\cdot\cdot}{:F:}}\ \overset{\overset{\cdot\cdot}{:F:}}{\underset{\underset{:\ddot{F}:}{S}}{\diagup\vert\diagdown}}\ :\ddot{F}:$$

The observed geometry can be explained by the hybridization of six of the sulfur orbitals (3s, $3p_x$, $3p_y$, $3p_z$, $3d_{x^2-y^2}$, and $3d_{z^2}$).

$$S^* \qquad \xrightarrow{\ 6:\ddot{F}\downarrow\ } \qquad SF_6$$

$$\begin{array}{ccc}
\underline{\uparrow} & \underline{\uparrow}\ \underline{\uparrow}\ \underline{\uparrow} & \underline{\uparrow}\ \underline{\uparrow}\ _\ _\ _ \\
3s & 3p & 3d
\end{array}
\qquad
\begin{array}{c}
\underline{\uparrow\downarrow}\ \underline{\uparrow\downarrow}\ \underline{\uparrow\downarrow}\ \underline{\uparrow\downarrow}\ \underline{\uparrow\downarrow} \\
d^2sp^3
\end{array}$$

Here S^* represents an electronically excited state of the sulfur atom in which the six atomic orbitals to be hybridized are singly occupied.

<u>Example 18.3</u>   Use hybrid orbitals to explain the bonding in C_2H_2, C_2H_4, and C_2H_6:

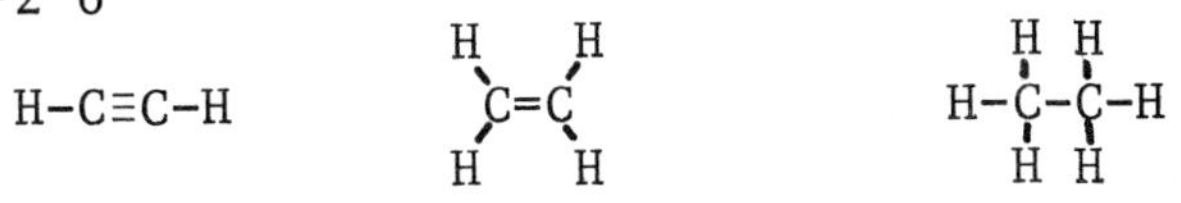

$$H-C\equiv C-H \qquad\qquad \begin{matrix} H & & H \\ & C=C & \\ H & & H \end{matrix} \qquad\qquad H-\overset{\displaystyle H}{\underset{\displaystyle H}{C}}-\overset{\displaystyle H}{\underset{\displaystyle H}{C}}-H$$

linear   planar   tetrahedral carbons

<u>Solution</u>   In ethane, C_2H_6, each carbon has 4 sigma (σ) bonds (AB_4). The C-H bonds result from the overlap between sp^3 hybrid orbitals on C and 1s atomic orbitals on H's. The C-C bond results from the overlap of sp^3 orbitals.

In ethylene, C_2H_4, each carbon has 3 sigma bonds (AB_3). The C-H bonds and the C-C sigma bond utilize the sp^2 hybrid orbitals on C and the 1s atomic orbital of H. The second bond in ethylene, a pi(π) bond, results from the overlap between the unhybridized p_z atomic orbitals on the C atoms. The p_z orbitals are perpendicular to the plane of the sp^2 hybrid orbitals. The six atoms in C_2H_4 must be coplanar for the two p_z atomic orbitals to overlap.

In acetylene, C_2H_2, each carbon has 2 σ bonds (AB_2). The C-H bonds and the C-C σ bond utilize the sp hybrid orbitals on C and the 1s atomic orbital on H. The two π bonds result from the overlap between the unhybridized p_y and p_z orbitals on the carbon atoms.

<u>Molecular Orbital Theory</u>

The bond length, bond dissociation energy and magnetic behavior of diatomic molecules are molecular properties that are easily correlated using molecular orbital theory. An energy level diagram of molecular orbitals for a diatomic molecule of second period atoms is shown to the right. Here x is the molecular axis. Each molecular orbital can accommodate two electrons. The asterisk indicates an antibonding orbital. The bond order is the number of electrons in bonding orbitals minus the number of antibonding orbitals, divided by two. A molecule or ion is paramagnetic if it contains one or more unpaired electrons; otherwise it is diamagnetic.

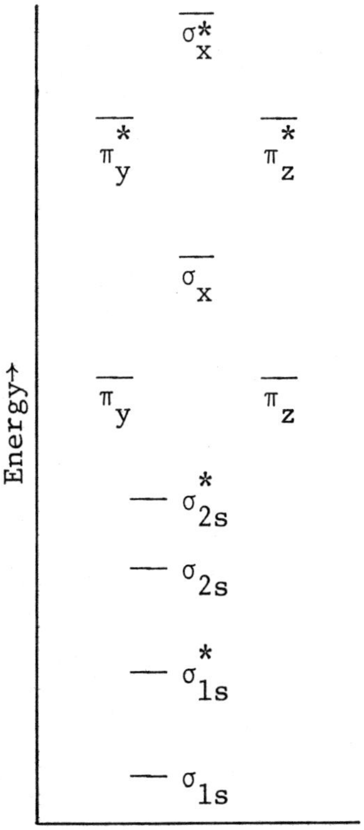

Energy →

σ_x^*

π_y^* π_z^*

σ_x

π_y π_z

σ_{2s}^*

σ_{2s}

σ_{1s}^*

σ_{1s}

These concepts are illustrated by the following examples.

Molecule	No. of electrons	Molecular Orbital Configuration	Bond Order	Magnetism
H_2	2	$(\sigma_{1s})^2$	1	Diamagnetic
B_2	10	$(\sigma_{1s})^2(\sigma_{1s}^*)^2(\sigma_{2s})^2(\sigma_{2s}^*)^2(\pi_y)^1(\pi_z)^1$	1	Paramagnetic
CO	14	$\ldots.(\pi_y)^2(\pi_z)^2(\sigma_x)^2$	3	Diamagnetic
O_2	16	$\ldots\ldots.(\pi_y^*)^1(\pi_z^*)^1$	2	Paramagnetic
O_2^{2-}	18	$\ldots\ldots.(\pi_y^*)^2(\pi_z^*)^2$	1	Diamagnetic

The first two molecular orbitals are frequently omitted (or abbreviated KK) when considering molecules with more than 4 electrons.

<u>Example 18.4</u>   Predict the bond order and magnetic behavior (paramagnetic or diamagnetic) for the following; NO^+, O_2^-, CN^-

<u>Solution</u> NO^+ and CN^- are isoelectronic with 10 valence electrons:

$$(\sigma_{2s})^2(\sigma_{2s}^*)^2(\pi_y)^2(\pi_z)^2(\sigma_x)^2$$

Both have bond order three and are diamagnetic.

O_2^-, the superoxide ion has 13 electrons:

$$(\sigma_{2s})^2(\sigma_{2s}^*)^2(\pi_y)^2(\pi_z)^2(\sigma_x)^2(\pi_{y,z}^*)^3$$

The bond order is (8-5)/2 or 1 1/2. The ion is paramagnetic

Questions

18.1 In which of the following is the octet rule violated?

    1. O_2^{2-} 2. SO_2 3. BrF_5

    a) 3
    b) 1 and 3
    c) 2 and 3
    d) All of them
    e) None of them

18.2 Which one of the following best describes the Lewis structure for
OCN^-?

    a)   one double bond, one or more single bonds, no unshared pairs of
electrons
    b)   one double bond, one or more single bonds, one or more unshared
pairs of electrons
    c)   three unshared pairs of electrons
    d)   one triple bond, one or more single bonds, one or more unshared
pairs of electrons
    e)   all single bonds, two unshared pairs of electrons

18.3* Draw the Lewis structure for each of the following molecules or ions.
(In case of resonance, one form is enough.) Deduce the formal
charges if any, on each atom. Which of the following assignments is
<u>wrong</u>?

    a) CO −1 on C
    b) CO_2 0 on C
    c) N_2O +1 on central N
    d) NO_2^- +1 on N
    e) NO_3^- +1 on N

18.4* In which of the following cases is the Lewis structure incomplete
unless resonance structures are included?

    1. HCO_2^- 2. NO_2^- 3. O_3

    a)   1 and 2
    b)   2 and 3
    c)   1 and 3
    d)   All of them
    e)   None of them

18.5 Which of the following molecules would be expected to have a dipole
moment?

    1. HI   2. $CHCl_3$ 3. SO_2F_2

    a)   1 and 2
    b)   1 and 3
    c)   2 and 3
    d)   All of them
    e)   None of them

150

18.6*Which of the following Lewis structures best represents the cyanogen
    molecule, C_2N_2?

    1. $:N=C=C=N:$ 2. $:N\equiv C-C\equiv N:$ 3. $:N-C\equiv C-N:$

    a) 1
    b) 2
    c) 3
    d) All of them in resonance
    e) 2 and 3 in resonance

18.7 Which of the following trends in bond angles is consistent with
    VSEPR theory?

                1. ✗ HNH: $NH_2^- < NH_3 < NH_4^+$
                2. ✗ OSO: $SO_4^{2-} < SO_2 < SO_3$
                3. ✗ ONO: $NO_2^- < NO_2 < NO_2^+$

    a) 1
    b) 1 and 2
    c) 2 and 3
    d) 1 and 3
    e) All of them

18.8*Which one of the following is <u>not</u> the geometry of the indicated
    molecule or ion based on the <u>valence shell electron pair repulsion</u>
    theory?

    a) BF_4^- =   regular tetrahedron

    b) SF_6 =   octahedral

    c) AlF_3 =   pyramidal

    d) ICl_4^- =   square planar

    e) $PbCl_2$ =   planar, bent

18.9 Which one of the following best describes the Cl-Pb-Cl bond angle
    in $PbCl_6^{2-}$?

    a) 120
    b) 109.5
    c) 109.5 < angle < 120
    d) 90 < angle < 109.5
    e) 90

18.10   The following statements are concerned with the concept of hybrid
orbitals in chemical bonding. Which one is wrong?

a) $BeCl_2$ is a linear molecule. The σ bonds between Be and Cl can
be considered to be the overlap between 3 equivalent sp^2 hybrid
orbitals on the Be atom and the appropriate wave function for
the chlorine atoms.

b)   Atomic orbitals not used to compose a hybrid set may be used in
forming π bonds to neighboring atoms.

c)   A σ bond using a hydrid orbital on one of the bonded atoms has
cylindrical symmetry with respect to the bond axis regardless
of the composition of the hybrid.

d)   A given element may utilize one of several different hybridized
orbital sets depending on the compound in which it exists.

e)   If a central atom forms N σ bonds and has M lone pairs of
electrons then the appropriate set of hybrid orbitals is obtained
by mixing (N + M) atomic orbitals.

18.11   Which one of the following gives the correct hybridization for the
central atom in the molecule or ion?

	GaF_3	NCO^-	HCN
a)	sp^2	sp	sp
b)	sp^2	sp^3	sp
c)	sp^3	sp^2	sp^2
d)	sp^2	sp^3	sp^3
e)	None of the above		

18.12   Use molecular orbital theory to predict which one of the following
diatomic molecules has the greatest bond dissociation energy.

a)   NF
b) H_2
c) Li_2
d) Cl_2
e)   All of the above species have the same bond dissociation energy.

18.13* Use molecular orbital theory to predict which one of the following
diatomic molecules or ions has the greatest bond length.

a) BO
b) NO^+
c) CO^+
d) O_2
e)   All of the above species have the same bond length.

152

18.14 How many net sigma and pi bonds are there in NF?

$$\begin{array}{ccc} & \sigma & \pi \\ \text{a)} & 1 & 1 \\ \text{b)} & 2 & 0 \\ \text{c)} & 0 & 2 \\ \text{d)} & 1 & 2 \end{array}$$

    e)   None of the above

18.15 Which of the following molecules or ions is (are) paramagnetic?

a) CO
b) CN^-
c) NO^+
d) O_2^{2-}
e)   All are diamagnetic

$$\text{CHAPTER 19. ACIDS AND BASES}$$

Introduction

Eight classes of acids and bases are considered here:

1. __Strong Acids__ $HX \xrightarrow{100\%} H^+ + X^-$

2. __Strong Bases__ $M(OH)_n \xrightarrow{100\%} M^{n+} + nOH^-$

3. __Weak Acids__ $HA \rightleftharpoons H^+ + A^-$ $K_a = \dfrac{[H^+][A^-]}{[HA]}$

4. __Weak Bases__ $B + H_2O \rightleftharpoons BH^+ + OH^-$ $K_b = \dfrac{[BH^+][OH^-]}{[B]}$

5. __Basic Salts__ $MA \xrightarrow{100\%} M^+ + A^-$

            $A^- + H_2O \rightleftharpoons HA + OH^-$ $K_h = \dfrac{[HA][OH^-]}{[A^-]} = \dfrac{K_w}{K_a}$

6. __Acidic Salts__ $BHX \xrightarrow{100\%} BH^+ + X^-$

            $BH^+ \rightleftharpoons B + H^+$ $K_a \text{ or } K_h = \dfrac{[B][H^+]}{[BH^+]} = \dfrac{K_w}{K_b}$

7. __Metal Ions__ $MX_n \xrightarrow{100\%} M^{n+} + nX^-$

            $M^{n+} + H_2O \rightleftharpoons MOH^{(n-1)+} + H^+$ $K_h = \dfrac{[MOH^{(n-1)+}][H^+]}{[M^{n+}]}$

8. __Polyprotic Acids__ (H_nA) For $n = 3$:

            $H_3A \rightleftharpoons H^+ + H_2A^-$ $K_1 = \dfrac{[H^+][H_2A^-]}{[H_3A]}$

            $H_2A^- \rightleftharpoons H^+ + HA^{2-}$ $K_2 = \dfrac{[H^+][HA^{2-}]}{[H_2A^-]}$

            $HA^{2-} \rightleftharpoons H^+ + A^{3-}$ $K_3 = \dfrac{[H^+][A^{3-}]}{[HA^{2-}]}$

Here the proton, H^+, is written for convenience. Reactions of H^+ actually involve the hydronium ion, H_3O^+. For example,

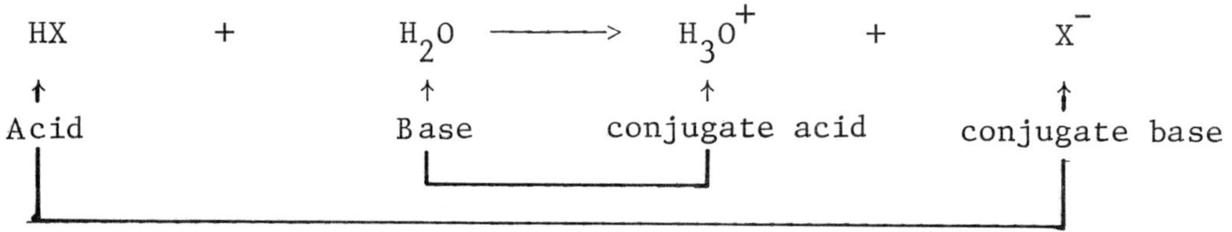

The difference between an acid and its conjugate base is the proton. For example,

Acid	HCl	HNO_2	HCOOH	H_3PO_4	$H_2PO_4^-$	HPO_4^{2-}	H_2O
Conjugate Base	Cl^-	NO_2^-	$HCOO^-$	$H_2PO_4^-$	HPO_4^{2-}	PO_4^{3-}	OH^-

Base	NH_3	OH^-	O^{2-}	H_2O	C_5H_5N	$(CH_3)_3N$	S^{2-}
Conjugate Acid	NH_4^+	H_2O	OH^-	H_3O^+	$C_5H_5NH^+$	$(CH_3)_3NH^+$	HS^-

Note that water is both an acid and a base.

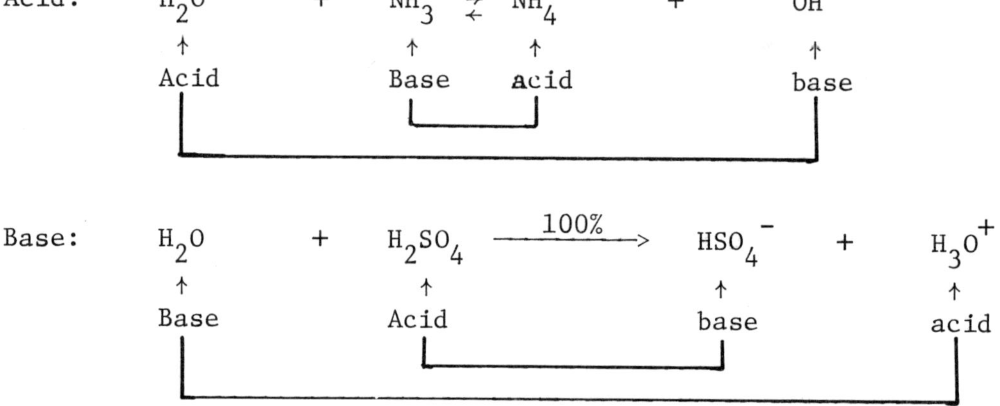

The autoionization (autoprotolysis) reaction of H_2O is

$$H_2O \rightleftarrows H^+ + OH^- \qquad\qquad K_w = [H^+][OH^-]$$

At 25°C, $K_w = 1.0 \times 10^{-14}$.   In pure H_2O,

$$[H^+] = [OH^-] = \sqrt{K_w} = 1.00 \times 10^{-7}$$

The equilibrium constants (K_a, K_b, K_w, etc). The $[H^+]$, and the $[OH^-]$ are often very small numbers. A logarithmetic scale is used to conveniently represent these numbers.

$$pH \equiv -\log [H^+] \qquad\qquad pOH \equiv \log [OH^-]$$

$$pK \equiv -\log K$$

For neutral water,

$$pH = pOH = -\log (1.00 \times 10^{-7}) = 7.00$$

The following equations are valid for any aqueous solution (acidic, neutral, or basic).

$$[H^+][OH^-] = K_w = 1.00 \times 10^{-14} \qquad pH + pOH = 14.00$$

Strong Acids and Bases

Strong acids and bases are completely dissociated in water,

$$HX \xrightarrow{\ 100\%\ } H^+ + X^-$$

$$M(OH)_n \xrightarrow{\ 100\%\ } M^{n-} + nOH^-$$

There are not many strong acids. The common laboratory reagents HCl, H_2SO_4, HNO_3 and $HClO_4$ are all strong acids. Also included in this category are HBr, HI, and a few other examples. Many metal hydroxides are strong bases, for example,

$$NaOH, \ KOH, \ RbOH, \ CsOH, \ Ca(OH)_2, \ Sr(OH)_2 \text{ and } Ba(OH)_2.$$

Most of the other metal hydroxides, for example, $Mg(OH)_2$ and $Al(OH)_3$ are very insoluble and so contribute little or no OH^- to solution.

<u>Example 19.1</u>   Consider a solution of 4.35 grams of $Ba(OH)_2$ in 475 ml. of solution. Calculate the pH and the pOH of this solution. The molecular weight of $Ba(OH)_2$ is 171 g/mole.

<u>Solution</u> $\qquad\qquad Ba(OH)_2(s) \xrightarrow{\ 100\%\ } Ba^{2+} + 2OH^-$

$$[OH^-] = \frac{2 \text{ moles } OH^-}{1 \text{ mole } Ba(OH)_2} \ x \ \frac{1 \text{ mole } Ba(OH)_2}{171 \text{ g}} \ x \ \frac{4.35 \text{ g}}{0.475\ell} = 0.107 \ \underline{M}$$

$$[H^+] = K_w/[OH^-] = 1.00 \times 10^{-14}/0.107 = 9.34 \times 10^{-14} \ \underline{M}$$

$$pH = -\log (9.34 \times 10^{-14}) = 13.03$$

$$pOH = -\log(0.107) = 0.970 = 14.00 - pH$$

<u>Example 19.2</u>   Calculate the $[H^+]$ and $[OH^-]$ in a solution of HCl for which the pH is 5.57.

<u>Solution</u>

$$pH = -\log [H^+]$$

$$[H^+] = 10^{-pH} = 10^{-5.57} = \text{antilog}(-5.57) = 2.7 \times 10^{-6}$$

$$[OH^-] = K_w/[H^+] = 1.0 \times 10^{-14}/2.7 \times 10^{-6} = 3.7 \times 10^{-9}$$

Alternatively, $[OH^-] = 10^{-pOH} = 10^{pH-14} = 10^{-8.43} = 3.7 \times 10^{-9}$

<u>Weak Acids</u>

Most acids are only partially dissociated in solution,

$$HA \rightleftharpoons H^+ + A^- \qquad\qquad K_a = \frac{[H^+][A^-]}{[HA]}$$

The larger the magnitude of K_a, the acid dissociation constant, the stronger the acid. Some examples of weak acids and the values of pK_a at 25°C are given below.

Name	nitrous	hydrofluoric	formic	acetic	hypochlorous	hydrocyanic
HA	HNO_2	HF	HCOOH	CH_3COOH	HOCl	HCN
pK_a	3.34	3.45	3.75	4.75	7.52	9.31

When the concentration of a weak acid solution is stated, as on a label, the initial concentration, $[HA]_0$, rather than the equilibrium concentration is implied. The addition of an appreciable amount ($>10^{-5}$ $\underline{M}$) of a weak acid will suppress the dissociation of H_2O:

$$HA \rightleftharpoons H^+ + A^- \qquad\qquad K_a \gg K_w$$

$$H_2O \longleftarrow H^+ + OH^- \qquad\qquad K_w = 1.0 \times 10^{-14}$$

<u>Example 19.3</u>   Calculate the pH of a 0.10 $\underline{M}$ HCN solution. Show that H_2O dissociation does not contribute significantly to the $[H^+]$. K_a for HCN is 4.9×10^{-10}.

<u>Solution</u>   Assume HCN is the only source of H^+. Let $[H^+] = x$, then at equilibrium, $[CN^-]$ is also x, and [HCN] is 0.1-x, as indicated below,

$$HCN \; \underset{\leftarrow}{\rightarrow} \; H^+ \; + \; CN^-$$

$$0.1-x \quad\quad x \quad\quad x$$

$$K_a = [H^+][CN^-]/[HCN] = (x)(x)/(0.1-x) = x^2/(0.1-x)$$

Rather than solving this quadratic equation for x, assume that since K_a is such a small number, x<<0.1. Then

$$\frac{x^2}{0.1-x} \;\simeq\; \frac{x^2}{0.1} \;=\; 4.9 \times 10^{-10}$$

$$x \simeq \sqrt{0.1 \times 4.9 \times 10^{-10}} = 7.0 \times 10^{-6} \; \underline{M} = [H^+] = [CN^-]$$

The assumption that x can be neglected compared to 0.1 is excellent, 0.000001 << 0.1. This assumption will be considered valid as long as x is less than 5% of initial concentration of the weak acid.

$$pH = -\log [H^+] = -\log (7.0 \times 10^{-6}) = 5.15.$$

$$[OH^-] = K_w/[H^+] = 1.0 \times 10^{-14}/7.0 \times 10^{-6} = 1.4 \times 10^{-9} \; \underline{M}$$

Since H_2O contributes H^+ and OH^- in equal amounts, the $[H^+]$ from water dissociation is also 1.4×10^{-9} $\underline{M}$. This is indeed negligible compared to the $[H^+]$ from HCN, 7.0×10^{-6} $\underline{M}$.

Example 19.4   A 0.073 $\underline{M}$ solution of a weak acid has a pH of 3.11. Calculate the dissociation constant of this weak acid and the percent dissociation at this concentration.

Solution $\quad\quad\quad HA \; \underset{\leftarrow}{\rightarrow} \; H^+ \; + \; A^-$

$$[H^+] = [A^-] = 10^{-3.11} = 7.8 \times 10^{-4}$$

$$[HA] = 0.073 - 7.8 \times 10^{-4} = 0.072$$

$$K_a = \frac{[H^+][A^-]}{[HA]} = \frac{(7.8 \times 10^{-4})^2}{0.072} = 8.4 \times 10^{-6}$$

The percent dissociation (% D) is

$$\% \; D = \frac{100 \times [H^+]}{[HA]_o} = \frac{100 \times 7.8 \times 10^{-4}}{0.073} = 1.1\%$$

Here $[HA]_o$ is the labelled concentration of weak acid, which does not take into account dissociation.

Weak Bases

The reaction of a weak base with water is

$$\underset{\text{Base}}{B} + \underset{\text{Acid}}{H_2O} \rightleftarrows \underset{\text{base}}{OH^-} + \underset{\text{acid}}{BH^+} \qquad K_b = \frac{[BH^+][OH^-]}{[B]}$$

This reaction is known as a hydrolysis (water breaking) reaction, or as the ionization of a weak base. Some examples of weak bases are:

B	CH_3NH_2	$(CH_3)_3N$	NH_3	N_2H_4	NH_2OH	C_5H_5N	$C_6H_5NH_2$
Name	methyl-amine	trimethyl-amine	ammonia	hydrazine	hydroxyl-amine	pyridine	aniline
pK_b	3.34	4.19	4.75	6.01	7.97	8.75	9.37

<u>Example 19.5</u>   Calculate the pH and percent hydrolysis (or percent ionization) of a 0.085 $\underline{M}$ solution of NH_2OH. K_b for NH_2OH is 1.1×10^{-8}.

<u>Solution</u> Let x be the $[OH^-]$ then

$$NH_2OH + H_2O \rightleftarrows NH_3OH^+ + OH^-$$

$$0.085-x \qquad\qquad x \qquad\qquad x$$

$$K_b = \frac{[NH_3OH^+][OH^-]}{[NH_2OH]} = \frac{x^2}{0.085-x} \simeq \frac{x^2}{0.085} = 1.1 \times 10^{-8}$$

$$x = [OH^-] \simeq \sqrt{0.085 \times 1.1 \times 10^{-8}} = 3.1 \times 10^{-5} \underline{M}$$

The assumption that x << 0.085 is justified (0.00003 << 0.085)

$$pH = 14.00 - pOH = 14.00 + \log(3.1 \times 10^{-5}) = 9.49$$

The percent hydrolysis is

$$\% \text{ hydrolysis} = \frac{100 \times [OH^-]}{[B]_o} = \frac{100 \times 3.1 \times 10^{-5}}{0.085} = 0.036\%$$

Basic Salts

Basic salts are those for which the anion is the conjugate base of a weak acid,

$$MA \xrightarrow{100\%} M^+ + A^-$$

$$A^- + H_2O \rightleftarrows OH^- + HA \qquad\qquad K_h = \frac{[HA][OH^-]}{[A^-]} = \frac{K_w}{K_a}$$

$$\underset{\text{Base}}{\uparrow} \quad \underset{\text{Acid}}{\uparrow} \quad \underset{\text{base}}{\uparrow} \quad \underset{\text{acid}}{\uparrow}$$

For example, solutions of the following salts will be basic: $NaNO_2$, CaF_2, KN_3, $NaCH_3COO$, $CsOCl$, and $RbCN$. These salts are all soluble, and completely ionized to M^+ and A^- in solution. The anions hydrolyze to form the conjugate acid (HA) and OH^-, resulting in a basic solution.

__Example 19.6__   Calculate the pH and the percent hydrolysis of a 0.23 __M__ solution of NaF. Ka for HF is 6.8×10^{-4}.

__Solution__ $NaF \xrightarrow{100\%} Na^+ + F^-$

$$\phantom{NaF \xrightarrow{100\%} }0.23 \quad 0.23$$

Let x be the $[OH^-]$. Then at equilibrium,

$$F^- + H_2O \rightleftarrows HF + OH^-$$

$$0.23{-}x x x$$

$$K_h = \frac{[HF][OH^-]}{[F^-]} = \frac{x^2}{0.23-x} \simeq \frac{x^2}{0.23} = \frac{1.0 \times 10^{-14}}{6.8 \times 10^{-4}} = 1.5 \times 10^{-11}$$

$$x = [OH^-] \simeq \sqrt{0.23 \times 1.5 \times 10^{-11}} = 1.8 \times 10^{-6}$$

The assumption that x is negligible compared to 0.23 is justified.

$$pH = 14.00 - pOH = 14.00 + \log [OH^-]$$

$$= 14.00 + \log (1.8 \times 10^{-6}) = 8.26$$

$$\% \text{ hydrolysis} = \frac{100 \times 1.8 \times 10^{-6}}{0.23} = 7.8 \times 10^{-4}\%$$

Acidic Salts

Acidic salts are those for which the cation is the conjugate acid of a weak base,

$$BHX \xrightarrow{100\%} BH^+ + X^-$$

$$BH^+ + H_2O \rightleftharpoons H_3O^+ + B \qquad K_a, \text{ also called } K_h$$

$$\text{acid} \qquad \text{base} \qquad \text{acid} \qquad \text{base}$$

or

$$BH^+ \rightleftharpoons B + H^+ \qquad K_a$$

Here

$$K_a = \frac{[B][H^+]}{[BH+]} = \frac{K_w}{K_b}$$

or

$$pH_a = 14.00 - pK_b$$

For example, solutions of the following salts will be acidic:

NH_4Cl - ammonium chloride NH_3OHClO_4 - hydroxylammonium perchlorate

$(CH_3)_3NHBr$ - trimethylammonium bromide C_5H_5NHI - pyridinium iodide

$N_2H_5NO_3$ - hydrazinium nitrate $C_6H_5NH_3Cl$ - anilinium chloride

These salts are all soluble and completely ionized to BH^+ and X^-. All these cations dissociate slightly to form the conjugate base and H^+. The reaction of an acidic salt with water is frequently termed a hydrolysis and the symbol K_h is often used rather than K_a.

<u>Example 19.7</u>   Calculate the pH of a 0.57 $\underline{M}$ $N_2H_5NO_3$ solution. K_b for N_2H_4 is 9.8×10^{-7}.

<u>Solution</u> $N_2H_5NO_3 \xrightarrow{100\%} N_2H_5^+ + NO_3^-$

Let x be the $[H^+]$

$$N_2H_5^+ \rightleftharpoons N_2H_4 + H^+$$

$$0.57-x \qquad\qquad x \qquad\qquad x$$

$$K_a = \frac{[N_2H_4][H^+]}{[N_2H_5^+]} = \frac{x^2}{0.57-x} \simeq \frac{x^2}{0.57} = \frac{1.0 \times 10^{-14}}{9.8 \times 10^{-7}} = 1.02 \times 10^{-8}$$

$$x = [H^+] \simeq \sqrt{0.57 \times 1.02 \times 10^{-8}} = 7.6 \times 10^{-5} \underline{M}$$

The assumption that x is negligible compared to 0.57 is justified.

$$pH = -\log(7.6 \times 10^{-5}) = 4.12$$

Hydrolysis By Metal Ions

Solutions containing salts with the general formula MX_n are often acidic due to the following reactions,

$$MX_n \xrightarrow{100\%} M^{n+} + nX^-$$

$$M^{n+} \quad + \quad H_2O \quad \rightleftarrows \quad H_3O^+ \quad + \quad MOH^{(n-1)+}$$

$$\uparrow \qquad\qquad \uparrow \qquad\qquad \uparrow \qquad\qquad \uparrow$$

$$\text{acid} \qquad\quad \text{base} \qquad\quad \text{acid} \qquad\quad \text{base}$$

Some examples of hydrolysis constants (as pK_a) for M^{n+} are:

M^{n+}	Zr^{4+}	Tl^{3+}	Bi^{3+}	Fe^{3+}	Hg^{2+}	Cr^{3+}	Al^{3+}	Be^{2+}
pK_h	0.22	1.15	1.58	2.19	3.70	4.01	5.14	6.50

__Example 19.8__   Calculate the pH of a 0.50 $\underline{M}$ $Be(NO_3)_2$ solution. K_h for Be^{2+} is 3.2×10^{-7}.

__Solution__ Let x be the $[H^+]$. Then

$$Be(NO_3)_2 \xrightarrow{100\%} Be^{2+} + 2NO_3^-$$

$$Be^{2+} + H_2O \rightleftarrows BeOH^+ + H^+$$

$$0.50-x \qquad\qquad\qquad x \qquad\quad x$$

$$K_h = \frac{[BeOH^+][H^+]}{[Be^{2+}]} = \frac{x^2}{0.50-x} \simeq \frac{x^2}{0.50} = 3.2 \times 10^{-7}$$

$$x = [H^+] \simeq \sqrt{0.50 \times 3.2 \times 10^{-7}} = 4.0 \times 10^{-4}$$

The assumption that x is negligible compared to 0.50 is justified

$$pH = -\log (4.0 \times 10^4) = 3.40$$

Polyprotic Acids

Many weak acids have more than one acidic proton. Some examples of diprotic (H_2A) and triprotic (H_3A) acids and the negative logs of their stepwise dissociation constants are given below.

H_2A	pK_1	pK_2		H_3A	pK_1	pK_2	pK_3
H_3AsO_3	9.22	13.52		H_3AsO_4	2.22	6.98	11.52
H_3PO_3	2.00	6.58		H_3PO_4	2.12	7.21	12.32
H_2SO_3	1.76	7.21		$C_6H_8O_7$	3.13	4.76	6.40
$H_2C_2O_4$	1.23	4.19					
H_2S	7.04	11.96					

Two types of acid-base reactions of polyprotic acids and their conjugate bases are considered here. For example, for the hypothetical diprotic H_2A,

1) dissociation $\qquad H_2A \rightleftharpoons H^+ + HA^- \qquad\qquad K_1$

$\qquad\qquad\qquad\qquad HA^- \rightleftharpoons H^+ + A^{2-} \qquad\qquad K_2$

2) hydrolysis $\qquad M_2A \xrightarrow{\ 100\%\ } 2M^+ + A^{2-}$

$\qquad\qquad\qquad\qquad A^{2-} + H_2O \rightleftharpoons HA^- + OH^- \qquad K_{h_1} = K_w/K_2$

$\qquad\qquad\qquad\qquad HA^- + H_2O \rightleftharpoons H_2A + OH^- \qquad K_{h_2} = K_w/K_1$

<u>Example 19.9</u> Calculate the pH, $[H_2SO_3]$, $[HSO_3^-]$, and $[SO_3^{2-}]$ of a 0.47 M solution of H_2SO_3. K_1 and K_2 for H_2SO_3 are 1.7×10^{-2} and 6.2×10^{-8} respectively.

<u>Solution</u>

$$H_2SO_3 \;\rightleftharpoons\; H^+ + HSO_3^- \qquad\qquad K_1 = 1.7 \times 10^{-2}$$

$$HSO_3^- \;\rightleftharpoons\; H^+ + SO_3^{2-} \qquad\qquad K_2 = 6.2 \times 10^{-8}$$

Since $K_2 \ll K_1$, assume the second dissociation does not contribute significantly to the $[H^+]$. Then

$$H_2SO_3 \;\rightleftharpoons\; H^+ + HSO_3^- \qquad\qquad K_1$$

$$\;0.47-x \qquad x \qquad\quad x$$

$$K_1 = \frac{[H^+][HSO_3^-]}{[H_2SO_3]} = \frac{x^2}{0.47-x} \simeq \frac{x^2}{0.47} = 1.7 \times 10^{-2}$$

$$x = [H^+] \simeq \sqrt{0.47 \times 1.7 \times 10^{-2}} = \sqrt{8.0 \times 10^{-3}} = 8.9 \times 10^{-2}$$

Here the assumption is <u>not</u> justified since $.089 > 0.05 \times 0.47$.

Using the quadratic formula (or the method of successive approximations),

$$x = [H^+] = [HSO_3^-] = 0.081 \; \underline{M}$$

$$pH = -\log\,(.081) = 1.09$$

The assumption that HSO_3^- does not contribute to the $[H^+]$ can now be checked. Let y be this additional $[H^+]$.

$$HSO_3^- \;\rightleftharpoons\; H^+ + SO_3^{2-} \qquad\qquad K_2$$

$$\;0.081-y \quad 0.081+y \qquad y$$

$$K_2 = \frac{[H^+][SO_3^{2-}]}{[HSO_3^-]} = \frac{(0.081+y)y}{0.081-y} \simeq \frac{0.081\,y}{0.081} = y$$

$$y = [SO_3^{2-}] = K_2 = 6.2 \times 10^{-8} \; \underline{M}$$

The assumption that y is negligible compared to 0.081 is justified. Since $6.2 \times 10^{-8} \ll 0.081$, the assumption that HSO_3^- contributes nothing to the pH is also justified. In summary:

$$pH = 1.09; \quad [H_2SO_3] = 0.47 - 0.081 = 0.39 \; \underline{M}; \quad [H^+] = [HSO_3^-] = 0.081 \; \underline{M};$$

$$\text{and} \quad [SO_3^{2-}] = 6.2 \times 10^{-8} \; \underline{M}$$

<u>Example 19.10</u> Calculate the pH of a 0.50 M Na_3PO_4 solution. Also calculate the equilibrium $[PO_4^{3-}]$, $[HPO_4^{2-}]$, $[H_2PO_4^-]$ and $[H_3PO_4]$. For H_3PO_4, $pK_1 = 2.12$, $pK_2 = 7.21$, and $pK_3 = 12.67$.

<u>Solution</u> Assume the dominant reaction is

$$PO_4^{3-} + H_2O \rightleftarrows HPO_4^{2-} + OH^- \qquad K_{h_1} = K_w/K_3 = 0.047$$

$$0.5-x \qquad\qquad\qquad x \qquad\quad x$$

Using the quadratic equation, or the method of successive approxima-
tions, $x = [OH^-] = [HPO_4^{2-}] = 0.13\ \underline{M}$. Then $[PO_4^{3-}] = 0.50 - 0.13 = 0.37\ \underline{M}$ and pH = 13.11.

This assumes the following successive hydrolysis reactions can
be ignored

$$HPO_4^{2-} + H_2O \rightleftarrows H_2PO_4^- + OH^- \qquad K_{h_2} = K_w/K_2 = 1.6 \times 10^{-7}$$

and

$$H_2PO_4^- + H_2O \rightleftarrows H_3PO_4 + OH^- \qquad K_{h_3} = K_w/K_1 = 1.3 \times 10^{-12}$$

Let y be the extent of the hydrolysis of HPO_4^{2-}. Then

$$HPO_4^{2-} + H_2O \rightleftarrows H_2PO_4^- + OH^-$$

$$0.13-y \qquad\qquad y \qquad\quad 0.13+y$$

$$y = 1.6 \times 10^{-7} \times \frac{(0.13-y)}{(0.13+y)} \simeq 1.6 \times 10^{-7} = [H_2PO_4^-]$$

Clearly, the hydrolysis reactions of HPO_4^{2-} (and $H_2PO_4^-$)
can be safely ignored.

$$[H_3PO_4] = \frac{[H^+][H_2PO_4^-]}{K_1} = \frac{(1.0 \times 10^{-14})(1.6 \times 10^{-7})}{(.13)(7.6 \times 10^{-3})} = 1.6 \times 10^{-18}\ M$$

<u>Questions</u>

19.1 The $[H^+]$ of an aqueous solution is 0.0977 $\underline{M}$. What is the $[OH^-]$?

    a) 1.02×10^{-13}
    b)   1.01
    c)   0.0977
    d) 1.16×10^{-13}
    e) 9.77×10^{-14}

19.2 The $[OH^-]$ of an aqueous solution is 0.0427 $\underline{M}$. What is the $[H^+]$?

    a)   1.37
    b)   0.0427
    c) 4.27×10^{-14}
    d) 2.34×10^{-13}
    e) 2.74×10^{-13}

19.3 The pH of an aqueous solution is 13.00. What is the pOH?

    a)   1.00
    b) 1.00×10^{-13}
    c)   13.0
    d)   3.00
    e)   1.17

19.4 The pOH of an aqueous solution is 9.93. What is the pH?

    a)   5.93
    b) 1.18×10^{-10}
    c)   4.69
    d)   9.93
    e)   4.07

19.5 The $[H^+]$ of an aqueous solution is 1.6×10^{-2} $\underline{M}$. What is the pH?

    a)   2.20
    b)   2.13
    c)   12.20
    d)   0.016
    e)   1.80

19.6 The $[OH^-]$ of an aqueous solution is 1.8×10^{-10} $\underline{M}$. What is the pOH?

    a) 1.8×10^{-10}
    b)   9.74
    c)   10.26
    d)   4.26
    e)   11.37

19.7 The pH of an aqueous solution is 9.54. What is the $[H^+]$?

    a) 2.9×10^{-10}
    b) 3.5×10^{-11}
    c)   9.54
    d) 3.3×10^{-10}
    e) 3.5×10^{-5}

19.8 The pOH of an aqueous solution is 5.98. What is the $[OH^-]$?

    a) 9.6×10^{-7}
    b) 9.4×10^{-7}
    c) 1.05×10^{-6}
    d) 9.6×10^{-9}
    e)   5.98

19.9 The $[H^+]$ of an aqueous solution is 5.37×10^{-11} $\underline{M}$. What is the pOH?

    a) 5.37×10^{-11}
    b)   10.27
    c)   3.73
    d)   4.13
    e)   4.27

19.10 The $[OH^-]$ of an aqueous solution is 2.75×10^{-4} $\underline{M}$. What is the pH?

    a) 2.75×10^{-4}
    b)   3.56
    c)   12.36
    d)   11.56
    e)   10.44

19.11* The pH of an aqueous solution is 3.640. What is the $[OH^-]$?

    a) 4.36×10^{-11}
    b) 4.95×10^{-11}
    c) 2.29×10^{-12}
    d)   3.64
    e) 2.29×10^{-4}

19.12 The pOH of an aqueous solution is 4.790. What is the $[H^+]$?

    a) 1.62×10^{-5}
    b) 1.62×10^{-11}
    c) 7.19×10^{-10}
    d)   4.79
    e) 6.17×10^{-10}

19.13 What is the $[H^+]$ of a 2.00×10^{-4} $\underline{M}$ HBr aqueous solution?

    a) 5.01×10^{-11}
    b) 2.00×10^{-4}
    c)   10.30
    d) 5.01×10^{-5}
    e) 2.30×10^{-4}

19.14 What is the $[OH^-]$ of a 0.0195 $\underline{M}$ LiOH aqueous solution?

    a) 5.13×10^{-13}
    b) 5.13×10^{-3}
    c)   0.0216
    d)   0.0195
    e)   12.29

19.15   What is the pH of a 8.51 x 10^{-4} $\underline{M}$ HNO_3 aqueous solution?

    a)   1.18 x 10^{-11}
    b)   2.621
    c)   10.930
    d)   3.070
    e)   4.930

19.16   What is the pOH of a 4.5 x 10^{-4} $\underline{M}$ CsOH aqueous solution?

    a)   4.65
    b)   2.22 x 10^{-11}
    c)   3.35
    d)   10.65
    e)   2.94

19.17   What is the $[H^+]$ of a 2.63 x 10^{-3} $\underline{M}$ CsOH aqueous solution?

    a)   3.80 x 10^{-12}
    b)   2.63 x 10^{-3}
    c)   2.63 x 10^{-13}
    d)   2.58
    e)   4.42 x 10^{-12}

19.18   What is the $[OH^-]$ of a .0120 $\underline{M}$ $HClO_4$ aqueous solution?

    a)   9.45 x 10^{-13}
    b)   8.33 x 10^{-13}
    c)   1.20 x 10^{-14}
    d)   1.92
    e)   1.20 x 10^{-2}

19.19   What is the pH of a 0.048 $\underline{M}$ CsOH aqueous solution?

    a)   10.24
    b)   13.32
    c)   12.68
    d)   1.32
    e)   0.048

19.20   What is the pOH of a 0.040 $\underline{M}$ HI aqueous solution?

    a)   13.40
    b)   0.040
    c)   1.40
    d)   12.60
    e)   14.52

19.21* What is the pH of a 2.8×10^{-8} $\underline{M}$ $HClO_4$ solution?

    a)   7.55
    b) 2.80×10^{-8}
    c)   7.89
    d)   6.94
    e)   7.06

19.22 Which of the following is(are) related as acid–conjugate base pairs?

    1. $H_2O–OH^-$ 2. $H_2PO_4^- – HPO_4^{2-}$ 3. $H_2SO_4 – H_2SO_3$

    a)   1 and 2
    b)   1 and 3
    c)   2 and 3
    d)   All of them
    e)   None of them

19.23* Which of the following salts, when dissolved in water, would give a basic solution (pH > 7)?

    1. NaF   2. NH_4NO_3 3. $Ca_3(PO_4)_2$

    a)   1 and 2
    b)   2 and 3
    c)   1 and 3
    d)   All of them
    e)   None of them

19.24 What is the pH of a 0.97 $\underline{M}$ formic acid solution? (K_a for formic acid, HCOOH is 1.7×10^{-4})

    a)   3.77
    b)   1.89
    c)   12.11
    d)   0.89
    e)   0.967

19.25 What is the pH of a 0.18 $\underline{M}$ pyridine solution? K_b for pyridine is 1.8×10^{-9}.

    a)   4.00
    b)   4.75
    c)   8.75
    d)   9.26
    e)   10.00

19.26* What initial concentration of HNO_2 gives a solution with a pH of 1.71? K_a for HNO_2 is 4.5×10^{-4}.

   a)   0.050 M
   b)   0.085 M
   c)   0.50 M
   d)   0.86 M
   e)   1.8 M

19.27 What initial concentration of methylamine, CH_3NH_2, gives a solution with a pH of 12.22? K_b for CH_3NH_2 is 3.7×10^{-4}.

   a)   1.8 M
   b)   0.050 M
   c)   0.50 M
   d)   0.76 M
   e)   0.092 M

19.28 A 0.66 M solution of a weak monoprotic acid has a pH of 2.81. Calculate the dissociation constant, K_a, of the weak acid.

   a) 2.4×10^{-6}
   b)   5.4
   c) 3.6×10^{-6}
   d) 2.4×10^{-3}
   e)   2.8

19.29* A 0.044 M solution of a weak base has a pH of 10.32. Calculate the ionization constant, K_b, of the weak base.

   a) 4.3×10^{-8}
   b)   210
   c) 4.7×10^{-3}
   d) 9.9×10^{-7}
   e)   6.0

19.30 A 0.185 M solution of a weak monoprotic acid is found by conductivity measurements to be 0.0369% dissociated at 25°C. What is the dissociation constant of the acid?

   a) 2.52×10^{-8}
   b) 7.33×10^{-3}
   c) 2.52×10^{-6}
   d) 3.69×10^{-4}
   e) 2.52×10^{-10}

19.31 A 0.267 M solution of a weak base is found by conductivity measurements to be 0.100% ionized at 25°C. What is the ionization constant
of the base?

a) 2.69×10^{-5}
b) 3.76×10^{-2}
c) 2.69×10^{-7}
d) 2.69×10^{-9}
e) 1.00×10^{-3}

19.32 The numerical value of K_a for hypochlorous acid, HClO, is
2.95×10^{-8}. What percentage of a 0.123 $\underline{M}$ solution of HClO is
dissociated?

a) 4.90×10^{-4}%
b) 4.90×10^{-2}%
c) 2.95×10^{-6}%
d) 3.63×10^{-7}%
e) 2.95×10^{-8}%

19.33 The numerical value of K_b for aniline, $C_6H_5NH_2$, is 4.60×10^{-10}.
What percentage of a 0.627 $\underline{M}$ solution of $C_6H_5NH_2$ is ionized?

a) 2.71×10^{-5}%
b) 2.89×10^{-10}%
c) 2.71×10^{-3}%
d) 2.89×10^{-8}%
e) 4.60×10^{-10}%

19.34* A 0.450 molal aqueous solution of a weak acid HA freezes at −0.855°C.
Estimate the dissociation constant of HA. The freezing point depression constant for water is −1.86 deg/molal.

a) 2×10^{-2}
b) 9×10^{-3}
c) 2×10^{-4}
d) 9×10^{-5}
e) 2×10^{-8}

19.35 A 0.155 molal aqueous solution of a weak base B freezes at −0.321°C.
Estimate the ionization constant of B. The freezing point depression
constant for water is −1.86 deg/molal.

a) 1.4×10^{-2}
b) 8×10^{-2}
c) 2×10^{-3}
d) 2×10^{-5}
e) 8×10^{-4}

19.36 What is the pH of a .594 molar sodium hypobromite solution (K_a for hypobromous acid, HBrO, is 2.09×10^{-9})?

    a)   4.45
    b)   7.00
    c)   8.68
    d) 11.23
    e)   9.55

19.37 * What is the pH of a .821 molar dimethylammonium chloride solution? (K_b for dimethylamine, $(CH_3)_2NH$, is 5.40×10^{-4}.) Dimethylammonium chloride is the salt formed from dimethylamine and HCl, $[(CH_3)_2NH_2]^+Cl^-$.

    a)   1.59
    b)   5.41
    c)   3.27
    d) 12.32
    e)   8.59

19.38 * For the reaction

$$Al^{3+} + H_2O \rightleftharpoons AlOH^{2+} + H^+ \qquad K_h = 7.9 \times 10^{-6}$$

What is the approximate pH of a 0.70 molar solution of $Al(ClO_4)_3$?

    a)   .700
    b) 11.37
    c)   2.63
    d)   5.10
    e)   2.07

19.39 What should the initial concentration of sodium acetate in solution be to give a pH of 9.38? K_a for acetic acid is 1.80×10^{-5}.

    a)   1.0 $\underline{M}$
    b)   0.56 $\underline{M}$
    c)   0.97 $\underline{M}$
    d) 2.4×10^{-5} $\underline{M}$
    e)   0.0010 $\underline{M}$

19.40 What should the initial concentration of ammonium chloride in solution be to give a pH of 4.88? K_b for ammonia is 1.75×10^{-5}.

    a) 1.3×10^{-5} $\underline{M}$
    b)   2.3 $\underline{M}$
    c)   0.43 $\underline{M}$
    d)   .30 $\underline{M}$
    e)   3.3 $\underline{M}$

19.41 What should the initial concentration of aluminum nitrate in
solution be to give a pH of 3.00? The hydrolysis constant
for Al^{3+} is 7.2×10^{-6}.

a) 7.2×10^{-6} $\underline{M}$
b) 1.0×10^{-3} $\underline{M}$
c) 0.14 $\underline{M}$
d) 3.0 $\underline{M}$
e) 7.2 $\underline{M}$

19.42* What is the pH of a solution that is 0.25 $\underline{M}$ in HNO_3 and 0.55 $\underline{M}$ in
HNO_2. K_a for HNO_2 is 4.6×10^{-4}.

a) -0.097
b) -0.60
c) 0.60
d) 2.99
e) 3.68

19.43 What is the pH of a solution that is 0.074 $\underline{M}$ in N_2H_4 and 0.068 $\underline{M}$
in NaOH? K_b for N_2H_4 is 9.8×10^{-7}.

a) 7.83
b) 8.15
c) 10.33
d) 12.67
e) 12.83

19.44 What is the pH of a solution that is 0.50 $\underline{M}$ in HCN and 0.50 $\underline{M}$ in
HF. The dissociation constants for HCN and HF are 4.9×10^{-10} and
3.5×10^{-4}, respectively.

a) 0.00
b) 0.30
c) 1.88
d) 3.34
e) 4.80

19.45 * Find the pH of a 1.94×10^{-3} M hydrofluoric acid, HF, solution.
K_a for HF is 6.76×10^{-4}.

a) 7.00
b) 11.29
c) 3.07
d) 2.71
e) 2.94

19.46 What is the pH of a 9.25×10^{-4} $\underline{M}$ ammonia solution? K_b for NH_3 is 1.85×10^{-5}.

    a)   10.12
    b)   10.09
    c)   7.00
    d)   3.91
    e)   3.03

19.47 * The stepwise acid dissociation constants for a hypothetical diprotic acid are:

$$H_2A \rightleftharpoons H^+ + HA^- \qquad\qquad K_1 = 1.71 \times 10^{-7}$$

$$HA^- \rightleftharpoons H^+ + A^{2-} \qquad\qquad K_2 = 2.61 \times 10^{-12}$$

What is the approximate molar concentration of A^{2-} in a .534 molar solution of H_2A?

    a) 1.71×10^{-7} molar
    b) 2.61×10^{-12} molar
    c) 3.02×10^{-4} molar
    d) 2.21×10^{-4} molar
    e) 2.07×10^{-3} molar

19.48 The stepwise acid dissociation constants for a hypothetical diprotic acid are:

$$H_2A \rightleftharpoons H^+ + HA^- \qquad\qquad K_1 = 2.78 \times 10^{-7}$$

$$HA^- \rightleftharpoons H^+ + A^{2-} \qquad\qquad K_2 = 6.91 \times 10^{-12}$$

What is the approximate molar concentration of HA^- in a 4.62 molar solution of Na_2A?

    a)   0.00145 molar
    b) 3.59×10^{-8} molar
    c)   0.176 molar
    d)   0.0818 molar
    e)   4.62 molar

19.49 The stepwise acid dissociation constants for a hypothetical diprotic acid are:

$$H_2A \rightleftharpoons H^+ + HA^- \qquad\qquad K_1 = 6.24 \times 10^{-5}$$

$$HA^- \rightleftharpoons H^+ + A^{2-} \qquad\qquad K_2 = 7.88 \times 10^{-9}$$

What is the approximate pH of a .440 molar solution of Na_2A?
    a)   11.68
    b)   10.87
    c)   2.28
    d)   6.15
    e)   3.30

19.50 * Phosphoric acid, H_3PO_4, is a triprotic acid for which K_1, K_2, and K_3 are: 7.50×10^{-3}, 6.20×10^{-8}, and 4.80×10^{-13} respectively. Calculate the pH of a 0.0907 M solution of H_3PO_4.

a)   2.35
b)   1.58
c)   12.42
d)   1.65
e)   1.28

19.51 Calculate the pH of a 0.050 $\underline{M}$ Na_3AsO_4 solution. For H_3AsO_4 $K_1 = 6.0 \times 10^{-3}$, $K_2 = 1.05 \times 10^{-7}$, and $K_3 = 3.0 \times 10^{-12}$.

a)   7.00
b)   11.52
c)   12.05
d)   12.11
e)   12.70

19.52 * What is the pH of a 0.10 $\underline{M}$ solution of NH_4NO_2?

$$K_a \text{ for } HNO_2 = 4.6 \times 10^{-4}$$

$$K_b \text{ for } NH_3 = 1.8 \times 10^{-5}$$

Hint: consider the equilibrium $NH_4^+ + NO_2^- \rightleftarrows NH_3 + HNO_2$

a)   2.26
b)   2.17
c)   7.30
d)   9.26
e)   6.30

19.53 * A hypothetical amino acid dissociates in aqueous solution as follows:

$$H_3N-R-COOH^+ \rightleftarrows H_3N-R-COO + H^+ \qquad pK_1 = 2.34$$

$$H_3N-R-COO \rightleftarrows NH_2-R-COO^- + H^+ \qquad pK_2 = 9.10$$

What is the pH at the isoelectric point? (At the isoelectric point the molar concentrations of $H_3N-R-COOH^+$ and $H_2N-R-COO^-$ are equal.)

a)   5.72
b)   11.44
c)   7.00
d)   3.38
e)   3.89

19.54 Calculate the pH obtained when 0.10 mole of H_3PO_4 and 0.10 mole of Na_3PO_4 are added to a volumetric flask and enough water is added to make the total volume 1.0 ℓ. For H_3PO_4 pK_1 is 2.12, $pK_2 = 7.21$, and $pK_3 = 12.67$. Hint: H_3PO_4 is an acid and PO_4^{3-} is a base.

a)   6.79
b)   7.00
c)   7.21
d)   7.39
e)   8.21

19.55 * The dissociation constants for the diprotic weak acid, H_2A, are $K_1 = 8.71 \times 10^{-6}$ and $K_2 = 8.40 \times 10^{-8}$. One millimole of H_2A is added to one liter of a buffer solution of pH = 6.85. What are the concentrations of the three species, H_2A, HA^-, and A^{2-}? (In a buffer solution the $[H^+]$ can be considered constant.)

	H_2A	HA^-	A^{2-}
a)	1.00×10^{-5}	6.20×10^{-4}	3.70×10^{-4}
b)	6.62×10^{-6}	4.09×10^{-4}	5.90×10^{-4}
c)	1.41×10^{-10}	1.23×10^{-15}	7.32×10^{-16}
d)	4.40×10^{-4}	4.24×10^{-4}	1.32×10^{-4}
e)	3.70×10^{-4}	6.20×10^{-4}	1.00×10^{-5}

CHAPTER 20.   BUFFERS AND TITRATIONS

Buffers

Buffer solutions contain appreciable concentrations of both a weak
acid and its conjugate base. The addition of a small amount of an acid or
base to neutral water or to an unbuffered solution changes the pH by
several units. Such an addition to a buffer results in a pH change of only
a fraction of a pH unit. The following solutions are buffers.

```
1. Add 1.0 mole of CH3COOH and 1.0 mole of CH3COONa to 1.0 liter of water
2.  "   0.5   "    "   NH3       "  0.25 "    "   NH4Cl   "  "    "     "    "
3.  "   0.1   "    "   NaH2PO4   "  0.2  "    "   Na2HPO4 "  "    "     "    "
4.  "   0.5   "    "   NaOH      "  1.0  "    "   HCN     "  "    "     "    "
5.  "   0.2   "    "   HNO3      "  0.5  "    "   N2H4    "  "    "     "    "
```

In cases 4 and 5 a neutralization reaction will occur upon mixing. In
case 4 this will result in a mixture of HCN (which was in excess) and its
conjugate base CN^- (which is the product of the neutralization). In case 5
there will be the excess N_2H_4 plus its conjugate acid $N_2H_5^+$ formed in the
neutralization.

The following solutions are not buffers.

```
 6. Add 1.0 mole of HCl and 1.0 mole of NaCl to 1.0 liter of H2O
 7.  "   0.5   "    "   NaOH  "   "     "   "   HClO4   "  "    "     "    "
 8.  "   0.1   "    "   H2S   "  0.1  "   "   Na2S    "  "    "     "    "
 9.  "   0.1   "    "   HF    "  0.1  "   "   NaOH    "  "    "     "    "
10.  "   0.2   "    "   NH3   "  0.3  "   "   HNO3    "  "    "     "    "
```

In cases 6 and 7 only the strong acids HCl and $HClO_4$ are encountered;
no undissociated acid is present. In cases 8 and 9 stoichiometrically
equivalent amounts of reactants are mixed resulting in case 8 in only HS^-,
and in case 9 only F^-, and no conjugates. In case 10 excess strong acid
(HNO_3) converts all the NH_3 to NH_4^+ leaving none of the conjugate base.

Example 20.1   Calculate the pH of a buffer solution which contains 0.125 M
NH_3 and 0.216 M NH_4^+. K_b for NH_3 is 1.8×10^{-5}

Solution The quantities $[NH_3]$, $[NH_4^+]$, and $[OH^-]$ are related by the K_b
for NH_3:

$$K_b = \frac{[NH_4^+][OH^-]}{[NH_3]} = 1.8 \times 10^{-5}$$

$$\text{or} \qquad [OH^-] = 1.8 \times 10^{-5} \times \frac{[NH_3]}{[NH_4^+]} = 1.8 \times 10^{-5} \times \frac{.125}{.216} = 1.04 \times 10^{-5}$$

$$pOH = -\log(1.04 \times 10^{-5}) = 4.98; \quad pH = 14 - pOH = \underline{9.02}$$

It was tacitly assumed in the above calculation that $[OH^-]$ was negligibly small compared to $[NH_3]$ and $[NH_4^+]$. This proved to be the case, as it will invariably be in buffer problems where large concentrations of buffering substances are used to control a very small concentration of H^+ or OH^-.

<u>Example 20.2</u>   Calculate the change in pH when:

a)   10.0 mls of 1.00 $\underline{M}$ HCl is added to 1.000 ℓ of H_2O.
b)   "   "   " "   " NaOH "   "   " "   "   "
c)   "   "   " "   " HCl   "   "   " "   " of a solution
    that is 0.25 $\underline{M}$ in HNO_2 and 0.75 $\underline{M}$ in $NaNO_2$.
d)   10.0 mls of 1.00 $\underline{M}$ NaOH added to solution that is 0.75 $\underline{M}$ in
    NH_4NO_3 and 0.25 $\underline{M}$ in NH_3.
    K_a for HNO_2 is 4.6×10^{-4} and K_b for NH_3 is 1.8×10^{-5}.

<u>Solution</u> a) The initial pH is 7.00. The total volume after the addition
    is 1.010 ℓ = 1010 ml

$$[H^+] = \frac{1.00 \times 0.010 \text{ mole}}{1.010 \; \ell} = .0099 \; \underline{M} \qquad pH = 2.00$$

The change in pH is $2.00 - 7.00 = -5.00$

b)   The initial pH is 7.00. After addition,

$$[OH^-] = \frac{1.00 \times 0.010 \text{ mole}}{1.010 \; \ell} = .0099 \; \underline{M}; \; pOH = 2.00$$

$$pH = 14.00 - pH = 14.00 - 2.00 = 12.00$$

The change in pH is $12.00 - 7.00 = +5.00$

c)   The initial pH is calculated as in example 20.1, assuming that
    the small $[H^+]$ at equilibrium is a negligible fraction of the
    $[HNO_2]$ and $[NO_2^-]$

$$K_a = \frac{[H^+][NO_2^-]}{[HNO_2]} ; \quad [H^+] = K_a \frac{[HNO_2]}{[NO_2^-]} = 4.6 \times 10^{-4} \times \frac{0.25}{0.75} =$$

$$1.53 \times 10^{-4} \; M$$

$$pH = -\log(1.53 \times 10^{-4}) = 3.81$$

When the HCl is added, the following reaction goes essentially to completion:

$$NO_2^- + H^+ \longrightarrow HNO_2$$

Initial Moles	0.75	0.010	0.25
Moles after reaction	0.74	-	0.26

The total volume is now 1.01 ℓ, so the concentrations of buffering substances are now:

$$[NO_2^-] = \frac{0.74}{1.01} \qquad [HNO_2] = \frac{0.26}{1.01}$$

$$[H^+] = K_a \frac{[HNO_2]}{[NO_2^-]} = 4.6 \times 10^{-4} \left\{ \frac{\frac{0.26}{1.01}}{\frac{0.74}{1.01}} \right\}$$

$$\simeq 4.6 \times 10^{-4} \left(\frac{0.26}{0.74}\right) = 1.62 \times 10^{-4} \underline{M}$$

$$pH = 3.79$$

$$\Delta pH = 3.79 - 3.81 = -0.02$$

d)   The initial pH can be calculated either as in example 20.1, or alternatively, from the equilibrium:

$$NH_4^+ \rightleftarrows NH_2 + H^+ \qquad K_a = K_w/K_b$$

$$[H^+] = K_a \frac{[NH_4^+]}{[NH_3]} = \frac{1.0 \times 10^{-14}}{1.8 \times 10^{-5}} \times \frac{0.75}{0.25} = 1.67 \times 10^{-9}$$

$$pH = 8.78$$

The addition of 0.010 mole of strong base converts 0.010 mole of the NH_4^+ to NH_3. As in part (c) the volume of solution cancels out,

$$[H^+] = \frac{K_a[NH_4^+]}{[NH_3]} = \frac{1.0 \times 10^{-14}}{1.8 \times 10^{-5}} \times \frac{0.74}{0.26} = 1.58 \times 10^{-9}$$

$$pH = 8.80$$

$$\Delta pH = 8.80 - 8.78 = 0.02$$

Note that the change in a and b was 5 pH units, whereas in c and d the pH changed 0.02 pH units. The solutions in c and d are buffer solutions.

<u>Example 20.3</u>   How many grams of NaOH should be added to 1.0 ℓ of 0.27 $\underline{M}$ CH_3COOH to produce a buffer solution with pH 4.89? K_a for CH_3COOH is 1.8×10^{-5}. The molecular weight of NaOH is 40.0 g/mole.

<u>Solution</u>

$$[H^+] = \frac{K_a[CH_3COOH]}{[CH_3COO^-]} = 10^{-4.89} = 1.3 \times 10^{-5} \ \underline{M}$$

Then $\dfrac{[CH_3COOH]}{[CH_3COO^-]} = \dfrac{[H^+]}{K_a} = \dfrac{1.3 \times 10^{-5}}{1.8 \times 10^{-5}} = 0.72$

Enough of the CH_3COOH must be neutralized to produce CH_3COO^- so that the ratio $\dfrac{[CH_3COOH]}{[CH_3COO^-]} = 0.72$

Let x = moles of NaOH added and assume the following reaction goes to completion.

$$CH_3COOH \ + \ OH^- \ \underset{\longleftarrow}{\longrightarrow} \ CH_3COO^- \ + \ H_2O$$

	CH_3COOH	OH^-	CH_3COO^-	H_2O
Initial moles	0.27	x	–	–
moles after reaction	0.27–x	–	x	–

$$\frac{0.27-x}{x} = 0.72$$

$$x = 0.27/1.72 = 0.16 \text{ mole}$$

$$\#g \ NaOH = \frac{40 \ g}{mole} \times 0.16 \text{ mole} = 6.3 \text{ g}$$

<u>pH Titration Curves</u>

Three cases are considered.

<u>Starting Solution</u>	<u>Titrant added</u>	<u>General Reaction</u>
1. Strong acid	– strong base	$H^+X^- + M^+OH^- \rightarrow H_2O + M^+X^-$
2. Weak acid	– strong base	$HA + M^+OH^- \rightarrow H_2O + M^+A^-$
3. Weak base	– strong acid	$B + H^+X^- \rightarrow (BH)^+X^-$

The equivalence point is the point on the titration curve that corresponds to stoichiometric amounts of the reactants. The pH at the equivalence point for these three titration cases is indicated below.

<u>Titration</u>	<u>Equivalence Point pH</u>	<u>Controlling Equilibrium at Equivalence Point</u>
1. $H^+X^- + M^+OH^-$	7.00	$H_2O \rightleftharpoons H^+ + OH^-$
2. $HA + M^+OH^-$	>7	$A^- + H_2O \rightleftharpoons HA + OH^-$
3. $B + H^+X^-$	<7	$BH^+ \rightleftharpoons B + H^+$

180

The pH can be calculated at any point on the titration curve. Four types of calculation are required. For example, for the titration of a weak acid with a strong base,

Type of Calculation	Controlling Equilibrium
1. Initial pH	$HA \rightleftharpoons H^+ + A^-$
2. Buffer region	$HA \rightleftharpoons H^+ + A^-$
3. Equivalence point	$A^- + H_2O \rightleftharpoons HA + OH^-$
4. Beyond the equivalence point.	$[OH^-]$ in excess

Example 20.4   Consider the titration of 50.00 ml of 0.10 $\underline{M}$ HOCl with 0.25 $\underline{M}$ NaOH. Calculate:

a)   The initial pH
b)   The pH at 50.00% of the equivalence point
c)   The pH at 90.00% of the equivalence point
d)   The pH at the equivalence point
e)   The pH at 150% of equivalence

The K_a for HOCl is 3.0 x 10^{-8}

Solution   The titration reaction is

$$HOCl + OH^- \longrightarrow H_2O + OCl^-$$

Let V = volume of NaOH at equivalence

$$V = \frac{50.00 \times 0.10}{0.25} = 20.00 \text{ ml.}$$

(a) The calculation of the initial pH is similar to example 19.3

$$[H^+] \simeq \sqrt{K_a [HOCl]} = \sqrt{0.10 \times 3.0 \times 10^{-8}} = 5.48 \times 10^{-5}$$

pH = 4.26

(b) The volume of NaOH added at 50.00% of titration is 10.00 ml. The total volume is 60.00 ml.

	HOCl	+	OH^-	$\longrightarrow$	H_2O	+	OCl^-
Initial milli-moles	5.0		2.5		–		–
Millimoles after reaction	2.5		–		–		2.5

This is a buffer solution, for which $[H^+]$ is calculated as in example 20.1 or 20.2c.

$$[H^+] = K_a \frac{[HOCl]}{[OCl^-]} = 3.0 \times 10^{-8} \times \frac{[\frac{2.5}{60}]}{[\frac{2.5}{60}]} = 3.0 \times 10^{-8}$$

$$pH = pK_a = 7.52$$

(c) At 90% of equivalence,

$$[H^+] = K_a \frac{[HOCl]}{[OCl^-]} = 3.0 \times 10^{-8} \times \frac{[\frac{0.5}{68}]}{[\frac{4.5}{68}]} = 3.33 \times 10^{-9}$$

$$pH = 8.47$$

(d) At the equivalence point, 5.00×10^{-3} mole of OCl^- have been formed, and the total volume is 70.00 ml, so:

$$[OCl^-] = \frac{5.00 \times 10^{-3}}{.070} = .0714 \underline{M}$$

However, the OCl^- will hydrolyze; let $X = [OH^-]$

$$OCl^- \; + \; H_2O \; \rightleftharpoons \; HOCl \; + \; OH^- \qquad\qquad K_h = \frac{K_w}{K_a}$$
$$.0714 - X \qquad\qquad\qquad\quad X \qquad\quad X$$

$$X = [OH^-] \simeq \sqrt{\frac{1.0 \times 10^{-14}}{3.0 \times 10^{-8}} \times .0714} \;\; = 1.54 \times 10^{-4}$$

$$pOH = 3.81; \quad pH = 14.00 - 3.81 = 10.19$$

(e) At 150% of titration the volume of NaOH added is 1.5 x 20.00 or 30.00 ml. The total volume is 80.00 ml

	HOCl	+	OH$^-$	$\longrightarrow$	H$_2$O	+	OCl$^-$
Initial mmoles	5.0		7.5		–		–
mmoles after reaction	–		2.5		–		5.0

The excess $[OH^-]$ is 2.5 mmole/80.0 ml or 0.031 $\underline{M}$
The contribution to the $[OH^-]$ from the hydrolysis of OCl^-
is negligible compared to 0.031 M.

$$pOH = -\log (.031) = 1.51$$
$$pH = 14.00 - 1.51 = 12.49$$

Indicators

Indicators are intensely colored dyes which are weak acids, and which change color upon neutralization,

$$HInd \rightleftharpoons H^+ + Ind^- \qquad K_a = \frac{[H^+][Ind^-]}{[HInd]}$$

Color A   Color B

The color of the indicator solution depends on the pH:

$$\frac{[Ind^-]}{[HInd]} = \frac{K_a}{[H^+]} = \frac{[Color\ B]}{[Color\ A]}$$

The solution is color A if $[H^+] \gg K_a$ as in a strongly acid solution, and color B if $[H^+] \ll K_a$ as in base. If $[H^+] \simeq K_a$ the color of the solution is the color of a mixture of A and B. The pH changes rapidly near the equivalence point in acid-base titrations. An indicator for which the pK_a is approximately equal to the pH at the equivalence point will change color when the equivalence point is reached. Actually, an interval of 1 to 2 pH units is required for the color to go from one extreme to the other.

Example 20.4   Which indicators should be used for the following titration:

a)   Titration of 50.00 mls of 0.10 $\underline{M}$ HCl with 0.10 $\underline{M}$ NaOH
b)   Titration of 50.00 mls of 0.10 $\underline{M}$ CH3COOH with 0.10 $\underline{M}$ NaOH
c)   Titration of 50.00 mls of 0.10 $\underline{M}$ NH3 with 0.10 $\underline{M}$ HCl

Indicator	pH Interval for Color Change	HInd Color	Ind^- Color
Methyl Red	4.2–6.3	Red	Yellow
Bromethymol Blue	6.0–7.6	Yellow	Blue
Phenolphthalein	8.3–10	Colorless	Pink

K_a for CH_3COOH is 1.8×10^{-5} and K_b for NH_3 is 1.8×10^{-5}.

Solution a) The equivalence point pH is 7.00. Use bromthymol blue.
   b) The equivalence point pH is the pH of a 0.050 $\underline{M}$ solution of $NaCH_3COO$, which is 8.72. Use phenolphthalein.
   c) The equivalence point pH is the pH of a 0.050 $\underline{M}$ solution of NH_4Cl, which is 5.28. Use methyl red.

Questions

20.1 What is the pH of a buffer solution for which the cyanic acid concentration is 1.35 molar and the cyanate concentration is .952 molar? (K_a for cyanic acid, HCNO, is 2.19×10^{-4}).

(see next)

a)   3.510
b) 1.548×10^{-4}
c) 3.093×10^{-4}
d)   10.19
e)   3.810

20.2*Calculate the pH of a solution that is .928 molar in HPO_4^{2-} and .586 molar in PO_4^{3-}. For phosphoric acid, H_3PO_4, K_1, K_2, and K_3 are: 7.50×10^{-3}, 6.20×10^{-8}, and 4.80×10^{-13} respectively.

a)   12.32
b)   12.12
c) 7.601×10^{-13}
d)   12.52
e)   1.881

20.3*A weak acid with pK_a of 3.27 dissociates according to the following reaction:

$$HA \; \underset{\leftarrow}{\rightarrow} \; H^+ \; + \; A^-$$

A buffer is prepared in which the concentration of HA is .216 M and the concentration of A^- is .230 M. If .0733 mole of NaOH is added to one liter of the solution, what is the pH of the resulting solution? (Assume no volume change.)

a)   3.60
b)   10.40
c)   10.70
d)   3.00
e)   3.30

20.4 A solution made by mixing .190 moles of sodium bromoacetate with 122 ml of .400 M bromoacetic acid has a pH of 5.28. What is the ionization constant, K_a, of bromoacetic acid?

a) 1.35×10^{-6}
b) 5.68×10^{-4}
c) 4.88×10^{-8}
d) 2.05×10^{-5}
e) 2.05×10^{-8}

20.5 Which of the following mixtures gives a buffer solution? (Acetic acid is a weak, monoprotic acid, $K_a = 1.76 \times 10^{-5}$.)

a)   45 ml of .291 molar sodium acetate plus 25 ml of .176 molar sodium chloride.
b)   26 ml of .189 molar sodium acetate plus 21 ml of .066 molar hydrochloric acid.
c)   26 ml of .141 molar acetic acid plus 40 ml of .170 molar sodium hydroxide.
d)   35 ml of .109 molar acetic acid plus 20 ml of .247 molar sodium chloride.   (over)

e)   None of the above

20.6*Which one of the following procedures will result in a solution that is <u>not</u> a buffer solution? (H_3A is a hypothetical triprotic acid).

a)   Mix 500 ml of 0.1 $\underline{M}$ H_3A and 500 ml of 0.1 $\underline{M}$ NaH_2A
b)   Mix 500 ml of 0.1 $\underline{M}$ Na_2HA and 500 ml of 0.1 $\underline{M}$ Na_3A
c)   Add 0.5 mole of NaOH to 1.0 liter of 1.0 $\underline{M}$ NaH_2A
d)   Add 0.5 mole of HCl to 1.0 liter of 1.0 $\underline{M}$ NaH_2A
e)   Add 0.5 mole of NaOH to 1.0 liter of 1.0 $\underline{M}$ Na_3A

20.7*Calculate the pH of a solution made by mixing 50 mls of 0.40 molar ammonia with 50 mls of 0.20 molar HCl. K_b for ammonia is 1.79×10^{-5}.

a)   4.75
b)   5.75
c)   10.25
d)   9.25
e)   8.73

20.8*How many grams of NaOH (MW = 40.0 g/mole) must be added to 1.00 liter of a .600 molar H_2A solution to prepare a buffer solution of pH = 9.95? (for H_2A, pK_1 = 5.65, pK_2 = 9.65)

a)   32.0
b)   40.0
c)   80.0
d)   1.00
e)   24.0

20.9 How many grams of NaOH (MW = 40.0 g/mole) must be added to 1.00 liter of a .800 molar H_3A solution to prepare a buffer solution of pH = 7.20? (For H_3A, pK_1 = 3.65, pK_2 = 7.05, pK_3 = 11.55)

a)   64.0
b)   1.27
c)   50.7
d)   32.0
e)   58.7

20.10 89.0 milliliters of .250 molar monoprotic weak acid are titrated with 0.250 molar NaOH. What is the pH at the equivalence point? (The dissociation constant of the weak acid is 9.28×10^{-6}.)

a)   9.21
b)   8.97
c)   13.70
d)   9.06
e)   5.03

20.11* 67.0 milliliters of a .350 molar hypothetical weak base, B, are titrated with 0.250 molar HCl. What is the pH at the equivalence point? (The dissociation constant, K_b, of the weak base is 3.47 x 10^{-6}.)

    a)   7.00
    b)   2.96
    c)   4.69
    d)   9.31
    e)   5.46

20.12* 60.0 milliliters of a .350 M solution of a weak acid are titrated with 0.250 molar NaOH. What is the pH when 57.0% of the acid has been titrated? (The dissociation constant of the weak acid is 8.87 x 10^{-8}.)

    a)   7.17
    b)   7.05
    c)   3.75
    d)   7.00
    e)   6.93

20.13 72.0 milliliters of a .105 molar weak base are titrated with 0.250 molar HCl. What is the pH when 21% of the base has been titrated? (The ionization constant, K_b, of the weak base is 4.10 x 10^{-6}.)

    a)   7.08
    b)   9.19
    c)   5.39
    d)   8.03
    e)   10.82

20.14* Given the following table of indicators for acid-base titrations:

Indicator #	1	2	3	4	5	6	7	8	9	10	11	12
pH range of color change	0-2	1-3	2-4	3-5	4-6	5-7	6-8	7-9	8-10	9-11	10-12	11-13

Which indicator should be chosen for a titration of 50.0 mls of a 0.100 molar solution of HA with 0.100 molar NaOH? The dissociation constant, K_a, for HA is 1.48 x 10^{-9}.

    a)   3
    b)   5
    c)   7
    d)   10
    e)   11

20.15 A solution of an unknown monoprotic weak acid was titrated with
standard base. The equivalence point was reached when 38.76 ml
of 0.125 $\underline{M}$ NaOH was added. Then 20.00 ml of 0.100 $\underline{M}$ HNO_3 was added
to the solution and the pH was 5.06. Calculate the K_a for the acid.

a) 2.11×10^{-6}
b) 3.60×10^{-6}
c) 6.12×10^{-6}
d) 8.71×10^{-6}
e) 1.24×10^{-5}

CHAPTER 21. SOLUBILITY

Introduction

The solubility of ionic compounds in water is represented by
a reversible reaction, for example,

$$CaF_2(s) \rightleftharpoons Ca^{2+}(aq) + 2F^-(aq) \qquad K_{sp} = [Ca^{2+}][F^-]^2$$

The equilibrium constant is called the "solubility product". Only the
species in solution are considered because the pure solid by definition
has unit thermodynamic activity.

Example 21.1 The solubility of CaF_2 at room temperature is 0.017 gram
CaF_2 per liter of solution. Calculate the solubility product of CaF_2.
The molecular weight of CaF_2 is 79.1 g/mole.

Solution   The equilibrium concentrations are

$$[Ca^{2+}] = \frac{0.017}{79.1} = 2.1 \times 10^{-4} \underline{M} \qquad\qquad [F^-] = 2\left(\frac{0.017}{79.1}\right) = 4.3 \times 10^{-4} \underline{M}$$

So $\qquad K_{sp} = [Ca^{2+}][F^-]^2 = (2.1 \times 10^{-4})(4.3 \times 10^{-4})^2 = 3.9 \times 10^{-11}$

Example 21.2 The solubility product of Ag_2S at room temperature is
5.5×10^{-51}. Calculate the solubility of Ag_2S in pure water and the
equilibrium concentrations of Ag^+ and S^{2-}.

Solution Let x be the solubility of Ag_2S.

$$Ag_2S(s) \rightleftharpoons 2Ag^+ + S^{2-}$$

Then $\qquad [Ag^+] = 2x$ and $[S^{2-}] = x$

$$K_{sp} = [Ag^+]^2[S^{2-}] = (2x)^2(x) = 4x^3 = 5.5 \times 10^{-51}$$

$$x = \left(\frac{5.5 \times 10^{-51}}{4}\right)^{1/3} = 1.1 \times 10^{-17} \underline{M}$$

So $\qquad [Ag^+] = 2.2 \times 10^{-17}\underline{M}$ and $[S^{2-}] = 1.1 \times 10^{-17}\underline{M}$

188

The solubility product depends on the temperature. In the following examples and questions it will be assumed that the temperature is 25°C and constant.

The Common Ion Effect

The solubility of an ionic compound in water is reduced if the solution contains either the same cation or anion as the compound. For example, the solubility of Ag_2S is less in a solution containing either $AgNO_3$ or Na_2S than it is in pure water. This is the common ion effect.

Example 21.3 Calculate the solubility of CaF_2 in a solution that is 0.065 $\underline{M}$ in $Ca(NO_3)_2$. Also calculate the equilibrium concentrations of Ca^{2+} and F^-. K_{sp} for CaF_2 is 3.9×10^{-11}.

Solution $Ca(NO_3)_2$ is a strong electrolyte:

$$Ca(NO_3)_2(s) \xrightarrow{\ 100\%\ } Ca^{2+} + 2NO_3^-$$

The $[Ca^{2+}]$ from $Ca(NO_3)_2$ is 0.065 $\underline{M}$. Let x be the solubility of CaF_2.

$$CaF_2(s) \rightleftharpoons Ca^{2+} + 2F^-$$

$$0.065 + x \qquad 2x$$

$$K_{sp} = [Ca^{2+}][F^-]^2 = (0.065 + x)(2x)^2 = 3.9 \times 10^{-11}$$

Assume x << 0.065. Then

$$K_{sp} = [Ca^{2+}][F^-]^2 \approx (0.065)(2x)^2 = 3.9 \times 10^{-11}$$

$$x = 1.2 \times 10^{-5} \underline{M}$$

The assumption that x is small compared to 0.065 is justified. The solubility of CaF_2 is 1.2×10^{-5} $\underline{M}$, $[Ca^{2+}] = 0.065$ $\underline{M}$, and $[F^-] = 2x = 2.4 \times 10^{-5}$ $\underline{M}$. Note that the solubility of CaF_2 is decreased by a factor of 20 in 0.065$\underline{M}$ $Ca(NO_3)_2$ compared to pure water (see example 21.1)

Example 21.4 Consider a solution prepared by mixing 75 ml of 0.050$\underline{M}$ K_2CrO_4 and 25 ml of 0.010$\underline{M}$ $AgNO_3$. Will Ag_2CrO_4 precipitate? Calculate the equilibrium concentrations of all ions in solution. K_{sp} for Ag_2CrO_4 is 1.1×10^{-12}.

<u>Solution</u> Let P be the "ion product", which is the same expression as for K_{sp} but in which the concentrations are not necessarily equilibrium values.

$$P = [Ag^+]^2[CrO_4^{2-}]$$

If P is less than K_{sp}, there will be no precipitate formation. If P exceeds K_{sp}, Ag_2CrO_4 will precipitate decreasing the concentrations of Ag^+ and CrO_4^{2-} until the ion product is equal to K_{sp}.

$$\text{mmoles } Ag^+ = 0.010 \times 25 = 0.25 \;;\; [Ag^+] = \frac{.25 \text{ mmoles}}{(25 + 75)\text{ml}} = 2.5 \times 10^{-3} \underline{M}$$

$$\text{mmoles } CrO_4^{2-} = 0.050 \times 75 = 3.75;\; [CrO_4^{2-}] = \frac{3.75 \text{ mmoles}}{(25 + 75)\text{ ml}} = 3.75 \times 10^{-2} \underline{M}$$

$$P = [Ag^{2+}]^2[CrO_4^{2-}] = (2.5 \times 10^{-3})^2(3.75 \times 10^{-2}) = 2.3 \times 10^{-7}$$

Since P is greater than K_{sp}, Ag_2CrO_4 will precipitate. To determine the equilibrium concentrations of Ag^+ and CrO_4^{2-} it is convenient to assume the precipitation reaction goes to completion and make the stoichiometric calculation:

$$2Ag^+ \quad + \quad CrO_4^{2-} \quad \xrightarrow{\;100\%\;} \quad Ag_2CrO_4(s)$$

	$2Ag^+$	CrO_4^{2-}	$Ag_2CrO_4(s)$
Initially (mmoles)	0.25	3.75	0
After (mmoles)	0	$3.75 - \dfrac{0.25}{2}$	$\dfrac{0.25}{2}$

The concentration of CrO_4^{2-} (the excess reagent) after precipitation is

$$[CrO_4^{2-}] = \frac{3.75 - \dfrac{0.25}{2}}{(25 + 75)} = 3.62 \times 10^{-2} \underline{M}$$

Now let x be the solubility of the Ag_2CrO_4. Then $[Ag^+] = 2x$, and $[CrO_4^{2-}] = 0.0362 + x$. Assuming that x is small compared to 0.0362,

$$K_{sp} = [Ag^+]^2[CrO_4^{2-}] = (2x)^2(0.0362 + x) \simeq (2x)^2(0.0362) = 1.1 \times 10^{-12}$$

$$x = 2.8 \times 10^{-6} \underline{M}.$$

$$[Ag^{2+}] = 2x = 5.6 \times 10^{-6} \underline{M}, \; [CrO_4^{2-}] = 0.0362 \underline{M},$$

$$[NO_3^-] = 2.5 \times 10^{-3} \underline{M} \text{ and } [K^+] = 2 \times 3.75 \times 10^{-2} = 7.5 \times 10^{-2}$$

Note that $\quad [Ag^+] + [K^+] = [NO_3^-] + 2[CrO_4^{2-}] = 0.075 \underline{M}$

Simultaneous Equilibria

When a system involving several reversible reactions is in equilibrium, all equilibrium constants must be satisfied simultaneously.

The reactions considered here include those that involve both solubility and acid-base reactions, or solubility and complexation reactions, for example,

$$AgCN(s) + H^+ \rightleftharpoons Ag^+ + HCN \qquad\qquad K_{eq} = K_{sp}/K_a$$
$$AgCl(s) + 2NH_3 \rightleftharpoons Ag(NH_3)_2^+ + Cl^- \qquad\qquad K_{eq} = K_{sp}K_f$$

Here $\qquad K_a = \dfrac{[H^+][CN^-]}{[HCN]} \qquad\qquad K_f = \dfrac{[Ag(NH_3)_2^+]}{[Ag^+][NH_3]^2} \quad ;$

$K_{sp} = [Ag^+][CN^-]$ in the first case, and $[Ag^+][Cl^-]$ in the second case.

These reactions indicate that the solubility of AgCN will be greater in an acidic solution than in pure water, and that the solubility of AgCl will be greater in the presence of NH_3 than in pure water.

<u>Example 21.5</u> Consider the conditions of the previous example except that the resultant solution is buffered at pH = 1.00. Will Ag_2CrO_4 precipitate? K_a for $HCrO_4^-$ is 3.0×10^{-7}.

<u>Solution</u> First calculate the $[H^+]$; $[H^+] = 10^{-1.00} = 0.10\underline{M}$. Since the solution is buffered this value is constant. Assume there is no precipitation of Ag2CrO4 initially. The chromate will react with H^+,

$$H^+ + CrO_4^{2-} \rightleftharpoons HCrO_4^- \qquad\qquad K_{eq} = \frac{1}{K_a} = 3.3 \times 10^6$$

Thus to a good approximation this reaction is quantitative, and $[HCrO_4^-] = 3.75 \times 10^{-2}\underline{M}$ (the value of CrO_4^{2-} in the previous problem). Then the equilibrium concentration of CrO_4^{2-} is

$$[CrO_4^{2-}] = \frac{K_a[HCrO_4^-]}{[H^+]} = \frac{(3.0 \times 10^{-7})(3.75 \times 10^{-2})}{(.10)} = 1.1 \times 10^{-7}$$

The ion product is

$$P = [Ag^+]^2[CrO_4^{2-}] = (2.5 \times 10^{-3})^2(1.1 \times 10^{-7}) = 6.9 \times 10^{-13}$$

Ag2CrO4 will <u>not</u> precipitate because P is less than the K_{sp}. The presence of acid in this case has kept the Ag_2CrO_4 in solution, in contrast to the previous problem.

Questions

21.1 The solubility of $PbCl_2$ is 8.1g $PbCl_2$ per liter of solution. What is the solubility product of $PbCl_2$? The molecular weight of $PbCl_2$ is 278 g/mole.

    a) $(8.1)^2$

    b) $4(8.1)^3$

    c) $(8.1/278)^2$

    d) $4(8.1/278)^3$

    e)   None of the above

21.2 The solubility product for CuCl is 3.2×10^{-7}. What is the molar solubility of CuCl in pure H_2O?

    a) 5.7×10^{-4}
    b) 3.2×10^{-7}
    c) 1.6×10^{-7}
    d) 6.8×10^{-3}
    e)   6.5

21.3 The solubility product for $PbCl_2$ is 1.0×10^{-4}. What is the molar solubility of $PbCl_2$ in pure H_2O? Neglect the hydrolysis of Pb^{2+}.

    a)   .046
    b)   4.0
    c)   .029
    d)   .010
    e) 1.0×10^{-4}

21.4*The solubility product for Ag_2CO_3 is 8.2×10^{-12}. What is the molar solubility of Ag_2CO_3 in pure H_2O? Neglect the hydrolysis of CO_3^{2-}.

    a)   0.00020
    b) 1.4×10^{-6}
    c)   0.00013
    d) 8.2×10^{-12}
    e) 2.9×10^{-6}

21.5 Find the molar solubility of CuSCN in a 1.2 M solution of $CuNO_3$. K_{sp} for CuSCN $= 4.0 \times 10^{-14}$.

    a)   1.2 molar
    b) 4.0×10^{-14} molar
    c) 1.8×10^{-7} molar
    d) 2.0×10^{-7} molar
    e) 3.3×10^{-14} molar

192

21.6 Calculate the molar solubility of Ag_2CrO_4 in a solution of 0.820
molar $AgNO_3$. K_{sp} for Ag_2CrO_4 = 1.1 x 10^{-12}; neglect the hydrolysis
of CrO_4^{2-}.

a)   1.6 x 10^{-12}
b)   1.3 x 10^{-6}
c)   6.5 x 10^{-5}
d)   1.3 x 10^{-12}
e)   1.1 x 10^{-12}

21.7*Consider a solution prepared by mixing 250 mℓ of 0.040$\underline{M}$ TlNO$_3$ and
750 mℓ of 0.040$\underline{M}$ Na$_2$C$_2$O$_4$. Calculate the equilibrium concentration
of Tl$^+$. K_{sp} for Tl$_2$C$_2$O$_4$ is 1.1 x 10^{-7}. Neglect the hydrolysis of
C$_2$O$_4^{2-}$.

a)   (1.1 x 10^{-7})/(0.025)

b)   [(1.1 x 10^{-7})/(0.025)]$^{1/2}$

c)   (1.1 x 10^{-7})/(0.010)

d)   [(1.1 x 10^{-7})/(0.010)]$^{1/2}$

e)   None of the above

21.8 What is the pH of a saturated solution of $Zn(OH)_2$? K_{sp} for
$Zn(OH)_2$ is 2.0 x 10^{-14}.

a)   4.47
b)   4.77
c)   9.23
d)   9.53
e)   13.70

21.9*What is the pH of a saturated solution of $Co(OH)_3$?
K_{sp} for $Co(OH)_3$ is 2.5 x 10^{-43}.

a)   2.99
b)   3.47
c)   10.53
d)   11.01
e)   Approximately 7

21.10*It is proposed to fluoridate a city's water supply by adding sodium
fluoride, NaF. However, the city's water is hard with a calcium ion
concentration of 5.6 x 10^{-4} molar. What is the maximum weight of
sodium fluoride that can be added per liter of water without precipitating CaF_2? (K_{sp} for CaF_2 is 3.9 x 10^{-11}; neglect the hydrolysis
of F$^-$)

a)   2.9 x 10^{-6} g
b)   5.5 x 10^{-3} g
c)   1.5 x 10^{-6} g
d)   3.5 x 10^{-2} g
e)   1.1 x 10^{-2} g

21.11*In an analytical procedure for SO_2 in stack gas the SO_2 is oxidized by H_2O_2 in aqueous solution to SO_4^{2-} which is then precipitated as $BaSO_4$ with excess $BaCl_2$ solution. If the precipitation is made from a solution of total volume 250 ml what should be the concentration of excess Ba^{2+} so that not more than 2.6×10^{-5} g of sulfur is lost as sulfate ion in the supernatant liquid? (K_{sp} for $BaSO_4$ is 1.5×10^{-9})

    a) 1.2×10^{-4} M
    b) 3.4×10^{-3} M
    c) 1.4×10^{-3} M
    d) 1.5×10^{-4} M
    e) 4.6×10^{-4} M

21.12*The waste solution from a silver plating operation contains 2.65×10^{-5} $\underline{M}$ Ag^+. How many grams of silver can be recovered per liter by adding Na_2CO_3 to precipitate Ag_2CO_3 until the excess CO_3^{2-} concentration reaches 2.5×10^{-2} M. K_{sp} for $Ag_2CO_3 = 8.2 \times 10^{-12}$.

    a) 9.10×10^{-4} g
    b)   0.00 g
    c) 8.40×10^{-6} g
    d) 1.96×10^{-3} g
    e) 2.96×10^{-3} g

21.13*The solubility product of AgCl is 1.7×10^{-10} and for AgI is 8.5×10^{-17}. If 0.015 moles of $AgNO_3$ is added to a one liter solution containing 0.0045 moles of NaCl and 0.0065 moles of NaI, what will be the result?

    a) Both $[Cl^-]$ and $[I^-]$ are very small and in the ratio $\dfrac{[Cl^-]}{[I^-]} = \dfrac{.0045}{.0065}$

    b) $[Cl^-] = [I^-] = 2.1 \times 10^{-14}$

    c) The ratio $\dfrac{[Cl^-]}{[I^-]} = \dfrac{.0045}{.0065} \times \dfrac{1.7 \times 10^{-10}}{8.5 \times 10^{-17}}$

    d) $[Cl^-] = 4.3 \times 10^{-8}$ M

    e) $[Cl^-] = 1.3 \times 10^{-5}$ M

21.14*A mixture of solid $BaSO_4$ and $BaCO_3$ is in equilibrium with a solution containing SO_4^{2-} and CO_3^{2-}. If $[SO_4^{2-}]$ is 1.0×10^{-2} $\underline{M}$, what is $[CO_3^{2-}]$. For $BaSO_4$, $K_{sp} = 1.0 \times 10^{-10}$; for $BaCO_3$, $K_{sp} = 8.0 \times 10^{-9}$.

    a) $1.0 \times 10^{-10} / 1.0 \times 10^{-2}$

    b) $8.0 \times 10^{-9} / 1.0 \times 10^{-2}$

    c) $\dfrac{1.0 \times 10^{-10} \times 1.0 \times 10^{-2}}{8.0 \times 10^{-9}}$

(see next)

194

d)
$$\frac{1.0 \times 10^{-10}}{8.0 \times 10^{-9} \times 1.0 \times 10^{-2}}$$

e)   None of the above.

21.15*A solution is to be analyzed for its chloride content. The
analyst adds 0.0050 M $CrO_4^=$ as an indicator and titrates it with
Ag^+. AgCl is less soluble and precipitates first, but eventually
the red compound Ag_2CrO_4, also precipitates. What is the $[Cl^-]$
when Ag_2CrO_4 first begins to form? (Assume the dilution factor
is negligible.)

$$K_{sp} \text{ for AgCl} = 1.7 \times 10^{-10}; \ K_{sp} \text{ for } Ag_2CrO_4 = 1.9 \times 10^{-12}$$

a)   0.45
b) 1.1×10^{-5}
c) 1.3×10^{-5}
d) 3.3×10^{-6}
e) 8.7×10^{-6}

21.16*A solution contains 0.010 M Fe^{2+} and 0.020 M Mn^{2+}. To what pH
should the solution be adjusted to precipitate as much of one
metal (as the hydroxide) as possible, without precipitating the
other metal hydroxide?

$$Mn(OH)_2 \rightleftharpoons Mn^{2+} + 2 \ OH^- \qquad K_{sp} = 4.5 \times 10^{-14}$$

$$Fe(OH)_2 \rightleftharpoons Fe^{2+} + 2 \ OH^- \qquad K_{sp} = 2.0 \times 10^{-15}$$

a)   7.65
b)   6.48
c)   8.18
d)   2.34
e)   11.66

21.17*A solution contains 0.010 M Fe^{2+} and 0.020 M Mn^{2+}. By adding OH^-
cautiously the less soluble metal can be precipitated as the hydroxide
while the other remains in solution. To what concentration can the
less soluble metal be reduced before the other metal starts to
precipitate? (See question 21.16 for K_{sp} values.)

a) 2.3×10^{-1} M
b) 7.9×10^{-6} M
c) 1.5×10^{-6} M
d) 8.9×10^{-4} M
e) 1.0×10^{-2} M

21.18 What is the value of the Cd^{2+} ion concentration that can exist in
a solution which is saturated with H_2S and buffered at a pH of 5.10?
(Assume that the solubility of H_2S is 0.10 mole per liter.)

(see next)

$$CdS \rightleftharpoons Cd^{2+} + S^{2-} \qquad K_{sp} = 1.0 \times 10^{-28}$$

$$H_2S \rightleftharpoons H^+ + HS^- \qquad K_1 = 9.1 \times 10^{-8}$$

$$HS^- \rightleftharpoons H^+ + S^{2-} \qquad K_2 = 1.1 \times 10^{-12}$$

a) 1.0×10^{-28} M
b) 1.0×10^{-27} M
c) 5.2×10^{-34} M
d) 1.0×10^{-14} M
e) 6.3×10^{-19} M

21.19 Calculate the molar solubility of AgCl in a 2.2 $\underline{M}$ NH_3 solution. K_{sp} for AgCl is 1.7×10^{-10}, and K_f for $Ag(NH_3)_2^+$ is 1.7×10^7.

a) 1.6×10^{-9}
b) 0.24
c) 1.3×10^{-5}
d) 0.014
e) 0.11

21.20*What is the solubility of AgCN in a solution buffered at pH = 6.48? For AgCN, $K_{sp} = 1.6 \times 10^{-14}$; for HCN, $K_a = 2.1 \times 10^{-9}$.

a) 1.3×10^{-7} $\underline{M}$
b) 1.6×10^{-14} $\underline{M}$
c) 1.6×10^{-6} $\underline{M}$
d) 14 $\underline{M}$
e) 0.0028 $\underline{M}$

CHAPTER 22. THERMODYNAMICS

ΔE and ΔH, Internal Energy and Ethalpy

The change in internal energy (ΔE) of a system is defined by the first law of thermodynamics,

$$\Delta E = q - w$$

Here q is the heat absorbed by the system and w is the work done by the system on the surroundings. In some texts the symbol U is used rather than E. ΔE for a chemical reaction is the observed heat of the reaction (q) when it is carried out at constant volume, as in a bomb calorimeter. Under these conditions the system can do no work, so w = o, and $\Delta E = q_V$, where q_V indicates the heat absorbed in a constant volume process

The enthalpy (H) is defined as follows,

$$H = E + PV$$

For a constant pressure process: $\Delta H = \Delta E + P\Delta V$

and it can be shown that $\Delta H = q_p$, where q_p is the heat absorbed in a constant pressure process.

The work term, $P\Delta V$, is usually small compared to ΔH and ΔE unless the reaction has a large volume change. The molar volume of a solid or liquid is negligible compared to the molar volume of a gas, so ΔV is not significant unless there is a change in the number of moles of gas in going from reactants to products.

Assume that all gases obey the ideal gas law, PV = nRT. Recall that the enthalpy of reaction is defined for a constant temperature process. Hence for P and T constant

$$P\Delta V = \Delta n_g RT$$

Here Δn_g is the change in the number of moles of gaseous reactants and products. For example,

Reaction	Δn_g
$2NH_3(g) \longrightarrow N_2(g) + 3H_2(g)$	+2
$CaCO_3(s) + SO_3(g) \longrightarrow CaSO_4(s) + CO_2(g)$	0
$SO_2(g) + H_2O(\ell) \longrightarrow H_2SO_3(aq)$	-1

The value of the ideal gas constant (R) in the appropriate units is 8.314 J/mole-K. To a good approximation ΔH and ΔE for a reaction are related by the equation

$$\Delta H = \Delta E + \Delta n_g RT$$

Example 22.1   The observed heat of the reaction

$$2H_2(g) + O_2(g) \longrightarrow 2H_2O(\ell)$$

measured using a **bomb** calorimeter, is 278.4 kJ of heat evolved per mole of O_2 consumed at 25°C. Calculate the enthalpy change of the reaction as written.

Solution $\Delta n_g = -3$

$$\Delta H = \Delta E + \Delta n_g RT$$

$$= -278.4 \text{ kJ} + (-3 \text{ mole})(8.314 \text{ J mole}^{-1}K^{-1})(298K) \frac{1kJ}{1000J}$$

$$= -278.4 - 7.4$$

$$= -285.8 \text{ kJ}$$

ΔS, ΔG, and K_{eq}, Entropy, Free Energy, and Equilibrium

It is a common experience that very exothermic reactions (ΔH very negative) tend to go to completion (for example combustion reactions). However there are some spontaneous processes that are endothermic (for example the dissolving of ammonium chloride). Thus, although a negative ΔH does favor a reaction it is not a sufficient criterion for spontaneity. Also to be considered is the change in randomness, or entropy (S) which occurs. Reactions are favored by an increase in the randomness, i.e. positive ΔS.

<u>Example 22.2</u>   For which of the following reactions is the increase in entropy the greatest?

a) $Fe(s) + 2HCl(g) \longrightarrow FeCl_2(s) + H_2(g)$

b) $H_2SO_4(\ell) + CaCO_3(s) \longrightarrow CaSO_4(s) + H_2O(\ell) + CO_2(g)$

c) $C_6H_{12}(g) \longrightarrow C_6H_6(g) + 3H_2(g)$

<u>Solution</u>   In general the entropy of a solid is less than that of a liquid, which is in turn less than that of a gas,

$$S_{solid} < S_{liquid} < S_{gas}$$

Thus for reactions which involve gases it is usually safe to concentrate on the net change in number of moles of gases.   In reaction (a) there is a decrease in entropy ($\Delta S < 0$), since there is a net decrease of one mole of gas.   Both reactions (b) and (c) will have an increase in entropy ($\Delta S > 0$), since there are more gaseous products than reactants in each case.   Reaction (c) has the greater increase in entropy since the increase in number of moles of gas is 3 for reaction (c) and only 1 for reaction (b).

It is observed that reactions accompanied by an increase in entropy ($\Delta S > 0$) tend to occur spontaneously.   The thermodymanic function that determines the spontaneity of a chemical reaction is the free energy, G.

$$G \equiv H - TS \qquad \text{and} \qquad \Delta G = \Delta H - T\Delta S$$

Under conditions of constant temperature and pressure, if the change in free energy of a reaction is negative ($\Delta G < 0$), then the reaction will occur spontaneously in the written direction.   If ΔG is greater than zero, then the reaction will not occur spontaneously in that forward direction, but will occur spontaneously in the reverse direction.   If ΔG is equal to zero the reaction is in equilibrium, that is, both forward and reverse reactions occur at equal rates, and there is no net change in composition.   The conditions that indicate the spontaneity of a reaction are indicated in the following table.

$$A + B \rightleftharpoons C + D$$

$$\Delta G = \Delta H - T\Delta S$$

ΔH	ΔS	ΔG	Direction of Spontaneous Reaction
<0	>0	<0	→
>0	<0	>0	←
<0	<0	?	?
>0	>0	?	?

In the latter two cases the direction will depend on which term, ΔH or $T\Delta S$, is numerically greater. In the last case, for example, if $|T\Delta S| > |\Delta H|$ the reaction is said to be entropy driven, despite the unfavorable enthalpy change.

For judging a chemical reaction one considers a mixture of standard states of reactants and products, and evaluates $\Delta G°$, the standard free energy change. One may calculate $\Delta G°$ for a reaction from tabulated values of $\Delta G_f°$, the standard free energies of formation of the compounds, in exactly the same way as $\Delta H°$ was calculated in Chapter 9. (There are no absolute G values just as there are no absolute H values). Alternatively $\Delta G°$ may be calculated from the relationship:

$$\Delta G° = \Delta H° - T\Delta S°$$

For many compounds absolute entropies are known and tabulated. S increases slowly with temperature, and the values tabulated at 298K and 1.00 atm are called standard entropies, $S°$. The absolute entropy scale is set by the third law of thermodynamics which states that the entropy of a perfect crystal is zero at the absolute zero of temperature.

Thermodynamic data are obtained experimentally in many ways. Direct methods include:

(a) ΔE and ΔH by calorimetry,
(b) ΔG by electrochemical potential (see Chapter 24) and by Keq measurement, and
(c) S by specific heat measurements down to low temperatures (not discussed in this text).

Furthermore there are indirect methods based upon thermodynamic relationships, such as obtaining ΔH by measuring Keq as a function of temperature (read ahead).

If $\Delta G°$ is negative a mixture of standard states will react in the direction of more product formation. The more negative $\Delta G°$ is the further toward completion the reaction will go. It can be shown that the equilibrium constant Keq is related to $\Delta G°$ as follows:

$$\Delta G° = -RT\ln Keq = -2.303 \; RT \; \log Keq$$

To use the above expression correctly Keq must by in the proper units, consistent with the definition of the standard states for which $\Delta G°$ is defined. For substances in solution the standard state is one molar (1.00 $\underline{M}$) and for gases it is 1.00 atmosphere. Hence for substances in solution Keq is expressed in molar concentrations, but for gases Keq is expressed in atmospheres. In general the units of Keq are those for which the standard state is defined as one unit.

<u>Example 22.3</u> Consider the following reaction and associated thermodynamic values:

$$NH_4HCO_3(aq) \; \rightleftharpoons \; NH_3(aq) + CO_2(aq) + H_2O(\ell)$$

Compound	NH_4HCO_3(aq)	NH_3(aq)	CO_2(aq)	$H_2O(\ell)$
ΔH_f° (kJ/mole)	−825	−79.5	−414	−284
S_{298}° (J/mole-K)	208	108	121	71.2

Will this reaction occur spontaneously if all reactants and products are present in their standard states? Calculate the equilibrium constant for the reaction.

Solution: First calculate ΔH° and ΔS° for the reaction

$$\Delta H^\circ = -79.5 - 414 - 284 - (-825) = 47.5 \text{ kJ}$$

$$\Delta S^\circ = 108 + 121 + 71.2 - 208 = 92.2 \text{ J/K}$$

To determine whether the reaction is spontaneous in the forward direction it is necessary to calculate ΔG°,

$$\Delta G^\circ = \Delta H^\circ - T\Delta S^\circ$$

$$= 47.5 - (298)(92.2)/1000$$

$$= 20.0 \text{ kJ}$$

Since $\Delta G^\circ > 0$, the reaction is <u>not</u> spontaneous in the foward direction. The equilibrium constant is calculated as follows:

$$\Delta G^\circ = -RT \ln K_{eq}$$

$$K_{eq} = \exp(-\Delta G^\circ/RT) = 10^{\frac{-\Delta G^\circ}{2.303 \, RT}}$$

$$= \exp(-20 \times 10^3/8.314 \times 298) = 10^{\frac{-20.0 \times 10^3}{2.303 \times 8.314 \times 298}}$$

$$= 3.12 \times 10^{-4} = \frac{[NH_3][CO_2]}{[NH_4HCO_3]}$$

Example 22.4 The following reaction has been suggested as a means of removing noxious NO(g) from a stream of furnace gases.

$$2 \text{ NO(g)} + 2 \text{ H}_2\text{(g)} \rightleftharpoons \text{N}_2\text{(g)} + 2 \text{ H}_2\text{O}(\ell)$$

Given the following standard free energies of formation, calculate the equilibrium constant for this reaction at 25°C.

Compound	NO(g)	H_2(g)	N_2(g)	$H_2O(\ell)$
ΔG_f° (kJ/mole)	86.69	0.0	0.0	−237.2

Solution

$$\Delta G^\circ = \Delta G_f^\circ (N_2) + 2\Delta G_f^\circ (H_2O) - \{2\Delta G_f^\circ (NO) + 2\Delta G_f^\circ (H_2)\}$$

$$= 0 + 2(-237.2) - (2 \times 86.69 + 0)$$

$$= -647.8 \text{ kJ}$$

$$\Delta G^\circ = -RT \ln Keq = -2.303 \log Keq$$

$$\log Keq = \frac{-\Delta G^\circ}{2.303\ RT} = \frac{647.8 \times 10^3}{2.303 \times 8.314 \times 298} = 113.53$$

$$Keq = 3.4 \times 10^{113} = \frac{P_{N_2}}{P_{NO}^2\ P_{H_2}^2}$$

Keq as a Function of Temperature

The standard enthalpy charge for a reaction determines the dependence of the equilibrium constant on temperature,

$$\Delta G^\circ = \Delta H^\circ - T\Delta S^\circ = -RT \ln Keq$$

$$\ln Keq = \frac{\Delta S^\circ}{R} - \frac{\Delta H^\circ}{RT}$$

If both ΔS° and ΔH° are constant over the temperature range $T_1 \leqslant T \leqslant T_2$, then

$$\ln\left(\frac{K_1}{K_2}\right) = \frac{\Delta H^\circ}{R}\left\{\frac{T_1 - T_2}{T_1 T_2}\right\}$$

or

$$\log\left(\frac{K_1}{K_2}\right) = \frac{\Delta H^\circ}{2.303\ R}\left\{\frac{T_1 - T_2}{T_1 T_2}\right\}$$

where K_1 and K_2 are the Keq values at T_1 and T_2 respectively. This equation is known as the van't Hoff equation. If the enthalpy charge for a reaction is known and can be assumed constant, and if the equilibrium constant is known at one temperature, then it is possible to estimate the equilibrium constant at another temperature. If the equilibrium constant is known at two temperatures, then it is possible to estimate the enthalpy change for the reaction, thus providing an important alternative to calorimetry for measuring ΔH°.

<u>Example 22.5</u> The solubility product of $Fe(OH)_3$ is 1.1×10^{-36} at $18°C$. Estimate its value at $43°C$ assuming the enthalpy charge for the solubility reaction is constant over this temperature range. The standard heats of formation are given below.

Compound or ion	$Fe(OH)_3$	Fe^{3+}	OH^-
ΔH_f° (kJ/mole)	−824.2	−47.7	−230.0

202

Solution

$$\Delta H^\circ = -47.7 + 3(-230.0) - (-824.2) = +86.5 \text{ kJ}$$

Let $T_1 = 43 + 273 = 316$ K and $T_2 = 18 + 273 = 291$ K

$K_1 = ?$ and $K_2 = 1.1 \times 10^{-36}$

then

$$\ln\left(\frac{K_1}{1.1 \times 10^{-36}}\right) = \frac{86.5 \times 10^3}{8.314}\left\{\frac{316 - 291}{316 \times 291}\right\} = 2.83$$

$$K_1 = 1.1 \times 10^{-36} \exp(2.83)$$

$$= 1.9 \times 10^{-35}$$

The solubility of $Fe(OH)_3$ increases with temperature. This is consistent with Le Chatelier's principle and the fact that the reaction is endothermic.

Example 22.6 NH_4Cl is a stable compound at room temperature,

$$NH_4Cl(s) \xrightarrow{\quad X \quad} NH_3(g) + HCl(g)$$

Estimate the temperature at which NH_4Cl will be in equilibrium with $NH_3(g)$ and $HCl(g)$, with both gases at 1.00 atm, given the following thermodynamic values,

Compound	$NH_4Cl(s)$	$NH_3(g)$	$HCl(g)$
ΔH°_f (kJ/mole)	-315.5	-46.0	-92.5
S°_{298} (J/mole-K)	94.6	192	187

Solution

$$\Delta H^\circ = \Delta H^\circ_f(NH_3) + \Delta H^\circ_f(HCl) - \Delta H^\circ_f(NH_4Cl)$$

$$= -46.0 + (-92.5) - (-315.5)$$

$$= 177 \text{ kJ}$$

$$\Delta S^\circ = 187 + 192 - 94.6 = 284 \text{ J/k}$$

$$\Delta G^\circ = \Delta H^\circ - T\Delta S^\circ$$

$$= 177 - 298 \times 284/1000$$

$$= 92.4 \text{ kJ}$$

The large positive value of ΔG° indicates that the equilibrium lies far to the left at 25°C. At the desired conditions $Keq = P_{NH_3} \times P_{HCl} = (1.00)(1.00)$; $\ln Keq = \ln 1 = 0$. The temperature at which the equilibrium constant is unity is calculated as follows:

$$\Delta G^\circ = -RT \, \ln \, K_{eq} = 0 = \Delta H^\circ - T\Delta S^\circ$$

$$T = \frac{\Delta H^\circ}{\Delta S^\circ} = \frac{177 \times 10^3}{284} = 623 \text{ K, or } 350^\circ\text{C}$$

Questions

22.1 Given the following table of standard enthalpies of formation of the solid compounds (kJ/mole at 298 K):

Compound	ΔH°_f
NaF	−569
MgS	−347
Na_2S	−373
MgF_2	−1102

Calculate the standard enthalpy change for the following reaction in kJ/mole MgF_2 formed.

$$NaF + MgS \rightarrow Na_2S + MgF_2$$

a) −686
b) 561
c) 10
d) −561
e) −10

22.2 Given the following table of standard entropies of the solid compounds (J/mole-K at 298 K):

Compound	S°
RbCl	96.2
BaS	72.0
Rb_2S	70.3
$BaCl_2$	125.5

Calculate the standard entropy change for the following reaction in J/K per mole $BaCl_2$ formed.

$$RbCl + BaS \rightarrow Rb_2S + BaCl_2$$

a) 27.6
b) −27.6
c) 68.6
d) −68.6
e) 364

22.3 Given the standard enthalpies of formation and third-law entropies
     tabulated below, calculate the standard free energy change (in kJ)
     at 298 K for the reaction:

$$SO_3(g) + H_2O(\ell) \rightarrow H_2SO_4(aq)$$

Compound	$SO_3(g)$	$H_2O(\ell)$	$H_2SO_4(aq)$
ΔH_f° kJ/mole	−395	−286	−908
S° J/(mole-K)	256	69.9	17.2

a) −135
b) 124
c) −1498
d) 135
e) 351

22.4* When the reaction below was carried out in a constant volume bomb
      calorimeter, 176 kJ of heat were evolved (on the basis of the
      number of moles indicated in the reaction as written). What is
      ΔH in kilojoules for this reaction? (Assume all steps are carried
      out at 25°C)

$$CO_2(g) + CaO(s) \rightarrow CaCO_3(s)$$

a) −178
b) −173
c) −176
d) +176
e) +178

22.5 In the current energy crisis methanol, CH_3OH, has been suggested as
     a fuel. The following reaction would be useful if it were thermo-
     dynamically feasible: (all substances are gases)

$$2H_2 + CO \rightarrow CH_3OH$$

To answer this question of feasibility, deduce ΔG° for the above
reaction, given the thermodynamic data for the following combustion
reaction: (all substances are gases and all units are kJ).

$$CH_3OH + 3/2\ O_2 \rightarrow CO_2 + 2H_2O \qquad \Delta G^\circ = -689.5$$
$$2H_2 + O_2 \rightarrow 2H_2O \qquad \Delta G^\circ = -456.9$$
$$CO + 1/2\ O_2 \rightarrow CO_2 \qquad \Delta G^\circ = -257.3$$

a) −889.1 kJ
b)   33.1 kJ
c) −1404 kJ
d) −24.7 kJ
e) −492.4 kJ

22.6 Given the values for the standard free energies of formation tabulated below, calculate the equilibrium constant, Keq, at 298 K for the reaction:

$$2 \text{ H}_2(g) + \text{CO}(g) \rightleftharpoons \text{CH}_3\text{OH}(g)$$

Compound	CO	CH_3OH
ΔG°_f kilojoules per mole	-137.3	-161.9

a) 2.1×10^4

b) 4.8×10^{-5}

c) 1.0

d) 3.1×10^{51}

e) .333

22.7*Which of the following reactions shows the <u>least</u> change in entropy?

a) $\text{Pt}^{4+}(aq) + 2 \text{ H}_2(g) \rightarrow \text{Pt}(s) + 4 \text{ H}^+(aq)$

b) $\text{NH}_4\text{NO}_2(s) \rightarrow 2 \text{ H}_2\text{O}(g) + \text{N}_2(g)$

c) $\text{C}_6\text{H}_{10}(g) \rightarrow \text{C}_6\text{H}_6(g) + 2 \text{ H}_2(g)$

d) $\text{HgCl}_2(s) + 2 \text{ Na}(s) \rightarrow 2 \text{ NaCl}(s) + \text{Hg}(\ell)$

e) $\text{Ag}_2\text{O}(s) + \text{H}_2(g) \rightarrow 2 \text{ Ag}(s) + \text{H}_2\text{O}(g)$

22.8 In the following reaction:

$$\text{Mg}(s) + 1/2 \text{ O}_2(g) \rightarrow \text{MgO}(s) \qquad \begin{aligned} \Delta H^\circ &= -602 \text{ kJ/mole} \\ \Delta G^\circ &= -569 \text{ kJ/mole} \end{aligned}$$

Which of the following is true?

a) The large enthalpy change insures that this reaction will spontaneously occur at any temperature.
b) The equilibrium is displaced toward the products because the unfavorable entropy change is more than overcome by the large favorable enthalpy change.
c) Since the spontaneity of this reaction is common knowledge, the entropy of the system must increase.
d) The reaction is not spontaneous because it is accompanied by a decrease in entropy.
e) The bright light emitted proves that the reaction is spontaneous.

22.9 Complete the statement below correctly. (ΔH = enthalpy change, ΔS = entropy change, Δn = change in number of moles.)

The enthalpy change in a chemical reaction:

a) is the heat obtained upon combustion in pure oxygen.
b) is the criterion for spontaneity of reaction.
c) can be predicted precisely from bond energies.
d) can be estimated from the internal energy change by a correction of $\Delta n_g RT$.
e) is equal to the voltage of the corresponding electrochemical cell times an appropriate conversion factor.

22.10*A certain substance exists in solution in two isomeric forms A and B for which the thermodynamic properties at 298 K are:

Isomer	ΔH°_f	S° J/mole-K
A	−114.50	175.03
B	−113.47	178.29

Consider the equilibrium between the two isomers:

$$A \rightleftharpoons B$$

(1) Which isomer is more stable when both are present in their standard states?

(2) Calculate the temperature at which an equilibrium mixture of the two isomers contains equal concentrations of A and B.

	(1)	(2)
a)	A	588 K
b)	A	125 K
c)	A	315 K
d)	B	65 K
e)	B	125 K

22.11 Given the following thermodynamic data at 298 K:

Compound	ΔH_f° (kJ/mole)	S° (J/mole-K)
N_2	0	191.50
O_2	0	205.02
N_2O	81.6	219.99
NO	90.4	210.62
N_2O_4	9.6	304.30

Which of the following reactions is endothermic but nevertheless spontaneous? (All substances are present in their standard states).

$$(1) \quad N_2 + O_2 \rightleftharpoons 2NO$$
$$(2) \quad N_2 + 2O_2 \rightleftharpoons N_2O_4$$
$$(3) \quad N_2 + 1/2\ O_2 \rightleftharpoons N_2O$$

a) (1) and (2)
b) (2) and (3)
c) (3)
d) All of them
e) None of them

22.12 Estimate the boiling point in °C of Br_2 given the following thermodynamic data:

	Standard Heat of Formation (kJ/mole)	Standard Entropy (joule/mole-K)
Br_2 (liq)	0.00	152.3
Br_2 (gas)	30.7	245.3

a) 57.0
b) 530
c) 76.0
d) 368
e) 95.0

22.13.* Given the following values for a hypothetical liquid at standard pressure:

Heat of vaporization = 23.0 kJ/mole
Boiling point = 148.0°C

Calculate the equilibrium vapor pressure of the liquid (in torr) at a temperature of 86.0°C.

a) 0.321
b) 284
c) 244
d) 79.6
e) 2.37×10^3

22.14 The atmospheric pressure at the top of a high mountain is 627 torr. Estimate the temperature in °C at which water will boil on this mountain top. The heat of vaporization of water is 40.38 kJ mole^{-1} at its normal boiling point of 100.0°C.

a) 82.5
b) 94.5
c) 20.0
d) 92.1
e) 367

22.15 The boiling point of a certain liquid at standard pressure (1.00 atm) is 109.0°C. The equilibrium vapor pressure of this liquid is 1020 torr at 123.8°C. Calculate the heat of vaporization of this liquid (in kJ/mole).

a) 39.83
b) 24.98
c) 34.52
d) 12.55
e) 5.94

22.16*The equilibrium constant Keq for the reaction

$$A + B \longrightleftharpoons C + D \qquad\qquad \Delta H = -4.85 \text{ kJ}$$

was found to be 12100 at 461 K. Calculate Keq at 450 K.

a) 1.25×10^4
b) 2.27×10^4
c) 1.21×10^4
d) 8.52×10^{-5}
e) 1.62×10^2

22.17 The equilibrium constant Keq for the reaction $A + B \longrightleftharpoons C + D$ was measured at two temperatures, with the results tabulated below.

Temperature, K	556	513
Equilibrium constant, Keq	3.08×10^{-4}	9.64×10^{-4}

Calculate ΔH in kilojoules for the reaction, assuming that ΔH and ΔS are constant in this temperature range.

a) .0682
b) 63.2
c) -6.32×10^4
d) 3008
e) -63.2

22.18 The solubility product of silver bromide at 25°C is 5.0×10^{-13}. What is it at 5°C?

$$AgBr(s) \rightleftharpoons Ag^+(aq) + Br^-(aq) \qquad \Delta H° = +84.5 \text{ kJ}$$

a) 5.8×10^{-12}
b) 2.3×10^{-13}
c) 4.3×10^{-14}
d) 4.7×10^{-13}
e) 3.9×10^{-14}

22.19*Pure water dissociates at 25°C:

$$H_2O \rightleftharpoons H^+ + OH^- \qquad K_w = 1.0 \times 10^{-14}; \ \Delta H° = 55.90 \text{ kJ}$$

Calculate the pH of pure water at 50°C.

a) 7.76
b) 6.24
c) 5.44
d) 7.38
e) 6.62

22.20*Which of the following statements is <u>not</u> consistent with the laws of thermodynamics?

a) The entropy change of a chemical reaction depends only on the initial and final states of the system and not on the exact path by which the reaction was carried out.
b) An endothermic reaction for which the entropy change is negative will become spontaneous as a sufficiently high temperature.
c) Energy is neither created nor destroyed in chemical reactions.
d) The free energy change of a chemical reaction is equal in magnitude but opposite in sign to the free energy change for the reverse reaction.
e) One cannot transfer heat from a cold object to a hot object unless one does some work.

CHAPTER 23. CHEMICAL EQUILIBRIUM

Introduction

The equilibrium constant for a chemical reaction is the quotient of
chemical activities, a. For example, for the simple case:

$$bB \; \rightleftharpoons \; cC$$

$$Keq = \frac{a_C^{\,c}}{a_B^{\,b}}$$

If the compound is an ideal gas, the activity is the partial pressure in
atmospheres. If the compound is a solute in an ideal solution, the
activity is the molar concentration. If the compound is a pure liquid
or solid (in a separate phase) the activity is unity.

If only concentrations are involved the symbol K_c is used and if
only partial pressures are involved the symbol K_p is used. In many texts
molar concentrations are used for gas phase reactions, in which case the
symbol K_c is used. (Note however that the Keq for gas reactions derived
from $\Delta G°$ must be K_p).

Some examples of equilibrium constants are given below:

$$N_2(g) \; + \; 3H_2(g) \; \rightleftharpoons \; 2NH_3(g) \qquad K_p = \frac{P_{NH_3}^2}{P_{N_2} \times P_{H_2}^3}$$

$$AgBr(s) \; + \; 2S_2O_3^{2-}(aq) \; \rightleftharpoons \; Ag(S_2O_3)_2^{3-}(aq) \; + \; Br^-(aq) \qquad K_c = \frac{[Ag(S_2O_3)_2^{3-}][Br^-]}{[S_2O_3^{2-}]^2}$$

$$Cl_2(g) \; + \; 2I^-(aq) \; \rightleftharpoons \; 2Cl^-(aq) \; + \; I_2(aq) \qquad Keq = \frac{[Cl^-]^2[I_2]}{P_{Cl_2}[I^-]^2}$$

The equilibrium constant for a reaction that is the algebraic sum
of two or more reactions can be calculated by multiplying the equilibrium
constants of the constituent reactions. Also if a reaction is written in
reverse Keq is the reciprocal of the forward Keq. If all coefficients
are multiplied by an integer n, Keq is raised to the n power.

$\underline{\text{Example 23.1}}$ Calculate K_p for the reaction

$$2\ SO_2(g) + O_2(g) \rightleftharpoons 2\ SO_3(g)$$

given:

$$S(s) + O_2(g) \rightleftharpoons SO_2(g) \qquad K_1 = 4.2 \times 10^{52}$$

$$2\ S(s) + 3\ O_2(g) \rightleftharpoons 2\ SO_3(g) \qquad K_2 = 5.9 \times 10^{129}$$

$\underline{\text{Solution}}$ It can be seen that the reaction of interest is the second reaction minus twice the first reaction, so write the first reaction in reverse, multiply by two, and add to the second.

$$2\ S(s) + 3\ O_2(g) \rightleftharpoons 2\ SO_3(g) \qquad K_2$$

$$2\ SO_2(g) \rightleftharpoons 2\ S(s) + 2\ O_2(g) \qquad 1/K_1^2$$

$$\rule{10cm}{0.4pt}$$

$$2\ SO_2(g) + O_2(g) \rightleftharpoons 2\ SO_3(g) \qquad K_p$$

$$K_p = \frac{P_{SO_3}^2}{P_{SO_2}^2\, P_{O_2}}$$

or

$$K_p = K_2 \times \frac{1}{K_1^2} = \frac{P_{SO_3}^2}{P_{O_2}^3} \times \frac{P_{O_2}^2}{P_{SO_2}^2} = \frac{P_{SO_3}^2}{P_{O_2}\, P_{SO_2}^2} \ , \ \text{as above}$$

$$K_p = \frac{5.9 \times 10^{129}}{(4.2 \times 10^{52})^2} = 3.3 \times 10^{24}$$

$$\rule{10cm}{0.4pt}$$

$$\underline{K_p \ \text{and} \ K_c \ \text{for Gas Phase Reactions}}$$

If all gases in a chemical reaction are ideal gases, then the partial pressure and the concentration are related as follows,

$$P = \frac{n}{V}RT = [\ ]\,RT$$

Here "[ ]" denotes the concentration in moles per liter. If [ ] RT is substituted for P in the expression for K_p the result is

$$K_p = K_c (RT)^{\Delta n_g}$$

where Δn_g is the number of moles of gaseous products minus reactants. The appropriate value of R is 0.0821 ℓ-atm/mole-K.

Example 23.2 At 450°C,

$$N_2(g) + 3\ H_2(g) \rightleftharpoons 2\ NH_3(g) \qquad Keq = K_p = 4.5 \times 10^{-5}$$

Calculate K_c for this reaction at 450°C.

Solution $\Delta n_g = 2-(1+3) = -2$, $T = 450 + 273 = 723K$

$$K_c = K_p(RT)^{-\Delta n}{}_g = K_p(RT)^2 = (4.5 \times 10^{-5})(0.0821 \times 723)^2$$
$$= 0.159$$

Measurement of Keq

The equilibrium constant is measured by determining the concentration of at least one compound in the reaction at equilibrium. If the starting concentrations are known the equilibrium concentrations of the remaining compounds can be related by stoichiometry. It is advisable to tabulate initial values, changes (which are stoichiometrically related), and final values, as shown in the following example.

Example 23.3 The equilibrium constant for the reaction

$$2\ NH_3(g) \rightleftharpoons N_2(g) + 3\ H_2(g)$$

is very small at 25°C. At this temperature a reaction vessel was filled with 2.79 atm of NH_3. It was then heated to 300°C and at equilibrium the total pressure was 9.37 atm. Calculate K_p for this reaction at 300°C.

Solution Assume that at 25°C, $P_{N_2} = P_{H_2} = 0.0$. Then the initial amount of NH_3 in atm at 300°C assuming no decomposition is given by the ideal gas law:

$$P_o = 2.79 \times \frac{300 + 273}{25 + 273} = 5.37 \text{ atm}$$

Since the observed total pressure is 9.37 atm, it follows that some of the NH_3 has dissociated at 300°C. Let x be the extent of the reaction in atm of N_2 formed.

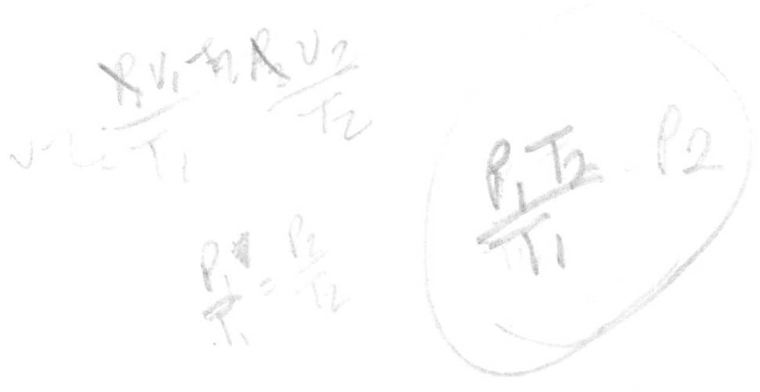

$$2\ NH_3(g) \rightleftharpoons N_2(g) + 3\ H_2(g)$$

	$2\ NH_3(g)$	$N_2(g)$	$3\ H_2(g)$
Initially	5.37	0	0
change	$-2x$	$+x$	$+3x$
at equilibrium	$5.37 - 2x$	x	$3x$

$$P_{tot} = P_{NH_3} + P_{N_2} + P_{H_2}$$

$$9.37 = 5.37 - 2x + x + 3x$$

$$x = 2.00\ atm = P_{N_2};\quad 3x = 6.00\ atm = P_{H_2}$$

$$5.37 - 2x = 5.37 - 2(2.00) = 1.37\ atm = P_{NH_3}$$

$$K_p = P_{N_2} P_{H_2}^3 / P_{NH_3}^2 = \frac{(2.00)(6.00)^3}{(1.37)^2} = 230$$

Equilibrium Calculations

The equilibrium concentrations of all reactions and products of a reaction can be calculated given the starting conditions and the equilibrium constant. In general these problems can become very complex algebraically but the examples and questions in this text have been selected to minimize such difficulties while illustrating the principles involved.

<u>Example 23.4</u> The equilibrium constant, Keq, for the following reaction is 14.7 at a particular temperature, T, (a high temperature).

$$H_2(g) + I_2(g) \rightleftharpoons 2\ HI(g)$$

If 2.00 moles of H_2 and 2.00 moles of I_2 are placed in a 4.50 liter reaction vessel and heated to temperature, T, what are the equilibrium concentrations of the three gases?

<u>Solution</u> For this reaction, $\Delta n_g = 0$, so

$$Keq = K_p = K_c$$

The initial concentrations are $[H_2]_0 = [I_2]_0 = 2.00/4.50 = 0.444\ \underline{M}$. Let x be the extent of reaction in moles per liter of H_2 decomposed. Then:

	$H_2(g)$	$I_2(g)$	$2\ HI(g)$
Initially	0.444	0.444	0
change	$-x$	$-x$	$+2x$
at equilibrium	$0.444-x$	$0.444-x$	$2x$

$$K_c = \frac{[HI]^2}{[H_2][I_2]} = \frac{(2x)^2}{(0.444-x)^2} = 14.7$$

214

The approximation $x \ll 0.444$ is not justified in this case. However, it is not necessary to solve the quadratic equation, because the algebraic expression is a perfect square. Taking the square root of both sides,

$$\sqrt{K_c} = \frac{2x}{0.444-x} = 3.83$$

$$x = 0.292 \text{ } \underline{M}$$

Then $\qquad [H_2] = [I_2] = .444 - .292 = .152 \text{ M}$

and $\qquad [HI] = 2(.292) = .584 \text{ M}$

This result can be checked by substituting the values into the equilibrium constant expression,

$$K_c = \frac{[HI]^2}{[H_2][I_2]} = \frac{(0.584)^2}{(0.152)^2} = 14.7$$

Le Chatelier's Principle

It is often useful to predict the effect of some imposed change upon the composition of a system at equilibrium, without making numerical calculations. Le Chatelier's principle states that a reversible reaction will always shift in that direction which tends to relieve the stress imposed.

Example 23.5 In each of the following equilibria predict in which direction there will be a net reaction in response to the change imposed.

(a) $CaSO_4(s) \rightleftharpoons Ca^{2+} + SO_4^{2-}$; add a solution of Na_2SO_4.

(b) $2 NH_3(g) \rightleftharpoons N_2(g) + 3 H_2(g)$; compress the mixture into a smaller volume.

(c) $2 NO(g) + O_2(g) \rightleftharpoons 2 NO_2(g) \qquad \Delta H° = -113 \text{ kJ}$; raise the temperature

Solution (a) Addition of SO_4^{2-} will cause the reaction to go to the left, because in this way SO_4^{2-} is consumed.

(b) Compression will cause the reaction to go to the left because in this way the number of moles of gas, hence the volume, is reduced.

(c) The reaction is exothermic as written but endothermic in the reverse direction. Raising the temperature will cause the reaction to go to the left because in this way the heat being supplied will be absorbed.

Questions

23.1 Given the following equilibria:

$$4\ NH_3(g) + 5\ O_2(g) \rightleftharpoons 4\ NO(g) + 6\ H_2O(g) \qquad K_1$$

$$2\ NO(g) + O_2(g) \rightleftharpoons 2\ NO_2(g) \qquad K_2$$

What is the equilibrium constant for the following reaction?

$$4\ NH_3(g) + 7\ O_2(g) \rightleftharpoons 4\ NO_2(g) + 6\ H_2O(g) \qquad K_3$$

a) $K_1 K_2$

b) $K_1 K_2^2$

c) K_1 / K_2

d) K_1 / K_2^2

e) None of the above

23.2 The equilibrium constant, Keq, for the reaction

$$SO_3(g) + NO(g) \rightleftharpoons SO_2(g) + NO_2(g)$$

is 7.1×10^{-7} at 25°C. What concentration of SO_2 will be in equilibrium with 0.15 $\underline{M}$ SO_3, 0.25 $\underline{M}$ NO, and 0.75 $\underline{M}$ NO_2?

a) $7.1 \times 10^{-7} \times 0.15 \times 0.25/0.75$

b) $7.1 \times 10^{-7} \times 0.75 \times 0.25/0.15$

c) $7.1 \times 10^{-7} \times 0.15 \times 0.75/0.25$

d) $0.15 \times 0.25/(0.75 \times 7.1 \times 10^{-7})$

e) None of the above

23.3 Consider the hydrogenation of benzene at 200°C.

$$C_6H_6(g) + 3\ H_2(g) \rightleftharpoons C_6H_{12}(g)$$

How are the equilibrium constants K_c and K_p related?

a) $K_p = K_c\ (0.0821 \times 473)^3$

b) $K_p = K_c\ (8.314 \times 473)^3$

c) $K_p = K_c / (0.0821 \times 473)^3$

d) $K_p = K_c$

e) None of the above

23.4 Given the reaction:

$$2\ XY(g) \rightleftharpoons X_2(g) + Y_2(g) \qquad K_p = 0.889 \text{ at } 243 \text{ K}$$

1.42 moles of XY are injected into a 7.32 liter container at 243 K.
What is the partial pressure of X_2 when equilibrium is reached?

a) 3.86 atm
b) 1.34 atm
c) 1.99 atm
d) 1.26 atm
e) More information is required

23.5*4.73 moles of XY(g) were injected into a 2.89 liter container at
298 K. The following reaction was allowed to reach equilibrium:

$$2\ XY(g) \rightleftharpoons X_2(g) + Y_2(g)$$

At equilibrium, 4.40 moles of XY remained. What is K_p for this
reaction at 298 K?

a) 6.19×10^{-3}
b) 5.63×10^{-3}
c) 1.41×10^{-3}
d) .0375
e) .0750

23.6*For the hypothetical chemical reaction

$$A(g) \rightleftharpoons B(g) + C(g)$$

The equilibrium constant, K_c, is 1.00 at 20.0°C. What is the
equilibrium concentration of A(g) if 67.5 grams of compound A are
injected into a 1.00 liter reaction vessel at 20.0°C? (The molecular
weight of A is 140 g/mole).

a) 0.874
b) 0.711
c) 0.126
d) 0.356
e) 1.36

23.7 Consider the equilibrium reaction:

$$2\ SO_3(g) \rightleftharpoons O_2(g) + 2\ SO_2(g)$$

1.0 mole of SO_3 is injected into a 1.0 liter reaction vessel at a
certain temperature. When equilibrium is reached, it is found that
37.5% of the SO_3 had dissociated. Calculate K_c for this reaction at
this temperature.

a) $(0.375)^3/(1-0.375)^2$
b) $(0.375/2)(0.375)^2/(1-0.375)^2$
c) $(0.375)(2{\times}0.375)^2/(1-0.375)^2$
d) $(0.375)(2{\times}0.375)^2/(1-2{\times}0.375)^2$
e) None of the above.

23.8* A vessel containing 2.10 atm of pure NH_3 at 25°C is heated to 230°C. At equilibrium, the pressure is 4.25 atm. Calculate K_p at 230°C for the reaction:

$$2\ NH_3(g) \rightleftharpoons N_2(g) + 3\ H_2(g)$$

    a) 18.6
    b) 0.154
    c) 0.134
    d) 0.054
    e) 0.027

23.9* At 250°C the equilibrium constant (K_p) for the reaction below is 1.81.

$$PCl_5(g) \rightleftharpoons PCl_3(g) + Cl_2(g)$$

Sufficient PCl5 is put into a vessel to give an initial pressure of 2.75 atmospheres at 250°C. What will be the final pressure at 250°C after the system has reached equilibrium?

    a) 4.25 atm
    b) 1.50 atm
    c) 1.25 atm
    d) 5.75 atm
    e) 2.95 atm

23.10 Consider the following gas phase equilibrium system:

$$PCl_5 \rightleftharpoons PCl_3 + Cl_2 \qquad \Delta H = 92.9\ kJ$$

Which of the following statements is <u>not</u> correct?

    a) Increasing the concentration of PCl_5 will cause the equilibrium to shift to the right.
    b) Increasing the temperature will shift the equilibrium to the right.
    c) A positive catalyst speeds up the approach to equilibrium and shifts the position of equilibrium to the right.
    d) When the total pressure of the system is decreased, the equilibrium is shifted to the right.
    e) The entropy change in this reaction favors the formation of products.

23.11 Consider the equilibrium:

$$4\ NH_3(g) + 3\ O_2(g) \rightleftharpoons 2\ N_2(g) + 5\ H_2O(g) \qquad \Delta H = -1268\ kJ$$

Which of the following changes will cause the reaction to shift to the right?

    a) increase the temperature
    b) selectively absorb the H_2O by reaction with sodium hydroxide solution
    c) add a catalyst
    d) selectively absorb the O_2 by reaction with an active metal surface
    e) none of the above

218

23.12*The density of an equilibrium mixture of N_2O_4 and NO_2 at 1.00 atm
and 50°C is 2.509 grams per liter. What is K_p at 50°C for the reaction

$$N_2O_4(g) \rightleftharpoons 2 \ NO_2(g)$$

(Hint: Assume ideal gas behavior and calculate the "average
molecular weight" of the gas mixture and then the mole fractions.)

a) .357
b) .692
c) .802
d) 1.16
e) 1.25

23.13 Consider the following dissociation equilibrium,

$$2 \ ClF_3(g) \rightleftharpoons Cl_2(g) + 3 \ F_2(g)$$

10.5 grams of ClF_3 were placed in a 1.00 liter reaction vessel and
heated to 1000 K. The equilibrium concentration of Cl_2 was
0.0056 $\underline{M}$. Calculate K_p and K_c for this reaction at 1000 K. The
molecular weight of ClF_3 is 92.4 g/mole.

	K_p	K_c
a)	0.017	2.5×10^{-6}
b)	6.3×10^{-4}	9.3×10^{-8}
c)	5.7×10^{-3}	8.4×10^{-7}
d)	1.8×10^{-3}	2.6×10^{-7}
e)	0.14	2.0×10^{-5}

23.14 The equilibrium constant, K_p, for the reaction

$$N_2O_4(g) \rightleftharpoons 2 \ NO_2(g)$$

is 0.114 at 25°C. What percentage of a sample of N_2O_4 is dissociated
if the total pressure is 1.00 atm?

a) 2.78
b) 10.2
c) 16.6
d) 33.2
e) 83.4

23.15*The equilibrium constant for the reaction

$$CO_2(g) + H_2(g) \rightleftarrows CO(g) + H_2O(g)$$

is 2.0 at 1000°C. In an equilibrium mixture of these gases the partial pressures are: $P_{CO_2} = 5.75$ atm, $P_{H_2} = 6.25$ atm, and $P_{H_2O} = 5.18$ atm. If H_2O is injected into the vessel, what must the final H_2O pressure be, in atm, in order to increase the partial pressure of H_2 to 10.0 atm?

a) 6.21
b) 10.0
c) 11.4
d) 13.6
e) 18.8

Voltaic Cells

The standard free energy change for the redox reaction

$$Cu^{2+} + Zn \;\rightleftharpoons\; Cu + Zn^{2+}$$

is -212 kJ. The equilibrium constant is 1.4×10^{37} at 25°C. This redox reaction can be used as a source of electrical energy by separating the oxidation and reduction half-reactions as shown below.

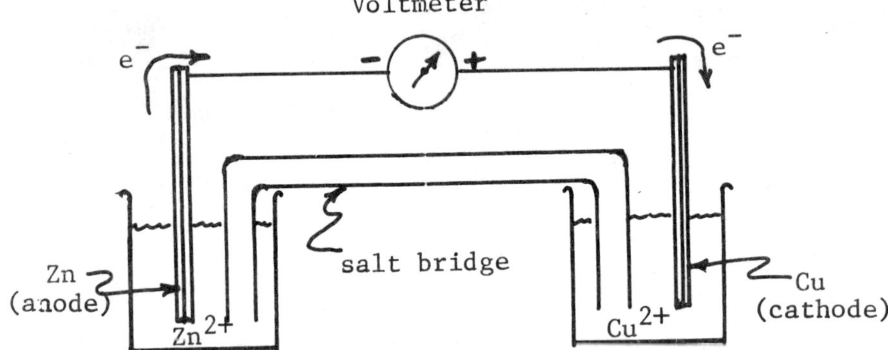

Oxidation occurs at the anode ($Zn \rightarrow Zn^{2+} + 2e^-$) and reduction occurs at the cathode ($Cu^{2+} + 2e^- \rightarrow Cu$). The negative terminal of the voltmeter is attached to the electron source. The salt bridge is a concentrated solution of an inert salt which provides counter-ions to maintain electrical neutrality. For example, the salt bridge might contain NaCl. Then chloride ions would migrate into the anode compartment and sodium ions would migrate into the cathode compartment. This arrangement is called a voltaic cell, or battery. The observed voltage of the cell depends on the temperature, the nature and concentration of the species in the anode and cathode compartments, the solvent, and other parameters of the cell. If the species are all present in their standard states, the cell potential is the "standard cell potential", E°. The relationship between the ∆G for the cell reaction and E (in volts) for the corresponding voltaic cell is

$$\Delta G = -nFE = -96.48 \; nE \qquad (kJ)$$

The following notation is used to represent voltaic cells,

$$anode \; || \; cathode$$

Here "||" represents the salt bridge. For example, the cell discussed above is represented:

$$Zn\,|\,Zn^{2+}\,||\,Cu^{2+}\,|\,Cu$$

The single vertical bar is used to represent a phase boundary.

Standard Redox Potentials

It is not possible to measure the potential of a half-cell. Only the potential difference between two half-cells can be measured. Consequently, the following half-cell is arbitrarily assigned the value of 0.00 volt, for reference purposes.

$$2H^+(1.00 \ \underline{M}) + 2e^- \ \rightleftharpoons \ H_2(g, \ 1.00 \ atm) \qquad E° = 0.00 \ volt$$

This assignment allows the electrochemist to order half-reactions according to standard half-cell potentials, measured experimentally. For example, the standard cell:

$$Pt \mid H_2, \ H^+ \mid\mid Cu^{2+} \mid Cu$$

Has a voltage of 0.34 volts and the Pt is the anode; the Cu is the positive terminal. Also the standard cell:

$$Li \mid Li^+ \mid\mid H^+, \ H_2 \mid Pt$$

has a voltage of 3.05 volts and the Li is the anode; the Li is the negative terminal.

Such measurements are tabulated in the form below, called "Standard Reduction Potentials".

Half-reaction	E° (volts)
$O_2(g) + 4H^+ + 4e^- \rightarrow 2H_2O$	1.23
$Cu^{2+} + 2e^- \rightarrow Cu(s)$	0.34
$2H^+ + 2e^- \rightarrow H_2(g)$	0.00
$Zn^{2+} + 2e^- \rightarrow Zn(s)$	-0.76
$Li^+ + e^- \rightarrow Li$	-3.05

<u>Example 24.1</u> Calculate the standard free energy change, and the equilibrium constant for the following reaction using the standard half-cell potentials given above. Devise a cell for which the overall reaction is the following, and calculate its standard potential.

$$4Li + O_2(g) + 4H^+ \ \rightleftharpoons \ 4Li^+ + 2H_2O$$

<u>Solution</u> Separate the overall reaction into two half reactions. These comprise the two half cells. Write the lithium reaction as an oxidation, corresponding to the overall reaction. The standard cell potential is the algebraic sum of the standard half-cell potentials.

$$\underline{\text{Reaction}} \qquad\qquad \underline{E^\circ}$$

$$\text{Li} \;\underset{\longleftarrow}{\longrightarrow}\; \text{Li}^+ + e^- \qquad\qquad 3.05$$

$$O_2(g) + 4H^+ + 4e^- \;\underset{\longleftarrow}{\longrightarrow}\; 2H_2O \qquad 1.23$$

$$4\text{Li} + O_2(g) + 4H^+ \;\underset{\longleftarrow}{\longrightarrow}\; 4\text{Li}^+ + 2H_2O \qquad 4.28$$

The cell is represented: $\text{Li}\,|\,\text{Li}^+\,||\,H^+, O_2\,|\,\text{Pt}$.

For the balanced reaction n = 4. Then

$$\Delta G^\circ = -nFE^\circ = (-4)(96.48)(4.28) \quad kJ$$

$$= -1652 \text{ kJ}$$

$$\log Keq = \frac{-\Delta G^\circ}{2.303 \; RT} = \frac{-1650 \times 1000}{2.303 \times 8.314 \times 298}$$

$$= 289.48$$

$$Keq = 3.0 \times 10^{289}$$

Example 24.2 Given the following standard reduction potentials,

$$\underline{\text{Reaction}} \qquad\qquad\qquad \underline{E^\circ \text{ (volts)}}$$

$$(1) \quad NO_3^- + 4H^+ + 3e^- \;\underset{\longleftarrow}{\longrightarrow}\; NO(g) + 2H_2O \qquad 0.96$$

$$(2) \quad MnO_4^- + 8H^+ + 5e^- \;\underset{\longleftarrow}{\longrightarrow}\; Mn^{2+} + 4H_2O \qquad 1.51$$

Will NO_3^- oxidize Mn^{2+} to MnO_4^- under standard state conditions? (Recall that standard $[H^+]$ is 1.00 M, i.e. the solution is strongly acidic.)

Solution The overall reaction under consideration is:

$$5NO_3^- + 3Mn^{2+} + 2H_2O \;\underset{\longleftarrow}{\longrightarrow}\; 5NO(g) + 3MnO_4^- + 4H^+$$

It can be separated into the half reactions: 5x reaction (1) above minus 3x reaction (2). The potential of the cell comprising these two half cells is accordingly 0.96 - 1.51 = -0.55 volts. The negative sign indicates that the overall reaction is not spontaneous as written. The same conclusion is in agreement with the sign of ΔG°, which may be calculated as follows:

$$\Delta G^\circ = -nFE^\circ = -(15)(96.48)(-0.55) = 796 \text{ kJ}$$

Example 24.3 Predict whether or not Hg_2^{2+} will disproportionate under standard state conditions. The disproportionation reaction is

$$Hg_2^{2+} \;\underset{\longleftarrow}{\longrightarrow}\; Hg^{2+} + Hg(\ell)$$

The relevant standard reduction potentials are:

Reaction	E° (volts)

(1) $2Hg^{2+} + 2e^- \rightleftharpoons Hg_2^{2+}$ 0.91

(2) $Hg_2^{2+} + 2e^- \rightleftharpoons 2Hg(\ell)$ 0.80

<u>Solution</u> Subtracting eq. (1) from eq. (2) gives twice the disproportionation reaction. The standard potential for the corresponding cell is

$$E° = 0.80 - 0.91 = -0.09 \text{ volt}$$

Since $E° < 0$, $\Delta G° > 0$, and the reaction will <u>not</u> occur spontaneously, as written.

You will note from the previous two examples that the following simple rule can be formulated regarding the direction of spontaneity of a pair of coupled half-reactions.

Spontaneously the half-reaction with the larger potential (algebraically) goes left to right and the one with the smaller potential (algebraically) goes in reverse.

The Nernst Equation

The Nernst equation relates the potential of a non-standard cell to the standard cell potential and to the concentration of reactants and products,

$$E = E° - \frac{RT}{nF}\ell n\ Q = E° - \frac{0.0592}{n} \log Q$$

Here R is the ideal gas constant, T is the absolute temperature, n is the number of electrons involved in the redox reaction, F is Faraday's constant, and Q is the "reaction quotient", a ratio of the same form as the equilibrium constant but containing the actual concentrations present in the cell, whatever they may be. The value of 2.303 RT/F is 0.0592 volts at 25°. When all reactants and products are in their standard states, Q is equal to 1.00. Since $\log 1 = 0$, $E = E°$. If the cell is allowed to run until equilibrium has been established Q is equal to Keq. At equilibrium, $E = 0$, hence (at 25°C):

$$E = E° - \frac{0.0592}{n} \log Q$$

$$0 = E° - \frac{0.0592}{n} \log Keq$$

$$Keq = nE°/0.0592$$

This result can also be derived using the following equations,

$$\Delta G° = -RT\ \ell n\ Keq = -nFE°$$

224

Example 24.4 Given the following standard potentials:

	Reaction	$E°$ (volts)
$Ni^{2+} + 2e^- \rightleftharpoons Ni(s)$		-0.25
$Cu^{2+} + 2e^- \rightleftharpoons Cu(s)$		0.34

Calculate the potential of the cell
$$Ni \mid Ni^{2+}(0.10 \ \underline{M}) \mid\mid Cu^{2+}(0.0020 \ \underline{M}) \mid Cu$$

Solution The cell reaction is
$$Ni(s) + Cu^{2+} \rightleftharpoons Ni^{2+} + Cu(s)$$

The standard cell potential is $E° = 0.34 - (-0.25) = 0.59$ volt

The Nernst equation is:
$$E = E° - \frac{0.0592}{2} \log \left\{ \frac{[Ni^{2+}]}{[Cu^{2+}]} \right\}$$

$$= 0.59 - \frac{0.0592}{2} \log \left\{ \frac{0.10}{0.0020} \right\}$$

$$= 0.59 - \frac{0.0592}{2} (1.7) = 0.59 - 0.05$$

$$= 0.54 \text{ volt}$$

Alternatively, using half-cell potentials,

$$Ni^{2+} + 2e^- \rightleftharpoons Ni(s) \qquad E_{Ni} = E°_{Ni} - \frac{0.0592}{2} \log \frac{1}{[Ni^{2+}]}$$

$$= -0.25 - \frac{0.0592}{2} \log \frac{1}{0.1} = -0.28$$

$$Cu^{2+} + 2e^- \rightleftharpoons Cu(s) \qquad E_{Cu} = E°_{Cu} - \frac{0.0592}{2} \log \frac{1}{[Cu^{2+}]}$$

$$= 0.34 - \frac{0.0592}{2} \log \frac{1}{0.002} = 0.26$$

For the cell reaction
$$Ni + Cu^{2+} \rightleftharpoons Ni^{2+} + Cu$$

$$E = E_{Cu} - E_{Ni} = 0.26 - (-0.28) = 0.54 \text{ volt}$$

Questions

24.1 In the redox reaction

$$Sn^{4+} + 2\ Fe^{2+} \rightarrow 2\ Fe^{3+} + Sn^{2+}$$

a) Sn^{4+} is the oxidizing agent and Fe^{2+} is the reducing agent.
b) Sn^{4+} is the reducing agent and Fe^{2+} is the oxidizing agent.
c) Sn^{4+} is the reducing agent and Fe^{3+} is the oxidizing agent.
d) Fe^{3+} is the oxidizing agent and Sn^{2+} is the reducing agent.
e) Sn^{2+} is the oxidizing agent and Fe^{3+} is the reducing agent.

24.2 Which statement correctly describes the following electrode reaction:

$$MnO_4^{2-} \rightarrow MnO_4^- + \varepsilon^-$$

a) An anion is oxidized at the anode.
b) An anion is reduced at the cathode.
c) A cation is reduced at the cathode.
d) An anion is oxidized at the cathode.
e) None of the above.

24.3 Consider a fuel cell for which the cell reaction is the combustion of cellulose,

$$[CH_2O](s) + O_2(g) \rightleftharpoons CO_2(g) + H_2O(\ell),$$

Here "$[CH_2O]$" represents the monomeric unit in cellulose. Calculate the initial voltage of this fuel cell operating under standard state conditions and determine which half-cell is the anode. The relevant half-reactions and standard <u>oxidation</u> potentials are:

	Reaction	$E°$ (volts)
(1)	$2\ H_2O(\ell) \rightleftharpoons O_2(g) + 4H^+ + 4e^-$	-1.23
(2)	$[CH_2O] + H_2O(\ell) \rightleftharpoons CO_2(g) + 4H^+ + 4e^-$	0.40

	Voltage	Anode
a)	$1.23 - 0.40$	(1)
b)	$1.23 - 0.40$	(2)
c)	$1.23 + 0.40$	(2)
d)	$1.23 + 0.40$	(1)
e)	None of the above	

226

24.4*Given the thermodynamic properties below, calculate the electrical
energy obtainable at 25°C from a fuel cell using the following
reactions. Assume all equipment is 100% efficient. All substances
are in their standard states in aqueous solution.

$$CH_3OH + 2OH^- \rightarrow CH_2O + 2\ H_2O + 2e^-$$

$$H_2O_2 + 2e^- \rightarrow 2\ OH^-$$

Compound	ΔH°_f kJ/mole	ΔG°_f kJ/mole	S° J/mole-K
CH_3OH	-246	-175	132
OH^-	-230	-157	-10.2
CH_2O	-158	-132	145
H_2O	-286	-237	69.9
H_2O_2	-191	-132	136

(Note: some values above have been obtained by interpolation).

a) 59.8 kJ per mole of H_2O_2 consumed.

b) 114 kJ per mole of CH_2O produced.

c) 292 kJ per mole of CH_2O produced.

d) 297 kJ per mole of CH_3OH consumed.

e) None, the reactions are not spontaneous.

24.5*Consider the following potential diagram in 1.0 M OH^-. The values
over the arrows are the standard <u>oxidation</u> potentials for going
from one species to the next.

$$I^- \xrightarrow{-0.535} I_2 \xrightarrow{-0.45} OI^- \xrightarrow{-0.14} IO_3^- \xrightarrow{-0.7} H_3IO_6^{2-}$$

List <u>all</u> of the iodine species which will <u>not</u> disproportionate
in 1.0 M OH^- ?

a) I^- and $H_3IO_6^{2-}$

b) I^-, $H_3IO_6^{2-}$ and IO_3^-

c) I^-, $H_3IO_6^{2-}$, IO_3^- and I_2

d) I^-, $H_3IO_6^{2-}$, IO_3^- and OI^-

e) I_2 and OI^-

24.6 Given the following hypothetical half-reactions and standard
     reduction potentials:

Reaction	E° (volts)
$M^{2+} + e^- \rightleftharpoons M^+$	-.663
$M^{3+} + e^- \rightleftharpoons M^{2+}$	.532

Which of the following is correct?

a) M^{2+} will not disproportionate, $E° = -1.195$ volts

b) M^{2+} will disproportionate, $E° = 1.195$ volts

c) M^{2+} will disproportionate, $E° = -0.131$ volts

d) M^{3+} will disproportionate, $E° = 1.195$ volts

e) More information is required.

24.7*Given the following standard reduction potentials in neutral
     solution:

Reaction	E° (volts)
$2\ H_2O + 2e^- \rightleftharpoons 2OH^- + H_2(g)$	-.414
$O_2(g) + 4H^+ + 4e^- \rightleftharpoons 2H_2O$	.816
$M^{3+} + e^- \rightleftharpoons M^{2+}$	.586

Which of the following statements is correct?

a) M^{3+} will oxidize H_2O to give M^{2+} and O_2 gas.

b) M^{2+} will oxidize H_2O to give M^{3+} and O_2 gas.

c) M^{2+} will oxidize H_2O to give M^{3+} and H_2 gas.

d) M^{3+} is stable in neutral solution, in the absence of H_2 gas.

e) M^{2+} will reduce H_2O to give M^{3+} and H_2 gas.

24.8 Given the following standard reduction potentials:

Reaction	E° (volts)
$Ce^{3+} \rightleftharpoons Ce$	-2.48
$Pd^{2+} \rightleftharpoons Pd$	0.987
$Tl^{3+} \rightleftharpoons Tl^{+}$	1.25

Which one of the reactions below will proceed spontaneously from left to right? The equations below are not necessarily balanced.

a) $Ce^{3+} + Tl^{3+} \rightarrow Ce + Tl^{+}$

b) $Ce^{3+} + Pd^{2+} \rightarrow Ce + Pd$

c) $Ce + Tl^{3+} \rightarrow Ce^{3+} + Tl^{+}$

d) $Tl^{+} + Pd^{2+} \rightarrow Tl^{3+} + Pd$

e) None of the above

24.9 What is the magnitude in volts of the potential of the following concentration cell?

$$Hg\,|\,Hg^{2+}(0.25 \text{ molar})\,||\,Hg^{2+}(0.45 \text{ molar})\,|\,Hg$$

The standard reduction potential of Hg^{2+} is 0.854 V.

a) 0.0076
b) 0.384
c) 0.846
d) 0.854
e) 0.015

24.10 Given the following standard reduction potentials:

Reaction	E° (volts)
$Al^{3+} + 3\,e^{-} \rightleftharpoons Al$	-1.66
$Fe^{2+} + 2\,e^{-} \rightleftharpoons Fe$	-0.41

Suppose the cell below is hooked up to deliver energy:

$$Al\,|\,Al^{3+}\,||\,Fe^{2+}\,|\,Fe$$

which one of the statements below is <u>incorrect</u>? Assume all substances are initially in the standard state.

a) The negative terminal of the voltmeter should be attached to Al and the positive to Fe.
b) The anode is Al.
c) If the standard hydrogen half cell were substituted for the Fe^{2+} half cell the voltage would be 1.66.
d) Anions in the salt bridge will flow in a direction from the Al compartment to the Fe compartment.
e) The concentration of Al^{3+} will increase.

24.11 Given the following standard reduction potentials:

Reaction	E° (volts)
$O_2(g) + 4H^+ + 4\ e^- \rightleftharpoons 2\ H_2O$	1.23
$I_2(g) + 2\ e^- \rightleftharpoons 2\ I^-$	0.54

Which one of the statements below is correct? Assume all substances are in the standard state.

a) H_2O will spontaneously oxidize $I_2(s)$ to form I^-.

b) I^- will spontaneously oxidize $O_2(g)$ to form H_2O.

c) $O_2(g)$ will spontaneously oxidize I^- to form $I_2(s)$.

d) H^+ will spontaneously reduce I^- to form $I_2(s)$.

e) $I_2(s)$ will spontaneously oxidize H_2O to form $O_2(g)$.

24.12* Given the following hypothetical half-reactions:

Reaction	E° (volts)
$M^{2+} + e^- \rightleftharpoons M^+$	-.758
$M^{3+} + e^- \rightleftharpoons M^{2+}$	-.248

Calculate the equilibrium constant at 25°C for the reaction

$$2\ M^{2+} \rightleftharpoons M^{3+} + M^+$$

a) log K = -4.32
b) log K = .116
c) log K = 4.32
d) log K = -.116
e) log K = -8.61

24.13* The pH of a solution can be determined by measuring the potential of the following cell:

$$Pt\,|\,H_2(g)(1.00\ atm)\,|\,H^+(?\ \underline{M})\,||\,Hg_2Cl_2(s)\,|\,Hg(\ell)\,|\,Cl^-(1.0\ \underline{M})\,|\,Pt$$

Given:

$$Hg_2Cl_2 + 2\ e^- \rightleftharpoons 2\ Hg(\ell) + 2\ Cl^- \qquad E° = 0.24\ volt$$

Calculate the pH of a solution for which the observed cell potential is 0.37 volts.

a) (.37-.24)/.0592
b) (.37+.24)/.0592
c) 2(.37-.24)/.0592
d) (.37-.24)/(2 x .0592)
e) None of the above

24.14*Given the following standard reduction potentials:

Reaction	$E°$ (volts)
$Co^{3+} + 3\ e^- \rightleftharpoons Co$	1.26
$Co(CN)_6^{3-} + 3\ e^- \rightleftharpoons Co + 6\ CN^-$	-0.13

Calculate the dissociation constant of $Co(CN)_6^{3-}$ at 25°C. The dissociation reaction is

$$Co(CN)_6^{3-} \rightleftharpoons Co^{3+} + 6\ CN^- \qquad K_d$$

a) $\log K_d = (3)(1.13)/.0592$

b) $\log K_d = -1.13/(0.0592)(3)$

c) $\log K_d = (3)(1.39)/.0592$

d) $\log K_d = -(3)(1.39)/.0592$

e) None of the above.

24.15 Given the following standard reduction potentials:

Reaction	$E°$ (volts)
$AgCN + e^- \rightleftharpoons Ag + CN^-$	0.106
$Ag^+ + e^- \rightleftharpoons Ag$	0.799

Calculate the solubility product of AgCN at 298 K.

a) 5.1×10^{-16}

b) 8.7×10^{-13}

c) $.821$

d) 4.0×10^{-24}

e) 2.0×10^{-12}

24.16*Given the following standard electrode potentials:

Reaction	$E°$ (volts)
$Sn^{4+} + 2\ e^- \rightleftharpoons Sn^{2+}$	0.150
$Tl^{3+} + 2\ e^- \rightleftharpoons Tl^+$	1.25

Calculate the potential of the following cell, given the tabulated molarities:

$$Pt\,|\,Sn^{2+},\ Sn^{4+}\,||\,Tl^{3+},\ Tl^+\,|\,Pt$$

$$[Sn^{2+}] = 9.00 \times 10^{-3}\ \underline{M} \qquad [Tl^{3+}] = 1.00 \times 10^{-3}\ \underline{M}$$
$$[Sn^{4+}] = .900\ M \qquad [Tl^+] = 2.00 \times 10^{-2}\ \underline{M}$$

a) 1.12
b) 1.30
c) 0.90
d) 1.00
e) 1.20

24.17*Given the following hypothetical half-reactions and standard reduction potentials:

Reaction	$E°$ (volts)
$2\ (M^{3+} + e^- \rightleftharpoons M^{2+})$	0.108
$M^{5+} + 2\ e^- \rightleftharpoons M^{3+}$	-0.444

Calculate the standard reduction potential for the half-cell reaction:

$$M^{5+} + 3\ e^- \rightleftharpoons M^{2+}$$

a) −.112
b) −.336
c) −.260
d) −.0769
e) .552

24.18* H_2O_2 is a strong oxidizing agent in 1.0 $\underline{M}$ H^+,

$$H_2O_2 + 2\ H^+ + 2\ e^- \rightleftharpoons 2\ H_2O \qquad E° = 1.77\ volts$$

What is the standard reduction potential in neutral solution?

a) $1.77 + (0.0592 \times 7)$

b) $1.77 - (0.0592 \times 7)$

c) $1.77 + \dfrac{0.0592 \times 7}{2}$

d) $1.77 - \dfrac{0.0592 \times 7}{2}$

e) None of the above

24.19*Given the following standard reduction potentials:

Reaction	E° (volts)
$Fe^{3+} + e^{-} \rightleftharpoons Fe^{2+}$	0.77
$8H^{+} + MnO_4^{-} + 5\ e^{-} \rightleftharpoons Mn^{2+} + 4H_2O$	1.52

Calculate the equilibrium molar concentration of Fe^{2+} if a stoichiometric quantity of $KMnO_4$ is added to 1.00 liter of 0.80 M Fe^{2+}. Assume the final $[H^{+}]$ is 1.0 M, the final volume is 1.0 liter, and the final temperature 25°C.

a) 4.3×10^{-10}
b) 2.2×10^{-11}
c) 4.4×10^{-12}
d) 9.3×10^{-34}
e) 4.7×10^{-33}

24.20*Consider the following half-reactions (not balanced) in 1.0 $\underline{M}$ H^{+}:

$Pb^{2+} \rightarrow Pb$ $U^{4+} \rightarrow U^{3+}$

$Mn^{3+} \rightarrow Mn^{2+}$ $H_3PO_3 \rightarrow H_3PO_2$

$Ag^{+} \rightarrow Ag$ $HNO_2 \rightarrow N_2O$

Given the following observations:

1) HNO_2 is a stronger oxidizing agent than H_3PO_3.
2) U^{3+} reduces both HNO_2 and H_3PO_3.
3) Neither HNO_2 nor H_3PO_3 oxidizes Mn^{2+} to Mn^{3+}.
4) Ag^{+} oxidized H_3PO_2 but does not oxidize N_2O.
5) Ag^{+} but not H_3PO_3 oxidizes Pb to Pb^{2+}.

Which of the following sequences is the correct order of decreasing strength of the metallic species as reducing agents?

a) $Mn^{2+} > Ag > Pb > U^{3+}$

b) $U^{3+} > Pb > Ag > Mn^{2+}$

c) $U^{3+} > Ag > Pb > Mn^{2+}$

d) $Mn^{2+} > Pb > Ag > U^{3+}$

e) None of the above

CHAPTER 25. KINETICS

The Rate Law

The rate law of a chemical reaction is an empirically determined
expression which relates the observed rate of the reaction to the
concentration of the reactants. The rate of the reaction is the change
in concentration per unit time, for example, moles per liter per second,
or atmospheres per minute.

For example,

Reaction	Observed Rate Law
$2\ N_2O_5(g) \rightarrow 4\ NO_2(g) + O_2(g)$	Rate $= k[N_2O_5]$
$H_2(g) + I_2(g) \rightarrow 2\ HI(g)$	Rate $= k[H_2][I_2]$
$2\ NO(g) + 2\ H_2(g) \rightarrow N_2(g) + 2\ H_2O(g)$	Rate $= k[NO][H_2]$
$S_2O_8^{2-} + 2\ I^- \rightarrow 2\ SO_4^{2-} + I_2$	Rate $= k[S_2O_8][I^-]^2$
$CO(g) + Cl_2(g) \rightarrow COCl_2(g)$	Rate $= k[CO][Cl_2]^{3/2}$

Note that there is <u>not</u> a one-to-one correlation between the exponents
in the rate law and the stoichiometric coefficients of the reaction. The
rate law contains the constant k, called the specific rate constant. The
value of k is characteristic of the particular reaction and also depends
on the temperature.

The rate law and the specific rate constant can be determined by
measuring the "initial rate" of the reaction. For example,

$$aA + bB \rightarrow \text{Products.}$$

$$\text{Rate}_o = k[A]_o^i[B]_o^j = \frac{\Delta[A]}{\Delta t}$$

Here $[A]_o$ and $[B]_o$ are the initial concentrations of the reactants A and
B, i is the order of the reaction with respect to A, j is the order of
the reaction with respect to B, k is the specific rate constant, and
$\Delta[A]$ is the change in concentration of A during a short time interval, Δt.
The ratio $\left(\dfrac{\Delta[A]}{\Delta t}\right)_o$ measured at the beginning of the reaction is the initial
rate. Alternatively $\left(\dfrac{\Delta[B]}{\Delta t}\right)_o$ or $\left(\dfrac{\Delta[\text{Products}]}{\Delta t}\right)_o$ could be measured instead of
$\left(\dfrac{\Delta[A]}{\Delta t}\right)_o$, depending on which is most practical experimentally.

234

Example 25.1 In the following gas reaction the rate may be observed easily by the fading of the brown color of NO_2, the only colored species present.

$$NO_2(g) + CO(g) \rightarrow NO(g) + CO_2(g)$$

The following initial rate data were collected at a particular temperature:

Experiment	$[NO_2]_o$ (M)	$[CO]_o$ (M)	Initial Rate (mole/ℓ-hr)
1	0.0021	0.0045	2.4×10^{-4}
2	0.0021	0.0090	4.8×10^{-4}
3	0.0063	0.0045	7.2×10^{-4}

Determine the rate law for the reaction and calculate the specific rate constant at this temperature.

Solution Let the rate law be

$$\text{Rate} = k[NO_2]^i[CO]^j$$

Consider experiments 1 and 2. The initial concentration of NO_2 is held constant, while that of CO is doubled. The initial rate of the reaction also doubled. Therefore the rate of the reaction must depend on the concentration of CO to the first power; i = 1. Similarly, in experiments 1 and 3, the initial concentration of CO is held constant, the initial concentration of NO_2 is tripled, and the observed rate also is tripled. Thus the reaction is also first order in NO_2; j = 1. The rate law is

$$\text{Rate} = k[NO_2][CO]$$

A more analytical approach will be necessary when the ratios are not so simple as the above. In a general case consider ratios of initial rate laws. For example the ratio of the initial rate in experiment 3 to that of experiment 1 is

$$\frac{(\text{Rate}_o)_3}{(\text{Rate}_o)_1} = \frac{k[A_o]_3^i[B_o]_3^j}{k[A_o]_1^i[B_o]_1^j} = \frac{(0.0063)^i(0.045)^j}{(0.0021)^i(0.0045)^j} = \frac{7.2 \times 10^{-4}}{2.4 \times 10^{-4}} =$$

$$3 = \left(\frac{0.0063}{0.0021}\right)^i = 3^i$$

$$i = 1$$

The value of k can be calculated from anyone of the three experiments. For example

$$(\text{Rate}_o)_2 = k[A_o]_2[B_o]_2$$

$$k = \frac{4.8 \times 10^{-4} \ (\text{mole}/\ell\text{-hr})}{(0.0021)(0.0090) \ \text{mole}^2/\ell^2} = 25 \ \ell/\text{mole-hr}$$

First-Order Reactions

If a reaction is first-order,

$$A \rightarrow \text{Products} \qquad\qquad \text{Rate} = k[A]$$

then the concentration of the reactant as a function of time can be readily derived (using calculus) and is given by the following equation:

$$\ln\left(\frac{[A]_o}{[A]}\right) = kt \qquad\text{or}\qquad \log\left(\frac{[A]_o}{[A]}\right) = \frac{kt}{2.303}$$

Here $[A]_o$ is the initial concentration (or partial pressure) of A and $[A]$ is the concentration of A remaining at time t. The "half-life", $t_{1/2}$ is the time required for exactly one-half of A to react. When $t = t_{1/2}$,

$$\ln\left(\frac{[A]_o}{[A]}\right) = \ln 2 = kt_{1/2}$$

so

$$t_{1/2} = \frac{\ln 2}{k} = \frac{0.693}{k}$$

Example 25.2 The decomposition of N_2O_5 in CCl_4 solution is first-order. At a certain temperature the half-life is 15.0 minutes. If a 0.855 $\underline{M}$ solution of N_2O_5 is held at this temperature for 42.0 minutes, what concentration of N_2O_5 will remain?

Solution

$$\ln\left(\frac{[A]_o}{[A]}\right) = kt = \frac{0.693\ t}{t_{1/2}}$$

$$\ln\left(\frac{0.855}{[A]}\right) = \frac{(0.693)(42.0)}{15.0} = 1.94$$

$$\frac{0.855}{[A]} = e^{1.94} = 6.96$$

$$[A] = 0.123\ \underline{M}$$

The Effect of Temperature on k

The rates of most reactions increase exponentially with temperature. This relationship is expressed by the Arrhenius equation,

$$k = Ae^{-Ea/RT}$$

where A, the "frequency factor", and Ea, the "activation energy", are experimentally determined constants. A convenient form of the Arrhenius equation which relates the observed specific rate constants at two temperatures to the activation energy is

$$\ln\left(\frac{k_1}{k_2}\right) = \frac{Ea}{R}\left\{\frac{T_1 - T_2}{T_1 T_2}\right\}$$

Ea can be interpreted as the minimum energy that two colliding reactant molecules must have in order for the collision to result in a chemical change rather than an elastic rebound.

Example 25.4 A certain reaction for which the activation energy is 82.1 kJ/mole proceeds very slowly at 25°C. To what temperature should the reaction vessel be heated to increase the rate by a factor of five?

Solution Let T_1 be the desired temperature. Then $T_2 = 273 + 25 = 298$ K, and $k_1/k_2 = 5.0$.

$$\ln\left(\frac{k_1}{k_2}\right) = \frac{Ea}{R}\left\{\frac{T_1 - T_2}{T_1 T_2}\right\}$$

$$\ln 5 = \frac{82.1 \times 1000\ (T_1 - 298)}{8.314 \times 298 \times T_1} = 1.61$$

$$T_1 = 313\ \text{K or } 40°C$$

Reaction Mechanisms

Most chemical reactions do not proceed as indicated by the overall equation but rather through a series of "elementary steps". For example, the hypothetical reaction

$$A + 2B \rightarrow AB_2$$

might actually proceed by two elementary steps,

$$A + B \rightarrow AB$$

$$AB + B \rightarrow AB_2$$

Such a sequence of elementary steps is called the mechanisms of the reaction. For elementary steps there is a one-to-one correspondence between the stoichiometry of the elementary reaction and the rate law for that step. For example,

Elementary Step	Rate Law
$A + B \rightarrow C$	Rate $= k[A][B]$
$2A \rightarrow B$	Rate $= k[A]^2$
$2A + B \rightarrow C$	Rate $= k[A]^2[B]$

If the rate of one of the elementary steps in the reaction mechanism is much slower than the others, this step is called the "rate determining" step. After proposing a mechanism the scientist must mathematically combine the rate laws for the elementary steps and deduce the overall rate law. For a mechanism to be acceptable the overall rate law predicted by the mechanism must be consistent with the observed rate law.

<u>Example 25.5</u> The observed rate law for the reaction

$$A + 2B \rightarrow AB_2$$

is

$$Rate = k[A][B].$$

Which of the following mechanisms is consistent with this rate law?

<u>Mechanism 1</u>	$A + B \xrightarrow{k_1} AB$	(slow)
	$AB + B \longrightarrow AB_2$	(fast)
<u>Mechanism 2</u>	$A + 2B \xrightarrow{k_2} AB_2$	(slow)
<u>Mechanism 3</u>	$A + B \underset{\longleftarrow}{\xrightarrow{Keq}} AB$	(rapid equilibrium)
	$AB + B \xrightarrow{k_3} AB_2$	(slow)
<u>Mechanism 4</u>	$2B \underset{\longleftarrow}{\xrightarrow{Keq}} B_2$	(rapid equilibrium)
	$A + B_2 \xrightarrow{k_4} AB_2$	(slow)

<u>Solution</u> The rate law predicted by Mechanism 1 is

$$Rate_1 = k_1[A][B]$$

This rate law is consistent with the observed rate law for the reaction, so Mechanism 1 is a plausible mechanism. The rate law predicted by Mechanism 2 is

$$Rate_2 = k_2[A][B]^2$$

This rate law is second-order in B, which is <u>not</u> consistent with the observed rate law for the reaction. Therefore this mechanism cannot be correct. The rate law predicted by Mechanims 3 is

$$Rate_3 = k_3[AB][B]$$

However [AB] is related to [A] and [B] by the equilibrium constant expression:

$$Keq = \frac{[AB]}{[A][B]}; \quad [AB] = Keq\,[A][B]$$

$$Rate_3 = k_3[AB][B] = k_3 Keq[A][B][B] = k_3 Keq[A][B]^2$$

This rate law does not agree with the observed rate law, so Mechanism 3 cannot be correct. The rate law predicted by Mechanism 4 is

$$\text{Rate}_4 = k_4 \text{Keq}[A][B]^2$$

The only mechanism that has a rate law consistent with the observed rate law is Mechanism 1.

Example 25.6 The overall reaction for the bromination of acetone is

$$Br_2 + (CH_3)_2CO \rightarrow CH_3COCH_2Br + H^+ + Br^-$$

This reaction proceeds rapidly in the presence of a strong acid, with the following rate law,

$$\text{Rate} = k[(CH_3)_2CO][H^+]$$

Which of the following mechanisms is consistent with this rate law?

Mechanism 1 $H^+ + (CH_3)_2CO \underset{\longleftarrow}{\overset{Keq}{\longrightarrow}} (CH_3)_2COH^+$ (rapid equilibrium)

$(CH_3)_2COH^+ + Br_2 \xrightarrow{k_1} CH_3COCH_2Br + H^+ + Br^-$ (slow)

Mechanism 2 $H^+ + (CH_3)_2CO \xrightarrow{k_2} (CH_3)_2COH^+$ (slow)

$(CH_3)_2COH^+ + Br_2 \longrightarrow CH_3COCH_2Br + H^+ + Br^-$ (fast)

Solution The rate law predicted by Mechanism 1 is

$$\text{Rate}_1 = k_1[(CH_3)_2COH^+][Br_2]$$

$$= k_1\text{Keq}[(CH_3)_2CO][H^+][Br_2]$$

The observed rate law is not first order in Br_2 but is independent of the Br_2 concentration, so Mechanism 1 is not possible. The rate law predicted by Mechanism 2 is

$$\text{Rate}_2 = k_2[H^+][(CH_3)_2CO]$$

This rate law is identical to the observed rate law, so Mechanism 2 is plausible.

Questions

25.1 For the hypothetical reaction A + B = C, Rate = k[A][B] with
$k = 9.83 \times 10^{-2}$ liter/(mole-sec) at 511°K. What is the initial
rate of the reaction at 511°K if the initial concentrations of A
and B are .850 and .700 mole per liter respectively?

a) 3.48×10^{-2} mole/liter-sec
b) 9.83×10^{-2} mole/liter-sec
c) .165 mole/liter-sec
d) 5.85×10^{-2} mole/liter-sec
e) 17.1 mole/liter-sec

25.2 Determine the rate law and specific rate constant for the
hypothetical reaction,

$$X + Y \rightarrow Z$$

given the following initial concentrations and initial rate data:

Initial [X]	Initial [Y]	Initial Rate of Formation of Z (mole/liter-sec)
.250	.250	0.469
.500	.500	3.75
.250	.500	1.88

Rate = $k[X]^i[Y]^j$

a) i = 0   j = 1   k = 120
b) i = 1   j = 2   k = 60.0
c) i = 1   j = 2   k = 30.0
d) i = 1   j = 0   k = 7.50
e) i = 2   j = 0   k = 120

25.3 The initial rate of formation of N_2O was measured in the following
experiments; all reactants and products are gases.

$$2\ NO + H_2 \rightarrow N_2O + H_2O$$

Exp. #	$[NO]_o$	$[H_2]_o$	Initial Rate of N_2O formation
1	.042 mole/liter	.042 mole/liter	0.10 (mole/liter)/min
2	.084 mole/liter	.042 mole/liter	0.40 (mole/liter)/min
3	.084 mole/liter	.084 mole/liter	0.40 (mole/liter)/min

Based on the above, select the one correct statement.

a) The rate constant has units of $\dfrac{(\text{liter/mole})^2}{\text{min}}$.

b) The reaction is second order overall.

c) Experiments 2 and 3 tell us the order with respect to NO.

d) Since all reactions are truly reversible this must be minus-one
order with respect to N_2O and also H_2O.

e) To determine a physically meaningful rate constant we must
convert from concentration to partial pressures in atmospheres.

25.4 In each of the gas phase experiments listed below the initial rate
was measured for the reaction

$$2\ NO + Cl_2 \rightarrow 2\ NOCl$$

Exp. #	P_{NO}	P_{Cl_2}	Rate of Pressure drop
1	0.50 atm	0.50 atm	2.3×10^{-4} atm/sec
2	1.00 atm	1.00 atm	1.84×10^{-3} atm/sec
3	0.50 atm	1.00 atm	4.6×10^{-4} atm/sec

Based on the above, select the one correct statement.

a) The order of the reaction with respect to NOCl is minus two.
b) The reaction is second order with respect to Cl_2.
c) Knowing the rate law the rate constant can only be determined
   from Experiment 2.
d) The rate of pressure decrease observed is the same as the rate
   of decrease of P_{Cl_2}.
e) The rate constant is in units of sec^{-1}.

25.5*The hypothetical decomposition reaction

$$AB \rightarrow A + B$$

is first order in AB and has specific rate constant k = 58.7 sec^{-1}
at 227°K. How long will it take for 72.8% of AB to decompose?

a) 4.18×10^{-3} sec
b) 2.22×10^{-2} sec
c) 14.4 sec
d) 9.63×10^{-3} sec
e) 7.30×10^{-2} sec

25.6 The decomposition of N_2O_5 follows first order kinetics,

$$N_2O_5(g) \rightarrow 2\ NO_2(g) + 1/2\ O_2(g)$$

What is the specific rate constant for this reaction at 25°C if it
takes 17.1 hours for 87.5% of a sample of $N_2O_5(g)$ to decompose?

(Note: $\dfrac{100}{12.5} = \dfrac{8}{1}$)

a) 0.0078. hr^{-1}
b) .0135 hr^{-1}
c) .0405 hr^{-1}
d) .0810 hr^{-1}
e) .122 hr^{-1}

25.7*The conversion from the cis to the trans isomer of a certain metal
complex occurs 24 times as fast as 474 K as it does at 438 K.
What is the energy of activation (in kJ/mole) for the process?

a) 150
b) -1.5×10^5
c) 2.2×10^{-7}
d) 2700
e) 110

25.8*The energy of activation of a certain chemical reaction is 68.6
kJ/mole. At what temperature (K) should it be run in order to
obtain a rate constant 3.00 times as great as it is at 619 K?

a) 657
b) 675
c) .537
d) 572
e) 1.48

25.9*The following rate constants were measured for the decomposition of
N_2O_5:

$$N_2O_5(g) \rightarrow 2 NO_2(g) + 1/2 O_2(g)$$

T (°C)	45°	100°
k (sec^{-1})	6.2×10^{-4}	0.20

Calculate the pre-exponential factor, A, and the activation
energy, E_a, for this reaction.

	A (sec^{-1})	E_a (kJ)
a)	7×10^{13}	104
b)	2×10^{15}	104
c)	2×10^{15}	95
d)	7×10^{13}	95
e)	4×10^6	52

25.10*The rate law for the reaction $\quad H_2O_2 + 2H^+ + 2I^- \rightarrow I_2 + 2H_2O$

    is $\quad$ Rate $= k[H_2O_2][I^-]$

Which of the following mechanisms is consistent with this rate law?

a) $H_2O_2 + 2I^- + 2H^+ \rightarrow I_2 + 2H_2O$

b) $H_2O_2 + I^- \rightarrow HOI + OH^-$ $\qquad\qquad$ (slow)

$\quad H^+ + OH^- \rightleftharpoons H_2O$ $\qquad\qquad$ (fast)

$\quad HOI + H^+ + I^- \rightarrow I_2 + H_2O$ $\qquad$ (fast)

c) $H_2O_2 + I^- \rightleftharpoons HOI + OH^-$ $\qquad\quad$ (fast)

$\quad H^+ + OH^- \rightleftharpoons H_2O$ $\qquad\qquad$ (fast)

$\quad HOI + H^+ + I^- \rightarrow I_2 + H_2O$ $\qquad$ (slow)

d) $H_2O_2 + I^- \rightleftharpoons HOI + OH^-$ $\qquad\quad$ (fast)

$\quad HOI + I^- \rightarrow I_2 + OH^-$ $\qquad\qquad$ (slow)

$\quad H^+ + OH^- \rightleftharpoons H_2O$ $\qquad\qquad$ (fast)

e) Both c and d.

CHAPTER 26. COORDINATION CHEMISTRY

Nomenclature

Coordination compounds are salts of Lewis acids and bases, for example,

$$Ag^+ + 2S_2O_3^{2-} \rightleftharpoons Ag(S_2O_3)_2^{3-}$$

Silver ion is the Lewis acid (electron pair acceptor) and the thiosulfate ion is the Lewis base (electron pair donor). The complex has the systematic name dithiosulfatoargentate(I). The Lewis bases are classified into ligands and chelates. Ligands donate one electron pair per molecule or ion. Chelates donate more than one electron pair per molecule or ion. The names and formulas of several ligands and chelates are tabulated below.

Ligands

Formula	Name	Formula	Name
F^-	fluoro	H_2O	aquo*
Cl^-	chloro	NH_3	ammine
Br^-	bromo	$S_2O_3^{2-}$	thiosulfato
I^-	iodo	CN^-	cyano
NO_2^-	nitro	SCN^-	thiocyanato
OH^-	hydroxo	CO_3^{2-}	carbonato

*The IUPAC approved name is aqua, but aquo is still in common use.

Chelates

Formula	Name	Formula	Name
CO_3^{2-}	carbonato	SO_4^{2-}	sulfato
$C_2O_4^{2-}$	oxalato	PO_4^{3-}	phosphato
$S_2O_3^{2-}$	thiosulfato	$C_5H_7O_2^-$	acetylacetonato
$NH_2CH_2CH_2NH_2$	ethylenediamine (abbreviated en)	$[CH_2N(CH_2COO)_2]_2^{4-}$	ethylenediamine-tetraacetato (abbreviated EDTA)

The prefixes di (2), tri (3), tetra (4), penta (5), and hexa (6)
are used to indicate the number of each ligand in the complex. When a
numerical prefix is part of the name of the ligand or chelate the name
is enclosed in parentheses and preceded by bis, tris, or tetrakis, in
place of di, tri, or tetra respectively. The metal is given the suffix
"ate" if the complex is anionic. Some examples are given below:

$[Pt(NH_3)_5Cl]Cl_3$	Chloropentaammineplatinum(IV) chloride
$[Pt(NH_3)_2Cl_4]$	Tetrachlorodiammineplatinum(IV)
$K[Pt(NH_3)Cl_5]$	Potassium pentachloromonoammineplatinate(IV)
$[Ni(H_2O)_6](ClO_4)_2$	Hexaaquonickel(II) perchlorate
$K_4[Fe(CN)_6]$	Potassium hexacyanoferrate(II)
$[Pt(en)_2Cl_2]Br_2$	Dichlorobis(ethylenediamine)platinum(IV) bromide
$[Pt(NH_3)_4][PtCl_4]$	Tetraammineplatinum(II) tetrachloroplatinate(II)

__Example 26.1__   Name the following complexes:

a) $Co(C_2O_4)_3^{3-}$

b) $Cr(en)_2(NO_2)_2^{+}$

c) $[Fe(H_2O)_4(SCN)_2]NO_2$

__Solution__

a)   tris(oxalato)cobaltate(III) ion

b)   dinitrobis(ethylenediamine)chromium(III) ion

c)   dithiocyanatotetraaquoiron(III) nitrite

Isomerism

Square planar and octahedral complexes exhibit cis-trans isomerism.
For example, for $[Ma_2b_2]$:

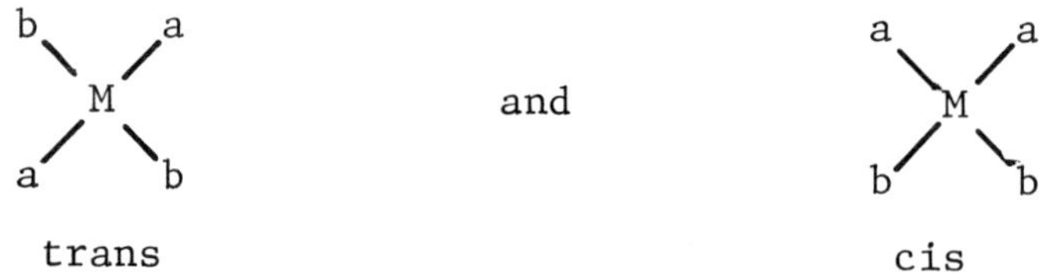

For $[Ma_4b_2]$:

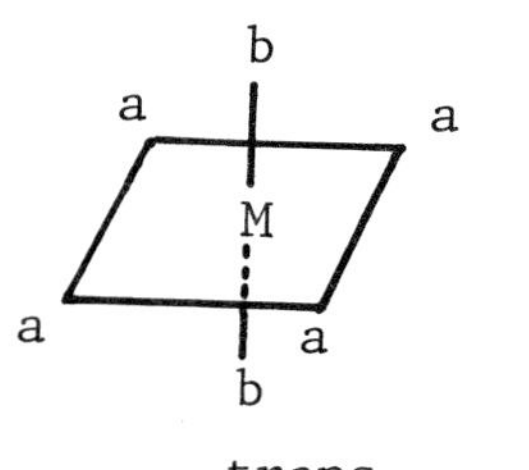

and

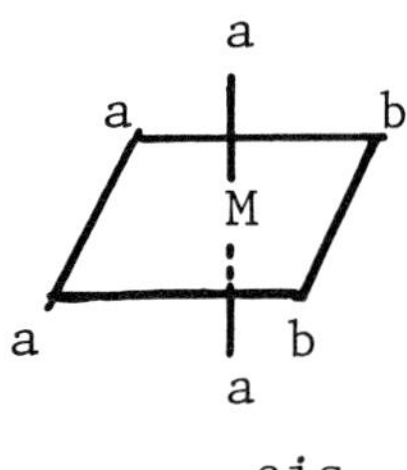

trans   cis

Octahedral complexes can also have optical isomers. For example the complex $[M(AA)_2b_2]$ where AA is a bidentate chelate and b is a ligand is optically active if the b's are cis:

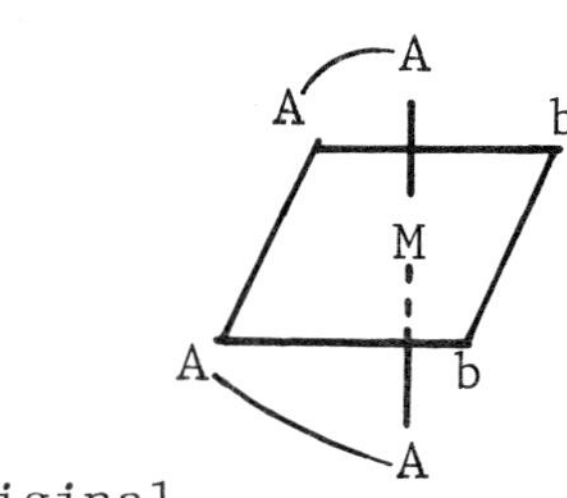

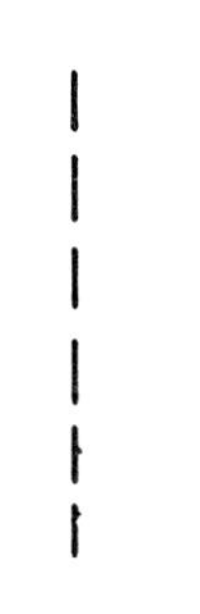

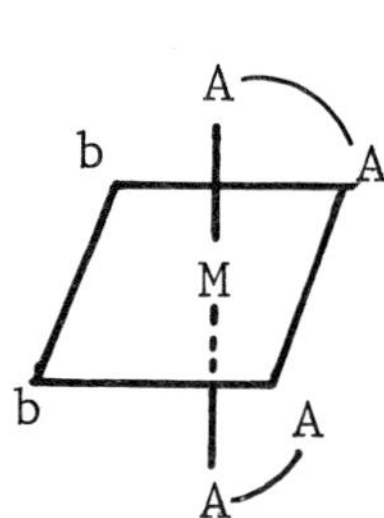

original   image

mirror

Here AA represents a chelate, like ethylenediamine. The image and the original are not superimposable and are optical isomers. Tetrahedral complexes in which all 4 ligands are different have two optical isomers.

<u>Example 26.2</u>   Which of the following complexes has geometrical isomers?

   a) $Ag(NH_3)_2^+$ b) $ZnCl_4^{2-}$

   c) $Fe(CN)_5Cl^{4-}$ d) $Co(NH_3)_4(CN)_2^+$

<u>Solution</u>   Cis and trans isomers are possible only for compound d.

<u>Example 26.3</u>   Which of the following complexes is optically active?

   a) $CuCl_4^{2-}$ b) $Cr(C_2O_4)_3^{3-}$

   c) $Co(NH_3)_4BrI^+$ d) $PtCNClBrI^{2-}$ (square planar)

<u>Solution</u>   Only compound b is optically active. Compound d would be optically active if it were tetrahedral rather than square planar.

Bonding

The following hybridization schemes explain the observed geometries of complexes.

Hybridization	Geometry	Examples
sp	linear	$AgCl_2^-$; $Au(CN)_2^-$, $HgCl_2$
sp^3	tetrahedral	MnO_4^-, $ZnCl_4^{2-}$, $Cd(NH_3)_4^{2+}$
dsp^2	square planar	$PtCl_4^{2-}$, $Pd(NH_3)_4^{2+}$, $Au(CN)_4^-$
d^2sp^3	octahedral	$Cr(H_2O)_6^{3+}$, $Fe(C_2O_4)_3^{3-}$, $Co(en)_3^{3+}$

In crystal field theory the metal is considered ionic and its magnetic and optical properties are related to its d electron configuration. In octahedral complexes, the five d orbitals are split into two groups, designated t_{2g} and e_g, by the electric field of the ligands. If the splitting is large the lower level will be filled completely before the upper level begins to fill. For example, for a d^5 configuration there are two possibilities:

$$\text{high spin} \qquad\qquad\qquad \text{and} \qquad\qquad\qquad \text{low spin}$$

The separation between the e_g and the t_{2g} energy levels is called 10 Dq. Ligands which cause large splittings are called strong field ligands and form low spin complexes. The following ligands are arranged according to their ability to split metal d orbitals:

$$CN^- > NH_2CH_2CH_2NH_2 > NH_3 > NCS^- > H_2O > F^- > Cl^- > Br^- > I^-$$

strong-field ligands	intermediate	weak-field ligands

Example 26.4 Predict the d electronic configuration of the following octahedral complexes using the designation $(t_{2g})^x(e_g)^y$:

a) $Cr(CN)_6^{3-}$

b) CoF_6^{3-}

c) $Co(CN)_6^{4-}$

d) $Fe(H_2O)_6^{3+}$

Solution   a) $(t_{2g})^3$

b) $(t_{2g})^4(e_g)^2$

c) $(t_{2g})^6(e_g)^1$

d) $(t_{2g})^3(e_g)^2$

The colors of coordination compounds can be explained in terms of the crystal field splitting. The greater the splitting of the d orbitals, the greater energy of a photon required to promote a t_{2g} electron to an e_g orbital. The following table correlates the observed colors with the wavelength of the absorbed photon.

Color observed	Color absorbed	Approximate wavelength (nm)	Approximate Energy (kJ/mole)
Colorless	Ultraviolet	<400	>299
Lemon yellow	Violet	410	292
Yellow	Indigo	430	277
Orange	Blue	480	249
Red	Blue-green	500	239
Purple	Green	530	226
Violet	Lemon yellow	560	214
Indigo	Yellow	580	207
Blue	Orange	610	196
Pale blue	Red	680	176
Pale blue	Deep red	720	166
Colorless	Infrared	>720	<166

Example 26.5   Given the following spectral data arrange the ligands H_2O, NH_3, NCS^-, Cl^- and CO_3^{2-} in order of increasing crystal field splitting.

Complex	Color Observed	Color Absorbed	Wavelength Absorbed (nm)
$Co(NH_3)_6^{3+}$	yellow	indigo	430
$Co(NH_3)_5H_2O^{3+}$	red	blue-green	500
$Co(NH_3)_5Cl^{2+}$	purple	green	530
$Co(NH_3)_5NCS^{2+}$	orange	blue	470
$Co(NH_3)_5CO_3^{+}$	pink	blue-green	500

Solution   For this set of related complexes the electronic transition has the following energy-wavelength relationship.

$$\Delta E = 10Dq = E_{e_g} - E_{t_{2g}} = h\nu = hc/\lambda$$

Large splittings are correlated with low values of λ.
The order of increasing 10Dq is:

$$Cl^- < H_2O = CO_3^{2-} < NCS^- < NH_3$$

weakest   strongest

Equilibria

Coordination compounds often exist in an equilibrium with the uncomplexed metal ion and the free ligand. The equilibrium reaction can be written either as a formation or a dissociation reaction. For example,

$$Cu^{2+} + 4NH_3 \rightleftharpoons Cu(NH_3)_4^{2+} \qquad K_f = \frac{[Cu(NH_3)_4^{2+}]}{[Cu^{2+}][NH_3]^4} = 1.0 \times 10^{12}$$

$$Cu(NH_3)_4^{2+} \rightleftharpoons Cu^{2+} + 4NH_3 \qquad K_d = 1/K_f = 1.0 \times 10^{-12}$$

Example 26.6 Calculate the equilibrium concentrations of Cu^{2+}, NH_3, and $Cu(NH_3)_4^{2+}$ in a solution prepared by dissolving 0.10 mole of $Cu(NO_3)_2$ in 1.0 liter of 0.75 $\underline{M}$ NH_3.

Assume the only complexation reaction is

$$Cu^{2+} + 4NH_3 \rightleftharpoons Cu(NH_3)_4^{2+} \qquad K_f = 1.0 \times 10^{12}$$

Assume the hydrolysis of NH_3 is negligible.

Solution This problem can be solved in two steps, the stoichiometry of the formation reaction assuming the reaction is quantitative ($K_f = \infty$), and then the equilibrium reaction using the dissociation reaction.

$$Cu^{2+} + 4NH_3 \longrightarrow Cu(NH_3)_4^{2+}$$

	Cu^{2+}	$4NH_3$	$Cu(NH_3)_4^{2+}$
Initially ($\underline{M}$)	0.1	0.75	0
After reaction	0	0.35	0.1

Let x be the equilibrium concentration of Cu^{2+}. Then

$$Cu(NH_3)_4^{2+} \rightleftharpoons Cu^{2+} + 4NH_3 \qquad K_d = 1/K_f$$

$$\phantom{Cu(NH_3)_4^{2+}} 0.1-x \qquad x \qquad 0.35 + 4x$$

$$K_d = \frac{[Cu^{2+}][NH_3]^4}{[Cu(NH_3)_4^{2+}]} = \frac{(x)(0.35+4x)^4}{0.1-x} = 1.0 \times 10^{-12}$$

Assume $x \ll 0.1$, $4x \ll 0.35$. Then

$$\frac{(x)(0.35)^4}{0.1} \simeq 1.0 \times 10^{-12}$$

$$x = 6.7 \times 10^{-12} \ \underline{M}$$

The assumption is excellent. The equilibrium concentrations are $[Cu^{2+}] = 6.7 \times 10^{-12}$ $\underline{M}$, $[NH_3] = 0.35$ $\underline{M}$ and $[Cu(NH_3)_4^{2+}] = 0.10$ $\underline{M}$.

Questions

26.1 Which of the following complexes is/are paramagnetic?

$$1. \quad Cr(H_2O)_6^{3+} \qquad 2. \quad CoF_6^{4-} \qquad 3. \quad Ag(CN)_2^-$$

a) 1
b) 1 and 2
c) 1 and 3
d) All of them
e) None of them

26.2 Which of the following complexes is correctly named?

1. $Co(NH_3)_5Cl^{2+}$ chloropentaamminecobalt(III) ion

2. $Fe(SCN)_4C_2O_4^{3-}$ oxalatotetrathiocyanatoferrate(III) anion

3. $Cr(H_2O)_5Br^{2+}$ bromopentaaquochromium(III) ion

a)   1
b)   1 and 2
c)   2 and 3
d)   1 and 3
e)   All of them

26.3 Which row in the following table contains an error?

Empirical formula	Number of ions per formula unit	Number of Cl^- ions per formula unit	Compound
a) $Pt(NH_3)_6Cl_4$	5	4	$[Pt(NH_3)_6]Cl_4$
b) $Pt(NH_3)_4Cl_4$	3	2	$[Pt(NH_3)_4Cl_2]Cl_2$
c) $Pt(NH_3)_2Cl_4$	0	0	$[Pt(NH_3)_2Cl_4]$
d) K_2PtCl_6	3	0	$K_2[PtCl_6]$

e) None of them

26.4 Which of the following complexes could be diamagnetic?

$$1. \quad Fe(CN)_6^{4-} \qquad 2. \quad Co(en)_3^{3+} \qquad 3. \quad Cr(NH_3)_6^{3+}$$

a)   1 and 2
b)   1 and 3
c)   2 and 3
d)   All of them
e)   None of them

26.5 The spin only formula for the magnetic moment of a coordination complex is

$$\mu = \sqrt{n(n+2)}$$

where n is the number of unpaired electrons. This gives:

n	1	2	3	4	5	
μ	1.73	2.83	3.88	4.90	5.92	Bohr magnetons

The observed magnetic moment of **hexacyanomanganate**(II) is **1.8 Bohr magnetons**. Indicate the d electron configuration of the **Mn**(II) complex.

	t_{2g}	e_g
a)	3	2
b)	2	3
c)	5	0
d)	0	5
e)	None of the above	

26.6*How many geometrical isomers are possible for a hypothetical octahedral complex M(AA)B_2C_2? (AA is a bidentate chelating agent and B and C are monodentate ligands)

a) 2
b) 3
c) 4
d) 6
e) None of the above

26.7*Which of the following complexes would exhibit optical isomerism?

1. cis-Co$(NH_3)_4(H_2O)_2^{3+}$ 2. trans-Cr(en)$_2Cl_2^{+}$ 3. Fe$(C_2O_4)_3^{3-}$

a) 1
b) 3
c) 1 and 3
d) 2 and 3
e) None of them

26.8 Which row in the following table contains an error? (↑ and ↓ denote metal electrons, x denotes a ligand electron).

Complex	Geometry	Hybridization	3d	4s	4p	4d
a) $CuCl_2^-$	linear	sp	↑↓ ↑↓ ↑↓ ↑↓ ↑↓	xx	xx __ __	__ __ __ __ __
b) $Ni(CN)_4^{2-}$	square planar	dsp^2	↑↓ ↑↓ ↑↓ ↑↓ ↑↓	xx	xx xx __	xx __ __ __ __
c) $Zn(NH_3)_4^{2+}$	tetrahedral	sp^3	↑↓ ↑↓ ↑↓ ↑↓ ↑↓	xx	xx xx xx	__ __ __ __ __
d) $Fe(CN)_6^{3-}$	octahedral	d^2sp^3	↑↓ ↑↓ ↑ xx xx	xx	xx xx xx	__ __ __ __ __
e) $Fe(H_2O)_6^{2+}$	octahedral	sp^3d^2	↑↓ ↑ ↑ ↑ ↑	xx	xx xx xx	xx xx __ __ __

26.9 Which of the following complexes is polar?

   1. trans-$[Pt(NH_3)_2Cl_2]$ 2. cis-$[Co(C_2O_4)_2Br_2]^{3-}$ 3. $Ag(S_2O_3)_2^{3-}$

  a) 1 and 3
  b) 2
  c) 2 and 3
  d) All of them
  e) None of them

26.10* The pH of a 0.10 $\underline{M}$ $NiCl_2$ solution is 5.0 because of the formation of $NiOH^+$. What is the formation constant K_f of the hydroxonickel(II) complex?

$$Ni^{2+} + OH^- \rightleftharpoons NiOH^+ \qquad K_f$$

  a) $K_f = 1 \times 10^8$

  b) $K_f = 1 \times 10^5$

  c) $K_f = 1 \times 10^6$

  d) $K_f = 1 \times 10^{-9}$

  e) $K_f = 1 \times 10^{23}$

26.11* One millimole of the soluble sodium salt of the complex anion MX_6^{3-} is dissolved in 50 milliliters of water. The concentration of uncomplexed M^{3+} was found to be 6.0×10^{-5} molar. The dissociation reaction is:

$$MX_6^{3-} \rightleftharpoons M^{3+} + 6 \ X^-$$

What is the dissociation constant for MX_6^{3-}?

a) 1.8×10^{-7}
b) 1.4×10^{-28}
c) 1.1×10^{-19}
d) 6.5×10^{-25}
e) 6.5×10^{-24}

26.12 * What is the concentration of Zn^{2+} in a solution that is 0.450 molar in OH^- and 0.050 molar in $Zn(OH)_4^{2-}$? The overall formation constant for $Zn(OH)_4^{2-}$ is 5.00×10^{14}. (Neglect all hydrolysis reactions).

a) 1.44×10^{-10}
b) 2.25×10^{-14}
c) 5.00×10^{-14}
d) 2.44×10^{-15}
e) 4.94×10^{-8}

26.13 What is the concentration of F^- in a solution that is 0.500 molar in Sn^{2+} and 0.350 molar in SnF_3^-? The overall formation constant for SnF_3^- is 8.0×10^9. (neglect all hydrolysis reactions).

a) 3.95×10^{-4}
b) 8.75×10^{-11}
c) 1.50
d) 5.63×10^{-4}
e) 4.44×10^{-4}

26.14 What is the equlibrium concentration of Fe^{2+} in a solution that is initially 0.450 molar in $Fe(OH)^+$? The overall formation constant for $Fe(OH)^+$ is 1.00×10^7. (Neglect all hydrolysis reactions.)

a) 2.12×10^{-4}
b) 1.86×10^{-4}
c) 4.50×10^{-8}
d) 1.28×10^{-4}
e) 1.06×10^{-4}

26.15 What is the equilibrium concentration of CN^- in a solution that is initially 0.950 molar in $CdCN^+$? The overall formation constant for $CdCN^+$ is 6.00×10^{18}. (Neglect all hydrolysis reactions.)

a) 3.54×10^{-10}
b) 1.58×10^{-19}
c) 7.96×10^{-10}
d) 1.99×10^{-10}
e) 3.98×10^{-10}

26.16*   What is the equilibrium concentration of Ni^{2+} in a solution that is initially 0.500 molar in Ni^{2+} and 0.550 molar in CN^-? The formation constant for $NiCN^+$ is 1.00×10^{30}. (Neglect all hydrolysis reactions.)

a) 5.50×10^{-29}
b) 1.00×10^{-30}
c) 1.10×10^{-29}
d) 1.00×10^{-29}
e) 3.16×10^{-15}

26.17   What is the equilibrium concentration of F^- in a solution that is initially 0.400 molar in Fe^{3+} and 0.300 molar in F^-? The overall formation constant for FeF_6^{3-} is 2.00×10^{15}. (Neglect all hydrolysis reactions.)

a) 6.00×10^{14}
b) 2.04×10^{-3}
c) 2.00×10^{15}
d) 4.51×10^{-2}
e) 3.38×10^{-3}

26.18*   What is the pH of a solution that is initially 0.10 $\underline{M}$ in the complex $Zn(OH)_4^{2-}$? The overall formation constant for the tetrahydroxozincate complex is 5.0×10^{14}.

a)   7.00
b)   3.02
c)   10.38
d)   10.98
e)   12.50

26.19*   Compare the solubility of AgI in pure water to its solubility in 1.0 $\underline{M}$ NH_3. The solubility product of AgI is 8.5×10^{-17}, and the formation constant of the diamminesilver(I) complex is 1.0×10^8 at 25°C.

	In H_2O	in 1.0M NH_3
a)	$\sqrt{8.5 \times 10^{-17}}$	$\sqrt{(8.5 \times 10^{-17})(1.0 \times 10^8)}$
b)	$\sqrt{8.5 \times 10^{-17}}$	$\sqrt{(8.5 \times 10^{-17})/(1.0 \times 10^8)}$
c)	8.5×10^{-17}	$\sqrt{(1.0 \times 10^8)/(8.5 \times 10^{-17})}$
d)	$\sqrt{8.5 \times 10^{-17}}$	$(8.5 \times 10^{-17})(1.0 \times 10^8)$

e)   None of the above

26.20   The color of some transition metal complexes can be explained
in terms of d-d electronic transitions. The following observations
were made for three octahedral complexes of the same metal:

Complex	Color of Complex	Spectral Color Absorbed	Wavelength Absorbed (Å)
MX_6^{3-}	purple	green	5300
MY_6^{3-}	green	purple-red	7200
MZ_6^{3-}	red	blue-green	5000

Which of the following is the correct sequence of ligands ordered
according to decreasing crystal field strength?

a)   X > Y > Z
b)   X > Z > Y
c)   Z > X > Y
d)   Y > Z > X
e)   None of the above

26.21*   The complex MY_6^{2+} is red in solution. If L has a greater crystal
field strength than Y, which of the following descriptions is most
likely for the spectral properties of ML_6^{2+}?

a)   deeper red in solution than MY_6^{2+}
b)   less red than MY_6^{2+}, perhaps even colorless, in solution
c)   blue to green in the solid state
d)   strongly absorbs yellow light
e)   absorbs near-infrared light

26.22   A 0.10 molal solution of the compound $Co(NH_3)_5Cl_3$ freezes at
-0.56°C. Which of the following is the correct designation of this
coordination compound? (The freezing point depression constant for
H_2O is 1.86°C/molal, and you may assume the solution behaves ideally.)

a) $[Co(NH_3)_5Cl_3]$
b) $[Co(NH_3)_5Cl_2]Cl$
c) $[Co(NH_3)_5Cl]Cl_2$
d) $[Co(NH_3)Cl_3]\cdot4NH_3$
e)   None of the above

26.23*   Given the following spectrochemical series:

$$CN^- > NO_2^- > en > NH_3 > SCN^- > H_2O > F^- > Cl^-$$

Which of the following statements is wrong?

a) F^- will cause less splitting than NH_3 among the d orbitals of a
metal ion.

b) $[Co(NH_3)_5Cl]^{2+}$ absorbs light at a longer wavelength than
$[Co(NH_3)_6]^{3+}$

c)   If M is a transition metal and $M(NH_3)_6^{2+}$ is yellow, it is
likely that $M(H_2O)_6^{2+}$ is blue.

d)   Since $[Co(NH_3)_6]^3$ is diamagnetic, $[Co(en)_3]^{3+}$ will also be diamagnetic

e)   You are more likely to form outer orbital complexes (in Valence
Bond theory) with NO_2^- than with Cl^-.

26.24*   A sample of the potassium salt of a hypothetical coordination
compound was dissolved in 1.0 liter of H_2O. The compound was
catalytically decomposed. Spectroscopic analysis indicated that
the solution was .0090 molar in M^{2+}.

The ammonia concentration was determined by titration with a standard
0.1 molar acid solution. The equivalence point was reached after
addition of 180 milliliters of the titrant.

The I^- concentration was determined gravimetrically by precipitation
with excess $AgNO_3$. This procedure yielded 8.46 grams of AgI.

What is the empirical formula of the complex?

a) $[M(NH_3)_3I_3]^-$

b) $[MNH_3I_5]^{3-}$

c) $[M(NH_3)_2I_4]^{2-}$

d) $[MNH_3I_3]^-$

e)   None of the above

26.25*   What is the crystal field stabilization energy (in Dq units)
of a hypothetical complex of Mn(VI) in an octahedral strong field?

a)   6 Dq
b)   18 Dq
c)   24 Dq
d)   4 Dq
e)   12 Dq

CHAPTER 27. NUCLEAR CHEMISTRY

Introduction

Although most chemical properties can be explained in terms of the
electronic structure of atoms and molecules, some phenomena of interest
to chemists are associated with atomic nuclei rather than the electrons
surrounding the nucleus. Isotopes are identified using the designation $^A_Z Q$,
for example $^3_1 H$, $^{12}_6 C$, $^{90}_{38} Sr$ and $^{238}_{92} U$. Q is the symbol of the element, Z is
the atomic number and A is the mass number. A is an integer approximately
equal to the atomic mass, within 0.1 amu, Z is the number of protons and
A−Z is the number of neutrons in the nucleus. Isotopes of an element
differ in the number of neutrons in the nucleus. For example, hydrogen
($^1_1 H$), deuterium ($^2_1 H$), and tritium ($^3_1 H$) atoms have one proton and zero, one,
and two neutrons in the nucleus, respectively.

The following fundamental particles are the ones usually encountered
in nuclear reactions:

Particle	Symbol	Mass (amu)	Charge
alpha	$^4_2 \alpha$	4.00205	+2
beta (electron)	$^0_{-1} e$	0.0005486	−1
positron	$^0_{+1} e$	0.0005486	+1
neutron	$^1_0 n$	1.008665	0
gamma ray	γ	0	0
proton	$^1_1 P$	1.007277	+1
hydrogen atom	$^1_1 H$	1.007825	
deuterium atom	$^2_1 H$	2.01410	
tritium atom	$^3_1 H$	3.01605	

Particle	Symbol	Mass (amu)	Charge
helium atom	$^{4}_{2}\text{He}$	4.00260	

In writing nuclear reactions it is common to use the symbol $^{4}_{2}\text{H}$ for an alpha particle, and $^{1}_{1}\text{H}$ for a proton.

The atomic mass unit is arbitrarily defined in terms of the $^{12}_{6}\text{C}$ isotope which is assigned mass 12.0000...atomic mass units (amu). One amu is 1.66×10^{-24} gram. The masses given above for $^{4}_{2}\text{He}$ and the three hydrogen isotopes are atomic masses, that is, the mass of the electron(s) is included.

Nuclear reactions must be balanced with respect to both atomic mass and nuclear charge. For example

$$^{14}_{7}\text{N} + {}^{4}_{2}\text{He} \longrightarrow {}^{17}_{8}\text{O} + {}^{1}_{1}\text{H}$$

The nuclear charge is 9 on both sides and the atomic mass is 18 on both sides. Gamma rays are frequently emitted but are not involved in the balancing, and so may or may not be written.

Example 27.1. Complete the following nuclear reactions:

a) $^{15}_{7}\text{N} + {}^{1}_{1}\text{H} \longrightarrow {}^{4}_{2}\text{He} + ?$

b) $^{235}_{92}\text{U} + {}^{1}_{0}\text{n} \longrightarrow {}^{95}_{38}\text{Sr} + {}^{139}_{54}\text{Xe} + ?$

c) $^{7}_{3}\text{Li} + {}^{1}_{1}\text{H} \longrightarrow 2\ ?$

d) $^{32}_{15}\text{P} \longrightarrow {}^{32}_{16}\text{S} + ?$

Solutions   a) $^{12}_{6}\text{C}$ b) $2\,{}^{1}_{0}\text{n}$ c) $^{4}_{2}\text{He}$ d) $^{0}_{-1}\text{e}$

Energy Changes in Nuclear Reactions

The protons and neutrons are held together in the nucleus by the binding energy. This can be calculated from the difference between the observed nuclear mass and the sum of the masses of the constituent particles. The observed mass is always less than it would be if mass were conserved. For example, the observed atomic mass (including the electrons) for $^{4}_{2}\text{He}$ is 4.00260 amu. The sum of the constituent particles is 2(mass of $^{1}_{1}\text{H}$ + mass of $^{1}_{0}\text{n}$ + mass of $^{0}_{-1}\text{e}$) or $2(1.007277 + 1.008665 + 0.0005486) = 4.03298$ amu. This difference, called the mass loss or mass defect, is -0.03038

amu for $_2^4$He. The "missing" mass has been liberated as energy. This energy, called the binding energy, would have to be supplied to separate a helium nucleus into its constituent particles. The relationship between energy and mass is given by Einstein's equation,

$$E = mc^2$$

where m is the mass in grams, c is the speed of light in cm/sec and E is the energy in ergs. The conversion factor used in nuclear chemistry is 931 MeV/amu, where MeV, million electron volts, is the unit of energy, and amu is the unit of mass. Thus the binding energy of $_2^4$He is 931 x 0.03038 or 28.3 MeV. A convenient measure of nuclear stability is the average binding energy per nucleon (proton or neutron). For $_2^4$He, this is 28.3/4 or 7.08 MeV/nucleon.

<u>Example 27.2</u>. Calculate the average binding energy per nucleon for $_{26}^{56}$Fe. The observed atomic mass of this isotope is 55.93493 amu.

<u>Solution</u>. The calculated mass of $_{26}^{56}$Fe is

$$26 \text{ mass of } _1^1p + 30 \text{ mass of } _0^1n + 26 \text{ mass of } _{-1}^0e$$

$$= 26 \,(1.007277) + 30 \,(1.008665) + 26 \,(0.0005486) = 56.46342$$

$$\frac{\text{Binding energy}}{\text{nucleon}} = \frac{931}{56}\,(56.46342 - 55.93493) = \frac{8.79 \text{ MeV}}{\text{nucleon}}$$

Enormous amounts of energy are involved in fission and fusion reactions. This energy, like the binding energy, is due to a mass loss during the reaction.

<u>Example 27.3</u>. Calculate the energy in MeV associated with the following reactions.

a) fission: $_{92}^{235}U + _0^1n \longrightarrow _{54}^{141}Xe + _{38}^{92}Sr + 3\,_0^1n$

                 235.04393   140.91   91.90

b) fusion: $_1^2H + _1^3H \longrightarrow _2^4He + _0^1n$

<u>Solution.</u>   a)   The change in mass, ΔM, is

$$\Delta M = 140.91 + 91.90 + 3 \times 1.008665 - (235.04393 + 1.008665)$$

$$= -0.22 \text{ amu}$$

The change in energy ΔE is

$$\Delta E = 931 \, (-0.22) = -205 \text{ MeV}$$

b) $\Delta M = 4.00260 + 1.008665 - (2.01410 + 3.01605) = -0.0189$

$$\Delta E = 931 \, (-0.0189) = -17.6 \text{ MeV}$$

<u>Radioactive Decay</u>

The majority of the known isotopes are radioactive. Four modes of radioactive decay are considered here: alpha; beta and positron emission; and electron capture. These processes and the corresponding energy changes are illustrated by the following examples. All masses are atomic masses.

1. <u>Alpha decay</u>

$$^{239}_{94}\text{Pu} \longrightarrow {}^{235}_{92}\text{U} + {}^{4}_{2}\text{He}$$

$$239.05216 \qquad 235.04393 \quad 4.00260$$

$$\Delta E = 931(235.04393 + 4.00260 - 239.05216)$$

$$= -5.24 \text{ MeV}$$

The alpha particle is actually the dipositive ion, He^{2+}.

The reaction, including electronic charge is

$$^{239}_{94}\text{Pu} \longrightarrow {}^{235}_{92}\text{U}^{2-} + {}^{4}_{2}\text{He}^{2+}$$

The total number of electrons on the right is the same as in the first equation so no error in total mass was made.

2. <u>Beta decay</u>

$$^{38}_{17}\text{Cl} \longrightarrow {}^{38}_{18}\text{Ar} + {}^{0}_{-1}\text{e}$$

$$37.965 \qquad\quad 37.963$$

Consider the symbols in the above equation to signify the nuclei. Then add 17 electrons to each side so that the mass of the left side is the mass of a chlorine-38 atom and the mass of the right side is the mass of an argon-38 atom.

$$\Delta E = 931(37.963 - 37.965) = -2 \text{ MeV}$$

3. <u>Positron decay</u>

$$^{15}_{8}O \longrightarrow ^{15}_{7}N + ^{0}_{+1}e$$

$$15.00307 \qquad 15.00011$$

If 8 electrons are added to each side the mass of the left side is the mass of the oxygen-15 atom. The left side now has the mass of the nitrogen-15 atom (the nucleus plus seven electrons) plus an additional electron and a positron (which have identical masses).

$$\Delta E = 931 \ (15.00011 + 2 \times 0.0005486 - 15.00307) = -1.73 \ \text{MeV}$$

4. <u>Electron capture</u> (EC)

In electron capture an orbital electron (1s) is absorbed by the nucleus.

$$^{7}_{4}Be + ^{0}_{-1}e \xrightarrow{\ EC\ } ^{7}_{3}Li$$

$$7.01693 \qquad 7.01601$$

If 3 electrons are added to each side, the mass of the left side will be that of a beryllium-7 atom and that on the right a lithium-7 atom.

$$\Delta E = 931 \ (7.01601 - 7.01693) = -0.86 \ \text{MeV}$$

<u>Example 27.4.</u> Calculate the energy changes associated with the following nuclear decay reactions:

a) $\quad ^{18}_{10}Ne \ (18.00572) \longrightarrow ^{18}_{9}F \ (18.00095) + ^{0}_{+1}e$

b) $\quad ^{244}_{98}Cf \ (244.06593) \longrightarrow ^{240}_{96}Cm \ (240.05550) + ^{4}_{2}He$

c) $\quad ^{37}_{18}Ar \ (36.96677) \xrightarrow{\ EC\ } ^{37}_{17}C\ell \ (36.96590)$

d) $\quad ^{60}_{27}Co \ (59.93381) \longrightarrow ^{60}_{28}Ni \ (59.93078) + ^{0}_{-1}e$

<u>Solutions.</u>

a) $\quad 931 \ (18.00095 + 2 \times 0.0005486 - 18.00572) = -3.42 \ \text{MeV}$

b) $\quad 931 \ (240.05550 + 4.00260 - 244.06593) = -7.29 \ \text{MeV}$

c) $\quad 931 \ (36.96590 - 36.96677) = -0.810 \ \text{MeV}$

d) $\quad 931 \ (59.93078 - 59.93381) = -2.82 \ \text{MeV}$

The calculation of energy changes in nuclear decay reactions may be easier if nuclear masses rather than atomic masses are used. For example, in positron emission,

$$^{15}_{8}O \longrightarrow ^{15}_{7}N + ^{0}_{+1}e$$

nuclear   14.99868 14.99627 0.0005486
mass

$$\Delta E = 931 \, (14.99627 + 0.0005486 - 14.99868) = -1.73 \text{ MeV}$$

Atomic masses are used in this chapter because the current nuclear chemistry references tabulate atomic rather than nuclear masses.

Rates of Radioactive Decay

Radioactive isotopes decay by a first order process, for example

$$^{14}_{6}C \longrightarrow ^{14}_{7}N + ^{0}_{-1}e$$

Let N be the number of $^{14}_{6}C$ atoms. Then

$$\frac{dN}{dt} = -kN = \frac{-0.693 \, N}{\tau}$$

$$N = N_o e^{-kt} = N_o e^{\frac{-0.693t}{\tau}}$$

$$\ln \frac{N_o}{N} = kt = \frac{0.693t}{\tau} = 2.303 \, \log \frac{N_o}{N}$$

Here N_o is the number of $^{14}_{6}C$ atoms at time (t) equal to zero, k is the specific rate constant for the first-order decay, and τ is the half-life of $^{14}_{6}C$.

Example 27.5. The half-life for the alpha decay of $^{254}_{100}Fm$ is 3.24 hours. How long will it take for a 1.00µg $(10^{-6}g)$ sample of the isotope to decay to 1.00 ng $(10^{-9}g)$?

Solution.

$$N = N_o e^{\frac{-0.693t}{\tau}}$$

$$\frac{N}{N_o} = \frac{10^{-9}}{10^{-6}} = e^{\frac{-0.683t}{3.24}}$$

$$\ln(10^{-3}) = \frac{-0.693t}{3.24} = -6.908$$

$$t = 32.3 \text{ hours}$$

Questions

27.1 Which of the following nuclear reactions is not balanced?

a) $^{68}_{30}Zn + ^{1}_{0}n \longrightarrow ^{65}_{28}Ni + ^{4}_{2}He$

b) $^{67}_{29}Cu + ^{2}_{1}H \longrightarrow ^{65}_{28}Ni + 2^{1}_{0}n$

c) $^{14}_{7}N + ^{1}_{0}n \longrightarrow ^{14}_{6}C + ^{1}_{1}H$

d) $^{14}_{7}N + ^{4}_{2}He \longrightarrow ^{17}_{8}O + ^{1}_{1}H$

e) $^{3}_{1}H + ^{2}_{1}H \longrightarrow ^{4}_{2}He + ^{1}_{0}n$

27.2 One of the natural radioactive decay series involves $^{232}_{90}Th$. This isotope decays with the emission of α and β^- particles to form, ultimately, $^{208}_{82}Pb$. How many α and β^- particles are emitted per atom of $^{208}_{82}Pb$ formed.

	α	β^-
a)	1	1
b)	4	6
c)	6	4
d)	8	6
e)	6	2

27.3 * Which one of the following is <u>not</u> the expected mode of decay of the indicated isotope?

a) $^{8}_{5}B \longrightarrow ^{8}_{4}Be + ^{0}_{+1}e$

b) $^{16}_{6}C \longrightarrow ^{16}_{7}N + ^{0}_{-1}e$

c) $^{7}_{4}Be + ^{0}_{-1}e \longrightarrow ^{7}_{3}Li$ (electron capture)

d) These are all the expected mode of decay.

e) None of these is the expected mode of decay.

27.4 An important parameter for judging the stability of a nuclide is the binding energy per nucleon. Calculate the binding energy per nucleon in MeV for the ^{15}O atom which has a mass of 15.00307 amu.

a) 108
b) 6.91
c) 7.19
d) 104
e) 7.46

27.5   Calculate the energy in MeV liberated in the following nuclear
        fission reaction:

$$^{235}U + neutron \longrightarrow {}^{91}Rb + {}^{142}Cs + 3\ neutrons$$

a) 2.04×10^{-4}

Isotope	^{235}U	^{91}Rb	^{142}Cs
Mass (amu)	235.04	90.91	141.92

b) 177
c) 3.94×10^4
d) 2.06×10^3
e) 206

27.6   In a present day nuclear reactor a typical reaction is:

$$^{235}U + neutron \longrightarrow {}^{93}Kr + {}^{140}Ba + 3\ neutrons$$

Select the one correct statement.

a)   The products have exceptionally low neutron/proton ratios.
b)   In the above reaction the number of nucleons and the charge
      are both conserved.
c)   Neutrons which escape the reactor will accumulate in the
      atmosphere and eventually threaten our environment.
d)   This reaction is called fusion.
e)   The above reaction is endothermic (ΔH is plus).

27.7*   Calculate the energy in MeV liberated in the following nuclear
        decay reaction:

$$^{36}Cl \longrightarrow {}^{36}Ar + \beta^-$$

a) 0.20

Isotope	^{36}Cl	^{36}Ar
Mass (amu)	35.96831	35.96755

b) 0.00082
c) 3.2
d) 0.0012
e) 0.71

27.8*   How many tons of coal will be required to generate as much heat as
        is generated when 1.00 pound of ^{4}He is formed according to the
        following fusion reaction?

        Assume coal is 100% carbon and the only combustion process is

$$C + O_2 \longrightarrow CO_2 \qquad \Delta H^\circ = -394\ kJ$$

1 ton = 2000 x 454 g and 1 eV/molecule = 96.5 kJ/mole

a)   5230 tons
b)   6470 tons
c)   11.5 tons
d)   49200 tons
e)   434 tons

$$^2_1H + {}^3_1H \longrightarrow {}^4_2He + {}^1_0n + 17.6\ MeV$$

27.9   The technique of radiocarbon dating relies on the following principles: (Select the one wrong statement.)

a)   Radioactive ^{14}C has been formed in the atmosphere at an almost constant rate for many thousands of years.
b)   Living organisms have always contained the same fraction of ^{14}C as the atmosphere.
c)   Old organic compounds build up easily detectable boron activity from the decay of ^{14}C.
d)   When ^{14}C decays it emits a beta particle which can be detected in an instrument such as a geiger counter.
e)   The specific ^{14}C activity, counts per minute per gram of carbon, is not affected by chemical decomposition and volatilization of the object.

27.10   The half-life of a radioactive isotope is 14.0 days. How many days will it take the activity of a given sample to be reduced to 61.0% of the original activity?

a) $\dfrac{0.693 \times 14.0 \times 100}{2.303 \times (100-61.0)}$

b) $\dfrac{2.303 \times 14.0 \times \log\ (100/61.0)}{0.693}$

c) $\dfrac{2.303 \times 14.0 \times 61.0}{0.693}$

d) $\dfrac{2.303 \times 14.0 \times \log(61.0/100)}{0.693}$

e)   None of the above.

27.11   If 71.0 percent of a radioactive isotope decays in 294 days, what is its half-life?

a) $(2)(294)\left(\dfrac{100 - 71.0}{100}\right)$ days

b) $(.5)\left(\dfrac{2.303}{.693}\right)\left(\dfrac{71.0}{294}\right)$ days

c) $\dfrac{2.303}{.693}\left\{\dfrac{294}{\log\left(\dfrac{100}{100 - 71.0}\right)}\right\}$ days

d) $\dfrac{.693}{2.303}\left\{\dfrac{294}{\log\left(\dfrac{100}{100 - 71.0}\right)}\right\}$ days

e)   None of the above.

27.12   The mass of a sample of a radioactive isotope is 4.55 grams.
Given a half-life of 54.1 yrs, what amount of the isotope will
be left after 83.1 yrs?

a)   2.96 g
b)   .0312 g
c)   .848 g
d)   1.57 g
e)   .892 g

27.13*   What is the approximate age of an artifact found by an archaeolo-
gist which has a specific activity due to carbon-14 of 10.2 counts
per minute per gram of carbon? Assume that living matter on the
earth had an activity of 15.3 counts per minute per gram of
carbon. The half-life of carbon-14 is 5770 years.

a)   8640 years
b)   1470 years
c)   3370 years
d)   1880 years
e)   7770 years

27.14*   The overall reaction for the decomposition of ^{232}Th is

$$^{232}Th \longrightarrow\ ^{208}Pb + 6\ \alpha + 4\ \beta$$

A geological sample contained a rock which had 1.97 grams of ^{208}Pb
for every 1.29 grams of ^{232}Th. Calculate the age of the rock in
years. (The overall half-life for ^{232}Th is 1.39×10^{10} years.)

a) 1.86×10^{10}
b) 2.36×10^{10}
c) 1.79×10^{10}
d) 2.00×10^{10}
e) 9.66×10^{9}

27.15   The half-lives of $^{235}_{92}U$ and $^{238}_{92}U$ are 7.1×10^{8} years and 4.5×10^{9}
years, respectively. The current composition of naturally
occurring uranium is 99.28% $^{238}_{92}U$ and 0.72% $^{235}_{92}U$. What will the
composition of naturally occurring uranium on the earth be after
another 7.1×10^{8} years have passed?

	$^{238}_{92}U(\%)$	$^{235}_{92}U(\%)$
a)	99.64	0.36
b)	99.60	0.40
c)	89.00	11.00
d)	99.99	0.01
e)	49.64	50.36

CHAPTER 28 ORGANIC CHEMISTRY

Nomenclature

It has been estimated that there are a million organic compounds
known. These range from the simplest stable hydrocarbon, methane (CH_4),
to the exceedingly complex bio-polymers found in living cells. The following classification scheme, which is based on functional groups, is
only an introduction to a systematic nomenclature of organic chemistry.

Class	Formula	Example
Alkane	C_nH_{2n+2}	C_3H_8 – propane
Alkene	C_nH_{2n}	C_4H_8 – butene
Alkyne	C_nH_{2n-2}	C_5H_8 – pentyne
Alcohol	ROH	$C_6H_{13}OH$ – hexanol
Aldehyde	RCHO	CH_3CHO – acetaldehyde
Ketone	$R-\overset{O}{\overset{\|}{C}}-R'$	$CH_3COC_2H_5$ – methyl ethyl ketone
Ether	R-O-R'	$(C_2H_5)_2O$ – diethylether
Acid	RCOOH	C_2H_5COOH – propionic acid
Amine	RNH_2	$C_3H_7NH_2$ – propyl amine
Amide	$RCONH_2$	CH_3CONH_2 – acetamide
Halide	RX	C_2H_5Br – ethyl bromide

Organic compounds can be further classified as cyclic or acyclic.
Some examples of cyclic compounds and their structural representations are
given below

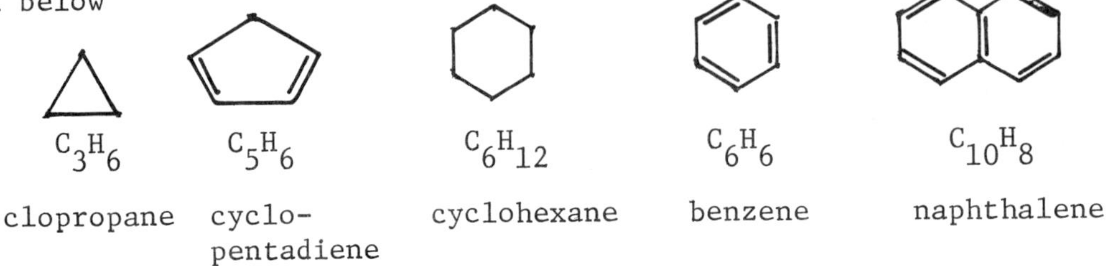

C_3H_6 C_5H_6 C_6H_{12} C_6H_6 $C_{10}H_8$

clopropane   cyclo-pentadiene   cyclohexane   benzene   naphthalene

Isomerism

Isomers are different substances which have the same chemical formula. The following examples demonstrate structural, geometric, and optical isomerism.

<u>Structural</u> – there are 5 isomers of C_6H_{14}

$$CH_3CH_2CH_2CH_2CH_2CH_3 \qquad CH_3\underset{\underset{CH_3}{|}}{C}HCH_2CH_2CH_3 \qquad CH_3CH_2\underset{\underset{CH_3}{|}}{C}HCH_2CH_3$$

$$CH_3\underset{\overset{|}{CH_3}}{C}H\,\underset{\overset{|}{CH_3}}{C}HCH_3 \qquad\qquad CH_3-\overset{\overset{CH_3}{|}}{\underset{\underset{CH_3}{|}}{C}}-CH_2CH_3$$

<u>Geometric</u> – There are two isomers of $H\overset{\overset{Cl}{|}}{C}=CHCH_3$. Because of the presence of the pi-bond the C=C bond cannot be rotated and so the cis and trans isomers cannot be superimposed.

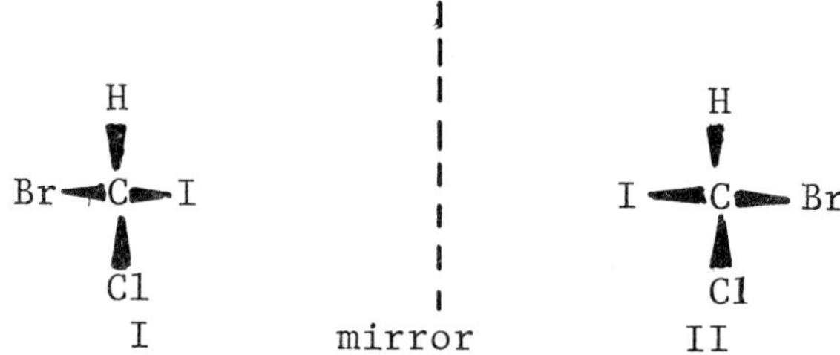

$$\text{cis} \qquad\qquad\qquad\qquad \text{trans}$$

<u>Optical</u> – There are two isomers (enantiomers) of CHClBrI

<pre>
          H   H
          ▲   ▲
  Br ◄─ C ─► I   I ◄─ C ─► Br
          ▲   ▲
          Cl   Cl
          I   mirror   II
</pre>

The molecule labelled I and its mirror image, II, are not super-imposable.

Bonding

The hybridization scheme involving the 2s and 2p atomic orbitals of carbon works very well in explaining the structures of organic compounds.

<u>Hybridization</u>	<u>Geometry</u>	<u>Examples</u>	
sp	linear	$H-C\equiv C-CH_3$	<HCC = 180°
sp^2	trigonal planar	$\overset{H}{\underset{H}{}}C=C\overset{Cl}{\underset{CH_3}{}}$	<HCC = 120°
sp^3	tetrahedral	$CH_3CH_2CH_3$	<HCC = 109.5°

268

Questions

28.1 Which one of the following statements is <u>incorrect</u>?

    a) $C_5H_{11}CHO$ is an aldehyde.

    b) $C_{11}H_{20}$ is an alkane.

    c) $C_5H_{11}OH$ is an alcohol.

    d) $C_5H_{11}NH_2$ is an amine.

    e)   None of the above.

28.2 Which of the following compounds is <u>incorrectly</u> named?

    a) C_5H_{12} pentane

    b) $C_3H_7NH_2$ propyl amine

    c) $CH_3CH=CHCH_3$ butene

    d) $CH_3OC_2H_5$ methyl ethyl ether

    e) C_4H_9COOH butyric acid

28.3*Which one of the following compounds has a dipole moment?

    a)   trans-1,2-dichloroethylene
    b)   cyanoacetylene
    c)   1,3,5-triiodobenzene
    d)   All of the above
    e)   None of the above

28.4 In which of the following molecules is the hybridization of the indicated carbon atom correctly assigned?

    1. $CCl_4 - sp^3$ 2. $C_6H_6 - sp^2$ 3. $CH_3CH_2\overset{*}{C}N - sp$

    a)   1 & 2
    b)   1 & 3
    c)   2 & 3
    d)   All of them
    e)   None of them

28.5*Which of the following bond angles is <u>incorrect</u>?

a) CH_2O < HCO $\simeq$ 120°

b) CH_3Cl < HCCl $\simeq$ 109°

c) $C_6H_{13}NH_2$ < HNH $\simeq$ 120°

d) CH_3COOH < OCO $\simeq$ 120°

e) CH_3COOH < COH $\simeq$ 109°

28.6*The equilibrium constant (K_c) for the hypothetical esterification reaction:

$$ROH \quad + \quad RCOOH \rightleftharpoons H_2O \; + \; RCOOR$$
$$\text{(alcohol)} \qquad \text{(acid)} \qquad\qquad \text{(ester)}$$

is 5.2×10^{-3} at 25°C.

Calculate the equilibrium concentration of RCOOR if the solution is initially 0.065 molar in both ROH and RCOOH. A non-aqueous solvent is used and water is absent initially.

a) .061
b) .0044
c) .0052
d) 3.4×10^{-4}
e) 2.2×10^{-5}

28.7*Which carbon atom in histidine is asymmetric (i.e., responsible for optical isomerism)?

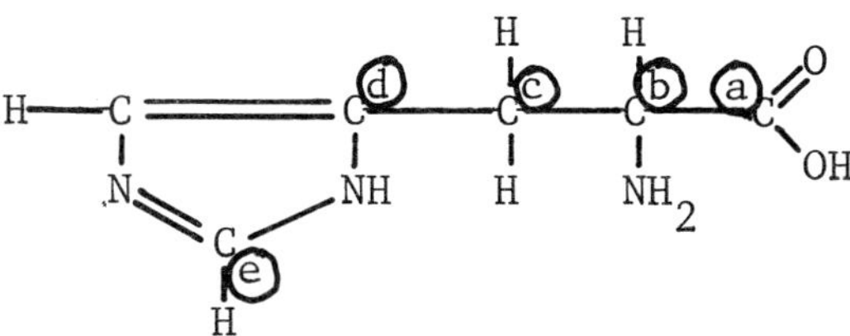

28.8 Which of the following compounds has an optical isomer?

1. $CH_3-CHCl-CH_2Cl$ 2. $CH_2=CHCl$ 3. $CH_3-CH=CHCl$

a) 1 & 3
b) 2
c) 1
d) All of them
e) None of them

270

28.9*Which of the following organic compounds has geometric (cis-trans) isomers?

      1. $CH_3CH=CClCH_3$ 2. $(CH_3)_2C=CCl_2$ 3. $CHCl=CHCH_3$

a) 1 & 2
b) 1 & 3
c) 2
d) All of them
e) None of them

28.10 Which one of the following organic compounds has neither geometric nor optical isomers?

a) $CH_3CH=CHCH_3$
b) $CH_2=CH-CHCH_3CH_2-CH_3$
c) $CH_3-CH_2-CH=CH_2$
d) $CH_2=CH-CHCH_3CH_2CH_3$
e) $CH_2Cl-CHCl-CH_3$

CHAPTER 29. DESCRIPTIVE CHEMISTRY

Introduction

The questions in this chapter are non-numeric and are grouped under the general heading "descriptive". The following concepts, among others, are included: classification of chemical reactions, classification of manner of occurrence of elements, solubility behavior, hydrogen bonding, atomic theory and environmental chemistry.

Questions

29.1*Heavy water has a molecular weight of 20 because it contains deuterium with mass number 2. Which of the following is the same regardless of whether the water is heavy water or ordinary water?

    a)   density of the gas produced at the cathode in electrolysis of the water

    b)   grams of gas formed when 0.100 mole of Ca (solid) is added to the water

    c)   grams of water decomposed per faraday. in the electrolysis of the water

    d)   density of the gas produced at the anode in electrolysis of the water

    e)   density of the gas produced when lithium is added to water

29.2*Which one of the following measurements or techniques can sometimes be used to find the atomic weight, or molecular weight, of a substance?

    a)   ionization potential
    b)   thermal conductivity
    c)   electrical conductivity
    d)   heat of combustion
    e)   specific heat

29.3*During the twentieth century classical atomic theory has been substantially revised by certain key experiments and by quantum theory. Which of the following statements is representative of classical atomic theory as opposed to modern atomic theory?

(see next)

   a)   All atoms have their mass concentrated in a very small fraction of their volume, called the nucleus.

   b)   The atom is composed of protons, neutrons and electrons.

   c)   Electrons in matter are more conveniently described as waves than as particles.

   d)   All atoms of the same element have the same mass.

   e)   The atomic number is the number of protons in the nucleus.

29.4 Listed below are a number of facts and phenomena which are understandable in terms of the ionic nature of one or more of the substances involved. Select the one statement which does not belong.

   a)   A solution of potassium nitrate in water conducts an electric current even though solid potassium nitrate does not.

   b)   Very pure acetic acid has a melting point of 17°C and one frequently finds it crystallized in the reagent bottle on a winter morning.

   c)   All the hydrated Cu(II) salts are about the same shade of blue.

   d)   If an aqueous solution of silver nitrate is added to an aqueous solution of potassium chromate a precipitate of silver chromate is formed instantly; there is no delay.

   e)   Molten sodium chloride conducts an electric current even though solid sodium chloride does not.

29.5 Which of the following categories best describes the manner of occurrence of the element Ca at the surface of the earth?

   a)   The element occurs as an ion in a soluble salt.

   b)   The element occurs either as: (1) a "hard" acid with oxide, fluoride, or an oxyanion, or if not (1) then (2) part of an oxyanion.

   c)   The element occurs as a "soft" acid in a binary sulfide or arsenide.

   d)   The free element exists in nature.

   e)   Both c and d.

29.6 Which one of the following statements is _not_ correct?

   a)   Metals characteristically have low electronegativities and low ionization potentials.

   b)   The following is an example of amphoteric behavior:

$$BeO(s) + 2 H^+ \rightarrow H_2O + Be^{2+}$$

$$BeO(s) + H_2O + 2 OH^- \rightarrow Be(OH)_4^{2-}$$

   c)   The following reaction occurs when calcium hydride is added to water:

$$CaH_2(s) + 2 H_2O \rightarrow Ca^{2+} + 2 H_2(g) + 2 OH^-$$

   d)   The following reaction occurs when barium oxide is added to basic solution:

$$BaO + H_2O + 2 OH^- \rightarrow Ba(OH)_4^{2-}$$ (see next)

e)   The following is an example of a Lewis acid-base reaction:

$$AlCl_3 + Cl^- \rightarrow AlCl_4^-$$

29.7*Which one of the following properties of water is attributable to
     hydrogen bonding?

a)   reaction with SO_2 to form sulfurous acid
b)   high dipole moment
c) $105°$ bond angle
d)   solvent for polymeric alcohols such as starch and cellulose
e)   self ionization (autoprotolysis)

29.8 Which of the following is an example of the indicated class of chemi-
     cal reactions?

1.   Oxidation-reduction: $N_2(g) + 3\ H_2(g) \longrightarrow 2\ NH_3(g)$

2.   Acid-Base: $BF_3 + NH_3 \longrightarrow F_3B\ NH_3$

3.   Metathesis: $BaCl_2 + Na_2SO_4 \longrightarrow BaSO_4 + 2\ NaCl$

a)   1 & 2
b)   1 & 3
c)   2 & 3
d)   All of them
e)   None of them

29.9*Which of the following would have the greatest tendency to dissolve
     in a polar solvent?

a)   methane, CH_4
b)   benzene, C_6H_6
c)   graphite, $C(s)$
d)   methanol, CH_3OH
e)   butadiene, C_4H_6

29.10 Select the most appropriate explanation of the following observation
       of solubility.

"Hydrogen chloride, HCl, is very soluble in water".

a)   Water promotes the ionization of many polar molecular substances.
b)   Opposites attract; that is, polar solutes dissolve in non-polar
      solvents and vice-versa.
c)   Water promotes the ionization of many ionic solids.
d)   Gases that have little interaction with solvents will dissolve
      freely.
e)   Low melting ionic crystals have low crystal energies and hence
      are readily soluble.

274

29.11* Which one of the following is <u>not</u> a strong electrolyte in aqueous
       solution?

       a)   ammonium nitrate, NH_4NO_3
       b)   cesium hydroxide, CsOH
       c)   hydrogen iodide, HI
       d)   phenol, C_6H_5OH
       e)   magnesium cyanide, $Mg(CN)_2$

29.12* What is the most harmful effect of the pollution of water by phos-
       phates?

       a)   They are non-biodegradable and so interfere with sewage plant
             operation.
       b)   They accumulate in the body and cause eventual nerve and brain
             damage.
       c)   They serve as nutrients for algae, and lead to eutrofication of
             lakes.
       d)   They inhibit bird life by causing thin egg shells.
       e)   They are easily reduced and tie up hemoglobin, especially in in-
             fants.

29.13* Drainage from abandoned coal mines is responsible for the pollution
       of water by:

       a)   nitrates
       b)   mercury
       c)   sulfuric acid
       d)   calcium bicarbonate
       e)   phosphates

29.14* Which of the following gases present in our atmosphere is used as a
       filler gas for incandescent lamps?

       a)   hydrogen
       b)   carbon monoxide
       c)   nitrogen
       d)   argon
       e)   nitrogen oxides

29.15* What is the difference between a soap and a detergent? Pick the one
       correct answer.

       a)   Detergents are carboxylates, soaps are sulfonates.
       b)   All soaps are biodegradable; some detergents, namely those with
             branched chains, are not.
       c)   Soaps are usually mixed with 50% or more of phosphates or other
             salts; detergents are usually sold as the pure compound.
       d)   Detergents are ionic; most soaps are not.
       e)   Detergents are precipitated by strong acids, soaps are not.

29.16 One may classify crystalline solids into five categories: ionic, polar molecular, nonpolar molecular, covalent network, and metallic. Which one of the following statements concerning these solids is correct?

    a)   All ionic solids are diamagnetic.

    b)   All polar molecular solids are semiconductors with electrical conductivity values between that of insulators and that of metals.

    c)   nonpolar molecular solids have high melting points because the molecules are packed in the crystal lattice very efficiently.

    d)   The transition metals are characterized by high melting points.

    e)   The atoms in a covalent network solid are joined by forces due to a sea of delocalized valence electrons.

29.17* Which set of examples below of the different classes of solids is correct?

	ionic	polar molecular	non-polar molecular	covalent network	metallic
a)	KI	C_6H_5Cl	I_2	SiC	Ba
b)	Na_2S	$SO_2(s)$	$Ar(s)$	GaAs	Si
c)	HF	H_2SO_3	CBr_4	SiO_2	V
d)	$CaCl_2$	H_2O	$POCl_3$	C(diamond)	Mg
e)	BaO	$NaHCO_3$	$CH_4(s)$	SrS	Pb

29.18 Which of the following batteries theoretically has the highest voltage?

    a)   hydrogen-oxygen fuel cell

    b)   lead storage battery (common automobile battery)

    c)   mercury battery (HgO-Zn)

    d)   lithium-chlorine

    e)   common dry cell (MnO_2-Zn)

29.19 In what respect is the mercury battery (HgO-Zn) outstanding compared to others now in use or under development?

    a)   cost

    b)   voltage constant with time

    c)   high voltage

    d)   ability to stand many charge-discharge cycles

    e)   low equivalent weight.

29.20 Which statement below best explains why galvanized (zinc coated)
       steel resists corrosion?

    a)   Zinc has a higher oxidation potential and so is oxidized in pre-
ference to iron.
    b)   Zinc forms an impervious layer over the iron.
    c)   Zinc rapidly forms an oxide coating which is resistant to further
attack.
    d)   The zinc-iron alloy formed has a lower oxidation potential than
either pure metal.
    e)   Electrons will flow from the iron to the zinc producing $H_2(g)$ at
the zinc surface.

The next five sentences consist of a statement and a reason. Answer using
the following key:

    TTR – both the statement and reason are true and are related as cause
and effect;
    TTN – both the statement and reason are true but are not related as
cause and effect;
    TF – the statement is true but the reason is false;
    FT – the statement is false but the reason is true; and
    FF – both the statement and reason are false.

29.21   The alkali metals are strong reducing agents because they have
low first ionization energies.

    a)   TTR
    b)   TTN
    c)   TF
    d)   FT
    e)   FF

29.22 HCl is a strong acid because hydrogen is more electronegative than
chlorine.

    a)   TTR
    b)   TTN
    c)   TF
    d)   FT
    e)   FF

29.23 A spontaneous process is always exothermic because the free energy
is always minimized when heat is released to the surroundings.

    a)   TTR
    b)   TTN
    c)   TF
    d)   FT
    e)   FF

29.24 The solubility of AgF in H_2O is less than that of AgBr because the
Ag-F bond is more polar than the Ag-Br bond.

    a)   TTR
    b)   TTN
    c)   TF
    d)   FT
    e)   FF

29.25 The boiling points of H_2O and HF are higher than the boiling points
of other hydrogen compounds of the characteristic elements in Groups
VI and VII because the liquid states of H_2O and HF are stabilized by
hydrogen bonding.

    a)   TTR
    b)   TTN
    c)   TF
    d)   FT
    e)   FF

PART II

SOLUTIONS TO SELECTED QUESTIONS

CHAPTER 2. UNITS AND CONVERSION FACTORS

2.8 Use the unit-factor method to convert kilometers to miles and liters to gallons. Then calculate the mileage in miles per gallon.

$$\# \text{ miles} = \frac{0.621 \text{ mile}}{\text{kilometer}} \times 427 \text{ kilometers} = 265 \text{ miles}$$

$$\# \text{ gallons} = \frac{0.264 \text{ gallon}}{\text{liter}} \times 37.9 \text{ liters} = 10.0 \text{ gallons}$$

$$\text{mileage} = \frac{\# \text{ miles}}{\# \text{ gallons}} = \frac{265 \text{ miles}}{10.0 \text{ gallons}} = 26.5 \text{ miles per gallon}$$

The answer is (b).

2.14 The relevant equations are $t_F = -t_c$ and

$$t_c = (t_F - 32)/1.8 \qquad \text{Then}$$

$$t_F = -t_c = -(t_F - 32)/1.8$$

$$1.8 \ t_F = t_F + 32$$

$$2.8 \ t_F = 32$$

$$t_F = 11.4$$

The answer is (b)

2.18 First convert 100 gallons to milliliters; then calculate the mass of this volume of Hg in grams; finally, convert this amount to tons

$$\# \text{ ml} = \frac{1000 \text{ ml}}{\text{liter}} \times \frac{3.79 \text{ liters}}{\text{gallon}} \times 100 \text{ gallons} = 3.79 \times 10^5 \text{ ml}$$

$$\# \text{ g} = \frac{13.6 \text{ g}}{\text{ml}} \times 3.79 \times 10^5 \text{ ml} = 5.15 \times 10^6 \text{ g}$$

$$\# \text{ tons} = \frac{1 \text{ ton}}{2000 \text{ lbs}} \times \frac{1 \text{ lb}}{454 \text{ g}} \times 5.15 \times 10^6 \text{ g} = 5.68 \text{ tons}$$

The answer is (a).

2.25 First calculate the mass of the solution in grams and then use the percentage by weight to calculate the mass of KCl.

$$\# \text{ g} = \frac{1.28 \text{ g}}{\text{ml}} \times 795 \text{ ml} = 992 \text{ g}$$

$$\text{\# g KCl} = \frac{35.5 \text{ g KCl}}{100 \text{ g sol'n}} \times 992 \text{ g sol'n} = 352 \text{ g KCl}$$

The answer is (c).

2.30 First use the conversion factor "1 g calcium carbonate/245 ml carbon
dioxide" to calculate the number of grams of calcium carbonate. Then
calculate the percentage calcium carbonate in the sample.

$$\text{\# g calcium carbonate} = \frac{1 \text{ g calcium carbonate}}{245 \text{ ml carbon dioxide}} \times 331 \text{ ml carbon dioxide}$$

$$= 1.35 \text{ g calcium carbonate}$$

$$\% \text{ calcium carbonate} = \frac{1.39 \times 100 \text{ g calcium carbonate}}{1.69 \text{ g sample}} = \underline{79.9\%}$$

The answer is (d).

CHAPTER 4. ATOMS AND MOLECULES

4.7 The formula indicates that there is one rubidium atom in each molecule
(or "formula unit", since we are discussing an ionic substance). Thus,
counting the number of molecules determines also the number of atoms.

$$389 \text{ g RbF} \times \frac{1 \text{ mole RbF}}{105 \text{ g RbF}} \times \frac{6.02 \times 10^{23} \text{ molecules}}{1 \text{ mole}} \times \frac{1 \text{ Rb Atom}}{1 \text{ RbF molecule}} =$$

$$2.23 \times 10^{24} \text{ Rb atoms.} \text{ The answer is c.}$$

The atomic weight of rubidium which was given was unnecessary information. Since the mass of RbF was given, it is appropriate to divide only
by the molecular weight of RbF.

4.14 Consider one molecule of compound (a) which weighs 140 amu and contains
six nitrogen atoms weighing 14.0 amu each.

$$\% \text{ N} = \frac{6 \times 14.0}{140} \times 100\% = 60\% \text{ nitrogen}$$

If you perform a similar calculation for the other compounds you will
find that (a) has the highest percentage.

4.16 The atomic weight of an element is the weighted average of the
naturally occurring isotopes. Let x be the percentage of the
heavier isotope. Then 100-x is the percentage of the lighter
isotope, and

$$47.17 = \frac{47.92 \text{ x} + 46.94(100-x)}{100}$$

Solving this equation for x gives x = 23.5, which is answer (e).

4.17 This solution illustrates the general method of obtaining empirical
formulas from percent composition. Assume as a basis of calculation
100 g of compound, which must contain 67.2 g of Xe and 32.8 g of O.

$$\text{moles of Xe} = \frac{67.2 \text{ g}}{131.3 \text{ g/mole}} = 0.512 \text{ moles Xe}$$

$$\text{moles of O} = \frac{32.8 \text{ g}}{16.0 \text{ g/mole}} = 2.05 \text{ moles O}$$

Since the ratio of moles of atoms is the same as the ratio of atoms we have;

$$\frac{2.05 \text{ atoms O}}{0.512 \text{ atoms Xe}} = \frac{4.01 \text{ atoms O}}{1.00 \text{ atom Xe}}$$

You should recognize the correct formula to be XeO_4, answer (e), and attribute the difference between 4.01 and 4.00 to experimental (or computational) error.

You are cautioned in these problems to preserve all significant figures and compute carefully, since a small numerical error could make it difficult to recognize the correct atomic ratio.

4.20 Since all of the calcium in the original $CaCl_2$ is present in the CaO weighed, there must have been 1 mole of $CaCl_2$ for each mole of CaO weighed.

$$\text{moles } CaCl_2 = \text{moles CaO} = 18.4 \text{ g CaO} \times \frac{1 \text{ mole CaO}}{56.1 \text{ g CaO}}$$

$$\text{g } CaCl_2 = \text{moles } CaCl_2 \times \frac{111 \text{ g } CaCl_2}{1 \text{ mole } CaCl_2}$$

$$\% \ CaCl_2 = \frac{\text{g } CaCl_2}{70.3 \text{ g mixture}} \times 100\%$$

Combining the above three ideas into one step.

$$\% \ CaCl_2 = 18.4 \text{ g CaO} \times \frac{1 \text{ mole CaO}}{56.1 \text{ g CaO}} \times \frac{111 \text{ g } CaCl_2}{1 \text{ mole } CaCl_2} \times \frac{100 \ \%}{70.3 \text{ g mixture}} \times$$

$$= 51.9\% \text{ which is answer c.}$$

In chemical analysis, problems like the above are very common and the factor $\frac{111}{56.1}$ is frequently called the "chemical factor" or "gravimetric factor" since it converts the mass of substance weighed to the mass of substance sought. In this case, the molecular weights appear in a ratio of 1:1, but that is not always the case.
You'll note that it is not necessary to find the mass of $CaCO_3$ to solve the problem. The mass of intermediate substance is not required as long as you can establish the molar ratio of initial to final substance. Think MOLES!

4.22 This problem is analogous to the problem of obtaining an empirical formula, except that here you are dealing in relative numbers of molecules rather than relative numbers of atoms.

$$323 \text{ g of hydrate minus } 149 \text{ g of } H_2O = 174 \text{ g of } BaO_2$$

$$\frac{149 \text{ g}}{18.0 \text{ g mole}^{-1}} = 8.28 \text{ moles } H_2O; \quad \frac{174 \text{ g}}{169 \text{ g mole}^{-1}} = 1.028 \text{ moles } BaO_2$$

$$\frac{\text{mole } H_2O}{\text{mole } BaO_2} = \frac{8.28}{1.028} = 8.05, \text{ very close to answer (b).}$$

4.24 Let the chlorides be MCl_x and MCl_y. Let A be the atomic weight of the metal.

100 grams of MCl_x contains 37.4 g Cl and 62.6 g M
100 grams of MCl_y contains 47.2 g Cl and 53.8 g M

MCl_x: (# moles Cl)/(# moles M) = X = (37.4/35.5)/(62.6/A)
MCl_y: (# moles Cl)/(# moles M) = Y = (47.2/35.5)/(52.8/A)

Then $\dfrac{x}{y} = \dfrac{52.8 \times 37.4}{47.2 \times 62.6} = 0.668$ or $\dfrac{2}{3}$

The empirical formulas are MCl_2 and MCl_3 or MCl_4 and MCl_6, etc. The answer is c.

CHAPTER 5.   CHEMICAL REACTIONS

5.6 The data enables one to calculate the atomic weight, AW, of the metal. From the formula:

$$\text{moles of M} = 2 \times \text{moles of S}$$

$$16.7/AW = 2 \times 2.48/32$$

$$AW = \frac{16.7 \times 32}{2 \times 2.48} = 107.5$$

This is close to AW of Ag, which is answer (c).

5.10 As in the previous problem, THINK MOLES, only in this case the given value is a gas volume rather than a mass. Accordingly, the first step is:

$$\text{moles } O_2 = 44.8 \text{ ℓ of } O_2 \times \frac{1 \text{ mole } O_2}{22.4 \text{ ℓ of } O_2} = 2.00 \text{ moles } O_2$$

Next, looking at the balanced equation:

$$2.00 \text{ moles } O_2 \times \frac{2 \text{ moles } KClO_3}{3 \text{ moles } O_2} = \frac{4.00}{3} \text{ moles } KClO_3$$

and to find the mass:

$$\frac{4.00}{3} \text{ moles of } KClO_3 \times \frac{123 \text{ g } KClO_3}{1 \text{ mole } KClO_3} = 164 \text{ g } KClO_3; \text{ the answer is a.}$$

Alternatively, setting it up in one step:

$$\text{g } KClO_3 = \frac{123 \text{ g } KClO_3}{1 \text{ mole } KClO_3} \times \frac{2 \text{ mole } KClO_3}{3 \text{ mole } O_2} \times \frac{1 \text{ mole } O_2}{22.4 \text{ ℓ } O_2} \times 44.8 \text{ ℓ } O_2 = 164 \text{ g}$$

5.12 The first step is write a balanced equation for the reaction which will provide the molar ratio for the substances of interest.

$$Zn + 2\,HCl \rightarrow ZnCl_2 + H_2.$$

Our first concern will be to convert the given amount, 1.05 g of Zn to moles:

$$1.05 \text{ g of Zn} \times \frac{1 \text{ mole of Zn}}{65.4 \text{ g of Zn}} = \frac{1.05}{65.4} \text{ moles of Zn.}$$

Next, we look to the equation to find the corresponding moles of HCl:

$$\frac{1.05}{65.4} \text{ moles of Zn} \times \frac{2 \text{ moles of HCl}}{1 \text{ mole of Zn}} = \frac{2 \times 1.05}{65.4} \text{ moles of HCl.}$$

Finally, we convert to the desired unit, grams:

$$\frac{2 \times 1.05}{65.4} \text{ moles of HCl} \times \frac{36.5 \text{ g of HCl}}{1 \text{ mole of HCl}} = 1.17 \text{ g HCl.}$$

The answer is c within the slide-rule accuracy.

Alternatively, we can do the problem in one step, utilizing the factor label method:

$$\text{g HCl} = \frac{36.5 \text{ g HCl}}{1 \text{ mole HCl}} \times \frac{2 \text{ mole HCl}}{1 \text{ mole Zn}} \times \frac{1 \text{ mole Zn}}{65.4 \text{ g Zn}} \times 1.05 \text{ g Zn}$$

5.15 The balanced equation is:

$$Ni + S + 2 O_2 \rightarrow NiSO_4$$

$$\text{\# moles Ni} = \frac{1 \text{ mole Ni}}{58.7 \text{ g Ni}} \times 12.0 \text{ g Ni} = 0.205 \text{ moles Ni}$$

Since the balanced equation requires as many moles of Ni as S, the .404 mole of S represents an excess, and answer c must be wrong.

$$\text{Moles } O_2 = 14.5 \text{ } \ell \text{ } O_2 \times \frac{1 \text{ mole } O_2}{22.4 \text{ } \ell \text{ } O_2} = 0.648 \text{ moles } O_2$$

$$\text{Moles } O_2 \text{ required} = 0.205 \text{ moles Ni} \times \frac{2 \text{ moles } O_2}{1 \text{ mole Ni}} = 0.410 \text{ moles } O_2$$

The 0.648 moles O_2 are also in excess and so the Ni is limiting.

$$\text{moles of NiSO}_4 = \text{moles of Ni} = 0.205 = \frac{12.0}{58.7}$$

Answer d is correct.

5.28 Let x = g of RbBr

17.7 - x = g of CsBr

AgBr from RbBr + AgBr from CsBr = total AgBr

Molecular Weights	
RbBr	165.4
CsBr	212.8
AgBr	187.8

$$\frac{187.8x}{165.4} + \frac{187.8(17.7-x)}{212.8} = 18.9$$

Solving, x = 13.06 $\quad \frac{13.06}{17.7} \times 100\% = 73.8$, answer e.

CHAPTER 6. REDOX REACTIONS

6.5 The balanced half-reactions are:

$$Cr_2O_7^{2-} + 14 \text{ H}^+ + 6 \text{ e}^- \rightarrow 2 \text{ Cr}^{3+} + 7 \text{ H}_2O \qquad (1)$$

$$H_2SO_3 + H_2O \rightarrow SO_4^{2-} + 4 \text{ H}^+ + 2 \text{ e}^- \qquad (2)$$

Reaction 2 must be multiplied by 3 so that there will be no net gain or loss of electrons. Multiply (2) by 3, add, and cancel terms which appear on both sides (H^+ and H_2O):

$$Cr_2O_7^{2-} + 3 \text{ H}_2SO_3 + 2 \text{ H}^+ \rightarrow 2 \text{ Cr}^{3+} + 3 \text{ SO}_4^{2-} + 4 \text{ H}_2O$$

The sum of coefficients is $1 + 3 + 2 + 2 + 3 + 4 = 15$; the answer is b.

6.6 The balanced half-reactions are

$$MnO_4^- + 3\ e^- + 2\ H_2O \rightarrow MnO_2 + 4\ OH^- \qquad (1)$$

$$I^- + 6\ OH^- \rightarrow IO_3^- + 6\ e^- + 3\ H_2O \qquad (2)$$

Reaction (2) plus two times (1) gives:

$$2\ MnO_4^- + I^- + H_2O \rightarrow 2\ MnO_2 + IO_3^- + 2\ OH^-$$

The sum of the coefficients is 9; the answer is c.

6.10 The balanced half-reactions are:

$$Br_2 + 2e^- \rightarrow 2\ Br^- \qquad (1)$$

$$Br_2 + 12\ OH^- \rightarrow 2\ BrO_3^- + 10e^- + t\ H_2O \qquad (2)$$

Reaction (2) + 5 times reaction (1) gives

$$6\ Br_2 + 12\ OH^- \rightarrow 10\ Br^- + 2\ BrO_3^- + 6\ H_2O$$

This equation can be simplified by dividing each stoichiometric coefficient by 2,

$$3\ Br_2 + 6\ OH^- \rightarrow 5\ Br^- + BrO_3^- + 3\ H_2O$$

The sum of the coefficients is 18, the answer is (a).

CHAPTER 7. FARADAY'S LAWS

7.8 In this problem you must first find out which reactant is limiting.

$$\text{Moles Na} = 14.5\ g \times \frac{1\ \text{mole}}{23.0\ g} = 0.631\ \text{moles}$$

$$\text{Moles Na}_2S_2 = 38.2\ g \times \frac{1\ \text{mole}}{110\ g} = 0.347\ \text{moles}$$

To balance the electron flow requires 2 moles of Na for every mole of Na_2S_2, or in this problem: $2 \times 0.347 = 0.694$ moles of Na.

Since you have only 0.631 moles, the Na is limiting.

Let t = time in seconds to consume the Na

$$\frac{(0.750\ t)}{(96,500)} = (1)(.631)$$

$$t = \frac{(.631)(96,500)}{(.750)} = 8.10 \times 10^4\ \text{sec, answer b.}$$

7.9 The electrolysis reaction is

$$2\ H_2O + 2\ NaCl \rightarrow Cl_2(g) + H_2(g) + 2\ NaOH$$

The number of electrons transferred is 2.
The number of faradays per hour, at 90% efficiency is

$$\frac{\#\ \text{faradays}}{\text{hour}} = \frac{1\ \text{faraday}}{96,500\ \text{coul}} \times \frac{80,000\ \text{coul}}{\text{sec}} \times \frac{3600\ \text{sec}}{\text{hr}} \times 0.9 = 2690\ \frac{\text{faradays}}{\text{hr}}$$

$$\frac{\text{\# kg NaOH}}{\text{hr}} = \frac{0.040 \text{ kg}}{\text{mole NaOH}} \times \frac{2 \text{ mole NaOH}}{2 \text{ faradays}} \times \frac{2690 \text{ faradays}}{\text{hour}} = 107 \text{ kg/hr}$$

$$\frac{\text{\# kg Cl}_2}{\text{hr}} = \frac{0.071 \text{ kg}}{\text{mole Cl}_2} \times \frac{1 \text{ mole Cl}_2}{2 \text{ faradays}} \times \frac{2690 \text{ faradays}}{\text{hr}} = 95 \text{ kg/hr}$$

$$\frac{\text{\# kg H}_2}{\text{hr}} = \frac{0.002 \text{ kg}}{\text{mole H}_2} \times \frac{1 \text{ mole H}_2}{2 \text{ faradays}} \times \frac{2690 \text{ faradays}}{\text{hr}} = 2.7 \text{ kg/hr}$$

The answer is (d).

7.10 The half-reaction corresponding to the deposition of Ag from a AgNO$_3$ solution is

$$Ag^+ + e^- \rightarrow Ag$$

One equivalent of electricity (96,500 coulombs) will deposit one equivalent (in this case, one mole or 108 grams) of Ag.

The number of equivalents in this problem is:

$$100 \text{ g} \times \frac{1 \text{ equivalent}}{108 \text{ g}} = \frac{100}{108} \text{ equivalents.}$$ Let z be the charge on the Hg cation. Then

$$Hg^{z+} + ze^- \rightarrow Hg$$

$$\text{\#g Hg} = \frac{201 \text{ g}}{\text{mole}} \times \frac{1 \text{ mole}}{z \text{ equivalents}} \times \frac{100}{108} \text{ equivalents} = 186 \text{ g}$$

Solving, z = 1.   The answer is b.

CHAPTER 8.   SOLUTION STOICHIOMETRY

8.8 In a titration, the number of moles of titrant used is stoichiometrically related to the moles of substance sought. This problem will be solved first in a stepwise manner and then it will be done in one step.

$$\text{moles of Fe}^{2+} = 37.3 \text{ ml} \times \frac{1 \text{ liter}}{1000 \text{ ml}} \times 0.149 \frac{\text{mole}}{\text{liter}} = 5.56 \times 10^{-3} \text{ moles.}$$

Since the balanced equation shows 2 chromium atoms for every six iron atoms

$$\text{moles Cr} = \frac{2}{6} \times 5.56 \times 10^{-3} = 1.85 \times 10^{-3} \text{ moles}$$

$$\text{mass of Cr} = 1.86 \times 10^{-3} \text{ mole} \times \frac{52.0 \text{ g}}{1 \text{ mole}} = 9.63 \times 10^{-2} \text{ g}$$

which is answer c.

The solution could be set down in one step:

$$\text{g of Cr} = \frac{52.0 \text{ g Cr}}{1 \text{ mole Cr}} \times \frac{2 \text{ mole Cr}}{6 \text{ mole Fe}} \times \frac{0.149 \text{ mole Fe}}{\text{liter}} \times .0373 \text{ liters}$$

$$= 9.63 \times 10^{-2} \text{ g Cr.}$$

8.11 Arbitrary amounts of acid and base were mixed, one of which was probably in excess.

$$\text{moles acid} = \frac{17.7 \text{ ml}}{1000 \text{ ml } \ell^{-1}} \times 1.90 \text{ mole } \ell^{-1} = 3.36 \times 10^{-2} \text{ moles}$$

$$\text{moles base} = \frac{39.8 \text{ ml}}{1000 \text{ ml } \ell^{-1}} \times 1.31 \text{ mole } \ell^{-1} = 5.21 \times 10^{-2} \text{ moles}$$

Since these reagents react mole for mole:

$$NaOH + HI \rightarrow NaI + H_2O$$

the base is in excess. After the 3.36×10^{-2} moles of acid has neutralized 3.36×10^{-2} moles of the base, the solution will contain 5.21×10^{-2} minus 3.36×10^{-2}, or 1.85×10^{-2} moles of excess base.

The total volume of solution is now 39.8 ml plus 17.7 ml, or 57.5 ml. The concentration of base (the OH^- ion) is:

$$\frac{1.85 \times 10^{-2} \text{ moles}}{57.5 \text{ ml}} \times 1000 \text{ ml } \ell^{-1} = .320 \text{ mole } \ell^{-1}, \text{ answer (b)}$$

8.14 Since the acid is monoprotic, the titration is ended when the moles of acid equals the moles of base.

Let MW = molecular weight of acid

$$2.80 \text{ g} \times \frac{1 \text{ mole}}{MW \text{ g}} = \frac{29.2 \text{ ml}}{1000 \text{ ml } \ell^{-1}} \times 0.500 \text{ mole } \ell^{-1}$$

$$MW = \frac{(1000)(2.80)}{(29.2)(.500)} = 192, \text{ answer c.}$$

8.17 Molality is defined as the number of moles of solute per kilogram of water. Take as a basis 1.000 liter of solution which weighs 1668 g.

$$\text{moles } HClO_4 = \frac{1668 \text{ g} \times 0.700}{100.5 \text{ g mole}^{-1}}$$

$$\text{kg of } H_2O = \frac{1668 \text{ g} \times 0.300}{1000 \text{ g kg}^{-1}}$$

$$\text{molality} = \frac{\text{Moles } HClO_4}{\text{kg of } H_2O} = \frac{1668 \times 0.700 \times 1000}{100.5 \times 1668 \times .300} = 23.2 \text{ molal}$$

Answer a is correct.

8.18 First calculate the molarity of the concentrated NaOH solution. Then solve the dilute problem. Consider 1.00 liter of the concentrated NaOH solution.

$$\frac{\# \text{ moles NaOH}}{\text{liter}} = \frac{1 \text{ mole NaOH}}{40.0 \text{ g NaOH}} \times \frac{50.0 \text{ g NaOH}}{100 \text{ g sol'n}} \times \frac{1.525 \text{ g}}{\text{ml}} \times \frac{1000 \text{ ml}}{\ell}$$

$$= 19.1 \text{ } \underline{M}$$

The dilution equation is $\quad M_1 U_1 = M_2 U_2$

$$V_1 = \frac{0.1 \text{ M} \times 3.5 \text{ } \ell}{19.1 \text{ } \underline{M}} = 0.0184 \text{ } \ell \text{ or } 18.4 \text{ ml}$$

The answer is (a).

8.21 The balanced equation is

$$5\ C_2O_4^{2-}\ +\ 2\ MnO_4^-\ +\ 16\ H^+\ \rightarrow\ 10\ CO_2 + 2\ Mn^{2+}\ +\ 8\ H_2O$$

The approach to this problem is as follows:

percentage Ca $\leftarrow$ # grams Ca $\leftarrow$ # moles Ca $\leftarrow$ # moles $C_2O_4^{2-}$ $\leftarrow$ # moles MnO_4^-

$$\text{\# moles } C_2O_4^{2-}\ =\ \frac{5}{2}\ \times\ \text{\# moles } MnO_4^-\ =\ \frac{5}{2} \times 0.123 \times \frac{35.7}{1000} = .01097$$

$$\text{\# moles } Ca^{2+}\ =\ \text{\# moles } C_2O_4^{2-}\qquad (CaC_2O_4)$$

$$\text{percentage Ca}\ =\ \frac{40.1 \times 0.01097}{1.06}\ \times 100 = 41.5.\quad \text{The answer is (c).}$$

8.22 This problem is an example of a method of analysis called iodometry.
Following the coefficients in the balanced reactions you see that two
$S_2O_3^=$ are required for each I_3^-, each derived from one I_2, which in turn
was formed from two I^- by a two-electron oxidation. Hence, one $S_2O_3^=$
is used for each electron gained by the oxidant. (The molarity then
is the same as the normality.) This is true in all iodometric
titrations.

Since the arsenic in H_3AsO_4 undergoes a two-electron change, its
equivalent weight is the atomic weight divided by two.

$$\text{Percent As}\ =\ \frac{(0.865)(35.97)}{(1000)}\ \times\ \frac{74.9}{2} \times \frac{100\%}{5.41}\ =\ 21.55\%,\ \text{answer b.}$$

8.24 This is an involved problem in which quantities are specified in
different ways. Convert everything to moles first to compare quantities
and figure excess amounts.
Incidentally, this reaction represents a commonly used laboratory
procedure for recovering copper from a solution.

$$\text{Moles } Cu^{2+}\ =\ \frac{400\ ml}{1000\ ml\ \ell^{-1}}\ \times\ 0.10\ \text{mole/}\ell\ =\ 0.040\ \text{moles}$$

Hence 0.400–0.040 = 0.360 moles of Zn will remain to react with the
acid. It will consume 2 x 0.360 = 0.720 moles of acid and produce
0.360 moles of $H_2(g)$. Answer a is OK.

$$\text{Concentration of } Zn^{++}\ =\ \frac{0.400\ \text{moles}}{0.400\ \ell}\ =\ 1.0\ \text{M; answer b is OK}$$

$$\text{Orig. Moles } H^+\ =\ \frac{400\ ml}{1000\ ml\ \ell^{-1}} \times\ 6.0\ \text{mole/}\ell\ =\ 2.40\ \text{moles}$$

$$\text{Moles } H^+ \text{ remaining} = 2.40 - 0.72 = 1.68\ \text{moles}$$

$$\text{Concentration of } H^+ = \frac{1.68\ \text{moles}}{0.40\ \ell}\ =\ 4.2\ \text{M; so answer c is}$$

incorrect and should be marked.

All the Cu^{++} (.040 moles) was converted to the metal so answer d is OK.

$$\text{Mass of Zn} = 0.400\ \text{moles} \times \frac{65.37\ g}{1\ \text{mole}}\ =\ 26.2\ g\ \text{so answer e is OK.}$$

8.25 In this problem reference is made to a normal solution (abbreviated $\underline{N}$). A 1.00 $\underline{N}$ solution contains one equivalent weight of solute per liter. In the case of an oxidizing agent an equivalent weight is the molecular weight divided by the number of electrons gained. Hence an equivalent weight has the capacity to gain just one mole of electrons.

The advantage of normality is that one can state for a titration:

Number of equivalent weights = number of equivalent weights
   of oxidizing agent   of reducing agent

For brevity chemists usually substitute "equivalents", abbreviated "eq", for "number of equivalent weights".

In acid-base work the concepts of normality and equivalents are used also, but hydrogen ions rather than electron change is counted.

In this problem since iron is titrated from Fe^{+2} to Fe^{+3} (change of 1 electron) its equivalent weight equals its atomic weight (divided by one).

$$\text{eq. of Fe} = \text{eq of oxidizing agent} = \frac{22.1 \text{ ml}}{1000 \text{ ml } \ell^{-1}} \times 0.0844 \text{ eq}/\ell$$

$$\text{Mass of Fe} = (\text{eq of Fe}) \times (\text{Eq. Wt. of Fe}) = \frac{22.1 \text{ ml}}{1000 \text{ ml } \ell^{-1}} \times \frac{0.0844 \text{ eq}}{\ell}$$

$$\times \frac{55.85 \text{ g}}{1 \text{ eq}} \times \frac{1000 \text{ mg}}{1 \text{ g}} = 104 \text{ mg, which is answer c.}$$

CHAPTER 9. HEATS OF REACTION

9.3 $\Delta H° = 4\Delta H_f°(CO_2) + 4\Delta H_f°(H_2O) - [\Delta H_f°(C_4H_8) + 6\Delta H_f°(O_2)]$

     $= 4(-394) + 4(-286) - [16 + 6(0)]$

     $= -2740 \text{ kJ}$

The answer is (a).

9.6 Let t = final temperature. Since there is so much more water than ice assume that t>0°C. The first two terms in the "heat-gained" side represent the warming of the ice to 0°C, and then its melting.

$$\text{Heat Gained} = \text{Heat Lost}$$

$$(75 \times 2.09 \times 25) + (75 \times 335) + 75 \times 4.18t = 600 \times 4.18(45 - t)$$

$$313.8 \, t + 2510.4t = 122968 - 3919 - 25125$$

$$t = 29.7°C \quad \text{Answer is c.}$$

Note that there are two other possible outcomes. If the amount of ice is much larger than the water, everything might freeze. At intermediate values you may end up with t=0 and a portion of the ice still not melted.

9.9 Let s be the specific heat of the metal in J/gram-degree. Equate
the heat lost by the metal to the heat gained by the water.

$$119(86.6 - 24.3)s = (390)(24.3 - 23.2)(4.184)$$

$$s = 0.242 \text{ J/gram-degree}$$

The answer is (b).

9.10 $\Delta H° = \Delta H°_f(SO_2Cl_2) - [\Delta H°_f(SO_2) + \Delta H°_f(Cl_2)]$

$$= -389 - (-297 + 0)$$

$$= -92 \text{ kJ}$$

92 kJ of heat are <u>evolved</u> when 1 mole of SO_2Cl_2 is formed. Thus
options a, c, and d are eliminated. When 91.0 g of SO_2Cl_2 are formed,

$$\# \text{ kJ} = \frac{-92 \text{ kJ}}{\text{mole}} \times \frac{1 \text{ mole}}{135 \text{ g}} \times 91 \text{ g} = -62.0 \text{ kJ}$$

The answer is (b).

9.13 In order to use Hess' Law, you must combine the given reactions in
such a way as to give the desired reaction. To obtain FeO on the left
side you must use 1/3 of reaction (3) in reverse (i.e., subtract it).
To obtain Fe on the right, use 1/2 of reaction (1). Then to eliminate
Fe_2O_3 subtract 1/6 of reaction (2). Try this scheme and see
how the other compounds come out.

$$
\begin{array}{lr}
 & \underline{\Delta H} \\
FeO + \tfrac{1}{3}CO_2 \rightarrow \tfrac{1}{3}CO + \tfrac{1}{3}Fe_3O_4 & -38.1/3 \\
\tfrac{1}{2}Fe_2O_3 + \tfrac{3}{2}CO \rightarrow Fe + \tfrac{3}{2}CO_2 & -27.6/2 \\
\tfrac{2}{6}Fe_3O_4 + \tfrac{1}{6}CO_2 \rightarrow \tfrac{1}{6}CO + \tfrac{3}{6}Fe_2O_3 & -(-58.6)/6
\end{array}
$$

If you carefully add the above equations (watch your fractions) you
obtain the desired equation, for which ΔH is the sum of ΔH's listed
above: $-38.1/3 + 58.6/6 - 27.6/2$, which is answer (b).

9.14 Start by drawing out the structural formulas.

$$\underset{H}{\overset{H}{>}}C=C\underset{H}{\overset{H}{<}} \;+\; Cl-Cl \longrightarrow H-\underset{\underset{Cl}{|}}{\overset{\overset{H}{|}}{C}}-\underset{\underset{Cl}{|}}{\overset{\overset{H}{|}}{C}}-H$$

$\Delta H = \Delta H(C=C) + \Delta H(Cl-Cl) - \{\Delta H(C-C) + 2\Delta H(C-Cl)\}$

$$= 619 + 243 - [347 + 2(326)] = -137, \text{ which is answer (b).}$$

9.15

$$H-C{\Large\{}H \; + \; H-O{\Large\}}O-H \longrightarrow H-C\cdots OH \; + \; H\cdots O-H$$

(with H above and below each C)

The bonds that must be broken are indicated by $\}$ and the ones which must be formed by $\cdots$.

$$\Delta H = -276 = \Delta H(C\text{-}H) + \Delta H(O\text{-}O) - \{\Delta H(O\text{-}H)\}$$

$$-276 = 414 + 138 - \Delta H(C\text{-}O) - 464$$

$$\Delta H(C\text{-}O) = 414 + 138 - 464 + 276 = 365 \text{ kJ, which is answer (b)}.$$

Note: Don't break (and remake) any more bonds than necessary. It isn't wrong, but it increases the chance of numerical error. In the above problem, for example, the fragment $H\text{-}C$ remained intact.

CHAPTER 10. COLLIGATIVE PROPERTIES

10.3 The freezing point depression, ΔT_f, is approximately proportional to the molality, m, of a non-electrolytic solute,

$$\Delta T_f = k_f m$$

Here k_f is a constant for a given solvent (−1.86 degrees per molal for H_2O). In this problem $\Delta T_f = -2.36 - 0.00 = -2.36$. The molality is the number of moles of solute divided by the number of kilograms of solvent. Let M be the molecular weight of the solute. Then

$$m = \frac{11.3}{M \times 0.105}$$

$$-2.36 = \frac{-1.86 \times 11.3}{M \times 0.105}$$

$$M = 84.8. \quad \text{The answer is d.}$$

10.8 The most fundamental of the colligative properties is the lowering of the vapor pressure, which is proportional to the mole fraction of solute.

$$\text{Moles } H_2O = \frac{850 \text{ g}}{18.0 \text{ g mole}^{-1}} = 47.2 \text{ moles}$$

$$\text{Moles solute} = \frac{449 \text{ g}}{54.0 \text{ g mole}^{-1}} = 8.32 \text{ moles}$$

$$\text{Total moles} = 47.2 + 8.32 = 55.5 \text{ moles}$$

$$\text{Vapor pressure lowering} = 71.0 \text{ torr} \times \frac{8.32 \text{ moles}}{55.5 \text{ moles}} = 10.6 \text{ torr}$$

$$\text{Vapor pressure of solution} = 71.0 - 10.6 = 60.4 \text{ torr, answer b.}$$

10.10 The most sensitive to measurement of the colligative properties is the osmotic pressure, which can be used to determine the molecular weights of very large molecules.

$$\text{The osmotic pressure,} \quad \pi = \frac{n}{V} RT$$

Let MW be the molecular weight of the protein.

$$\text{Then n} = 2.00 \text{ g} \times \frac{1 \text{ mole}}{\text{MW g}} = \frac{2.00}{\text{MW}} \text{ moles}$$

$$10.5 \text{ torr} = \frac{2.00 \text{ mole}}{\text{MW}} \times \frac{1}{.150 \text{ } \ell} \times \frac{.0821 \text{ } \ell\text{-atm}}{\text{mole-deg}} \times 300 \text{ deg} \times \frac{760 \text{ torr}}{1 \text{ atm}}$$

$$\text{MW} = \frac{(2.00)(.0821)(300)(760)}{(10.5)(.150)} = 23,800, \text{ answer d.}$$

CHAPTER 11. MIXED STOICHIOMETRY

11.1 You must convert all the amounts to a common unit; moles is the best common unit in chemistry.

a) $\dfrac{40.5 \text{ g}}{201 \text{ g mole}^{-1}} = 0.201 \text{ mole}$

b) given directly as 0.460 mole

c) $\dfrac{1.83 \times 10^{23} \text{ atoms}}{6.02 \times 10^{23} \text{ atoms mole}^{-1}} = .304 \text{ mole}$

d) $\dfrac{6.0 \text{ } \ell}{22.4 \text{ } \ell \text{ mole}^{-1}} \times \dfrac{1 \text{ mole Hg}}{1 \text{ mole Hg(CH}_3)_2} = .267 \text{ mole}$

e) $5.0 \text{ } \ell \times .020 \text{ mole } \ell^{-1} \times \dfrac{1 \text{ mole Hg}}{1 \text{ mole Hg(NO}_3)_2} = .100 \text{ mole}$

Now you can see that (b) is the answer.

11.2 From the anode reaction you see that 2 Faradays removes 2 moles of Cl^-, so 0.10 faraday removes 0.10 mole of Cl^-.

The original solution contained

$$\frac{500 \text{ ml}}{1000 \text{ ml } \ell^{-1}} \times 1.0 \text{ mole } \ell^{-1} = .500 \text{ mole Cl}^-$$

The Cl^- remaining is .500 − .10 = .40 moles which is answer (a).

You should verify that the other choices are all wrong.

b) Cl_2 produced = .100 mole Cl^- removed $\times \dfrac{1 \text{ mole Cl}_2}{2 \text{ mole Cl}^-} = .050$ mole

   mass of Cl_2 = .050 mole × 71.0 g mole^{-1} = 3.55 g

c) According to the cathode reaction only 0.10 mole of OH^- is produced by 0.10 faraday.

d) The concentration of OH^- is

   $[OH^-] = \dfrac{.10 \text{ mole}}{.500 \text{ } \ell} = .200 \text{ mole } \ell^{-1} = .200 \text{ M}$

e) Using the information in (b)

   volume of Cl_2 (STP) = .050 mole × 22.4 ℓ mole^{-1} = 1.12 ℓ

292

11.3 Convert both quantities to moles so that you can determine the limiting reagent.

$$\frac{7.48 \text{ g}}{23 \text{ g mole}^{-1}} = .325 \text{ mole Na}$$

$$\frac{2.96}{22.4 \text{ } \ell \text{ mole}^{-1}} = .1321 \text{ mole Cl}_2$$

According to the balanced equation, the Cl_2 required is:

$$.325 \text{ mole Na} \times \frac{1 \text{ mole Cl}_2}{2 \text{ mole Na}} = .1625 \text{ mole}$$

Hence the Na is in excess and <u>all</u> the Cl_2 is converted to NaCl.

mass of Cl_2 = .1321 mole $\times$ 70.90 g mole^{-1} = 9.36 g

total mass of solids = 7.48 g Na + 9.36 g Cl = 16.84 g, answer e.

11.4 Convert to moles so that you can determine the limiting reagent.

$$\text{moles Ca(HCO}_3)_2 = \frac{10.6 \text{ g}}{162 \text{ g mole}^{-1}} = 0.0654 \text{ moles}$$

moles HCl required to react with all $CaCO_3$ = 2 $\times$ .0654 = .1308 moles

moles HCl = 0.131 ℓ $\times$ 0.117 mole/ℓ = .0153 moles

Since there is insufficient HCl the HCl is limiting, and we can compute the volume of CO_2 from the moles of HCl.

$$\text{Vol. of CO}_2 = \frac{22.4 \text{ } \ell \text{ CO}_2}{1 \text{ mole CO}_2} \times \frac{2 \text{ mole CO}_2}{2 \text{ mole HCl}} \times 0.0153 \text{ mole HCl}$$

$$= .343 \text{ } \ell \text{ which is answer c.}$$

11.5 Moles Cl_2 = $\dfrac{11.2 \text{ } \ell}{22.4 \text{ } \ell \text{ mole}^{-1}} = 0.50 \text{ mole}$

Moles OH$^-$ required = 2 $\times$ 0.50 = 1.00 mole

Moles NaOH = moles OH$^-$ = 1.0 ℓ $\times$ 6.0 mole ℓ^{-1} = 6.0 moles

Since there is more than sufficient NaOH all the Cl_2 will be absorbed

Answer c is OK

Na$^+$ is not involved in the reaction, [Na$^+$] remains unchanged at 6.00 M, so answer (a) is OK.

Moles NaOCl = moles OCl$^-$ = moles Cl_2 = 0.50, so answer b is incorrect and is the one to select.

Moles Cl$^-$ = moles Cl_2 = 0.50, [Cl$^-$] = $\dfrac{.500 \text{ mole}}{1.00 \text{ } \ell}$ = .500 M, so d is OK.

Moles OCl$^-$ = moles Cl$^-$, so [OCl$^-$] = [Cl$^-$] = .500 M, so e is OK

11.7 In this problem you must first write and balance the equations in order to establish the molar ratio between Zn and Fe_3O_4.

$$Zn \; + \; H_2SO_4 \; \rightarrow \; H_2 \; + \; ZnSO_4$$

$$4 \; H_2 \; + \; Fe_3O_4 \; \rightarrow \; 3 \; Fe \; + \; 4 \; H_2O$$

$$g \; Zn = \frac{65.37 \; g \; Zn}{1 \; mole \; Zn} \; x \; \frac{1 \; mole \; Zn}{1 \; mole \; H_2} \; x \; \frac{4 \; moles \; H_2}{1 \; mole \; Fe_3O_4} x \; \frac{1 \; mole \; Fe_3O_4}{231.6 \; g \; Fe_3O_4} x \; 314g \; Fe_3O_4$$

$$= 356 \; g \; Zn \; \text{which is answer d.}$$

11.13 Let x be the number of kilograms of alloy #1 and y be the number of kilograms of alloy #2. Then, x + y = 6.25. The number of kilograms of Co in the x kilograms of alloy #1, in the y kilograms of alloy #2, and in the 6.25 kilograms of the third alloy are 0.49x, 0.72y and (0.59) (6.25) respectively. Then

$$0.49x + 0.72y = (0.55)(6.25)$$

Since y = 6.25-x, you can solve for x.   x = 3.53, y = 2.72,

which is answer b.

11.14 The reactions are: $Be \; + \; 2 \; H^+ \; \rightarrow \; H_{2(g)} \; + \; Be^{2+}$

$$Sc \; + \; 3 \; H^+ \; \rightarrow \; 3/2 \; H_{2(g)} \; + \; Sc^{3+}$$

Let x = # of grams of Be and y = # grams of Sc. The volume of H_2 due to Be is 22.4x/9.01. The volume of H_2 due to Sc is

$$\frac{3}{2} \; x \; \frac{22.4y}{45} = \frac{33.6y}{45} \; .   \text{Then}   x \; + \; y = 4.60$$

and $\dfrac{22.4x}{9.01} + \dfrac{33.6y}{45} = 7.95$. Solving the simultaneous equations,

x = 2.59,   % Be = $\dfrac{2.59}{4.60}$ x 100% = 56.4%, which is answer d.

11.15 The balanced equations for the reactions with Sn^{2+} are:

$$Hg_2^{2+} \; + \; Sn^{2+} \; \rightarrow \; Sn^{4+} \; + \; 2 \; Hg$$

$$Hg^{2+} \; + \; Sn^{2+} \; \rightarrow \; Sn^{4+} \; + \; Hg$$

Let x = conc. of Hg_2^{2+} which was oxidized

.100 − x = conc. of Hg_2^{2+} which remained

2x = conc. of Hg^{2+} which formed

moles of $Sn^{2+} = \dfrac{(45.8)(.100)}{(1000)} = \dfrac{40(.100-x)}{1000} + \dfrac{40(2x)}{1000}$

Solving, x = .0145 Hg_2^{2+} = .100 − .0145 = .0855

$$Hg^{2+} = 2(.0145) = 0.0290$$

Answer is a.

CHAPTER 12.   LABORATORY STOICHIOMETRY

12.1 a) You need the molecular weight to compute the number of moles. From
the dimensions of the film, you can count the number of molecules.
Avogadro's number = number of molecules/number of moles.
   b) Yes, you must assume that all the oleic acid remains in the film.
   c) Yes, in order to compute the volume of the film and hence the
number of molecules.
   d) Yes, in order to compute the volume and the mass of oleic acid
dropped on the water.
   e) No; this is the answer to mark. You could deduce Avogadro's number
from m_e (but this was not your experiment) as follows: from the
charge/mass ratio of J. J. Thomson calculate the charge on the
electron and from the Faraday unit (which is the charge on a mole of
electrons) compute Avogadro's number.

12.2 a) The student's percent was too high, meaning the weight loss was too
great. If something fell into the crucible the weight <u>loss</u> would
be too small.
   b) If this happened the weight <u>loss</u> would be too small.
   c) As long as <u>all</u> weighings were on this balance they were relatively
correct and no error would result.
   d) Again the weight <u>loss</u> was too small.
   e) If it were wet, the "wetness" water would be lost along with the
hydrate water, and the weight loss would be excessive. Mark this
answer.

12.3 a) It would be necessary in this case to use an excessive volume of
NaCl to reach maximum $PbCl_2$ yield and the students would have
computed an excessive ratio of Cl to Pb. This is a possible source
of error; mark this answer.
   b) In this case all $PbCl_2$ yields would be equally high but the peak
yield would still occur at the correct molar ratio.
   c) As above, except that all yields are low.
   d) & e) According to the lab procedure both Pb^{++} and Cl^- are dispensed
from burets. Although the absolute volumes might be wrong the
relative quantities are the same as if the burets were perfect.

12.4 a) The total mass of metal, which exceeded the available iron, was
thought to be iron. This could explain the high yield.
   b) If the Fe_2O_3 were in excess, not all of the possible iron would
appear as metal and the yield would be low. This can <u>not</u> explain
the high yield and is the answer to mark.
   c) The weight of iron (including material thought to be iron) was high.
Valid explanation.
   d) Fe_3O_4, thought to be iron because of its attraction to the magnet,
weighs more than its equivalent amount of iron. Again, a high
weight of iron is reported.
   e) You get more iron out of Fe_3O_4 than you get from the same weight of
Fe_2O_3, so your actual yield will exceed your anticipated yield.

12.5 Since no precipitate forms with HCl, group I is absent eliminating (c) Hg_2^{2+} and (e) Ag^+. The precipitate formed with H_2S in acid indicates the unknown is from group II, eliminating (b) Fe^{3+}. The NaOH test tells you nothing more, nor does the HNO_3 test. Since Cd^{2+} forms a soluble complex with NH_3 you know that it is not the unknown, eliminating (a). However, Bi^{3+} precipitates as the hydroxide upon NH_3 addition, confirming that (d) is the answer.

CHAPTER 13. THE GAS LAWS

13.4 The reaction is: $2 H_2(g) + O_2(g) \rightarrow 2 H_2O(g)$

Since at constant temperature and pressure the volume of any gas is proportional to the number of moles, you may use volumes in place of moles to work the problem.

According to the balanced equation 17.3 ℓ of O_2 require (2)(17.3) = 34.6 ℓ of H_2. The actual volume of H_2 is in excess by 3.2 ℓ. The final volume will be the 34.6 ℓ of H_2O formed by the reaction and the 3.2 ℓ of excess H_2, for a total of 37.8 ℓ. Answer (a).

13.9 Let y = the molecular weight of the compound

$$n = 11.8 \text{ g}/y \text{ g mole}^{-1} = 11.8/y \text{ mole}$$

$$\frac{(630 \text{ torr})(1.31 \text{ ℓ})}{(760 \text{ torr atm}^{-1})} = \frac{(11.8 \text{ mole})(.082 \text{ ℓ atm deg}^{-1} \text{ mole}^{-1})(312 \text{ deg})}{y}$$

Solving, y = 279, answer (d).

13.13 First, determine the empirical formula from the composition, then determine the molecular weight from the gas data.

$$\overset{C}{\frac{49.0}{12.0}} = 4.08 \qquad \overset{H}{\frac{2.00}{1.00}} = 2.00 \qquad \overset{O}{\frac{49.0}{16.0}} = 3.06$$

$$\approx 4 \qquad\qquad\qquad \approx 2 \qquad\qquad\qquad \approx 3$$

Empirical formula is $C_4H_2O_3$, formula wt. = 98

$$n = \frac{(670.3)(740)}{(1000)(760)(.0821)(426)} = 1.87 \times 10^{-2} \text{ moles.}$$

$$\text{Mol. wt.} = \frac{1.83 \text{ g}}{1.87 \times 10^{-2} \text{ mole}} = 97.4 \text{ g mole}^{-1}$$

There is just one formula unit per molecule, answer (e).

13.15 $PV = nRT = \frac{m}{MW} RT$, where m is the mass of the gas

$$\text{density} = p = \frac{m}{V} = \frac{(P)(MW)}{RT} = \text{constant} \left(\frac{P}{T}\right)$$

comparing two sets of conditions, $\dfrac{\rho_2}{\rho_1} = \dfrac{P_2T_1}{P_1T_2}$

Condition 2 in this problem is STP

$$\rho_2 = 4.38 \text{ g } \ell^{-1} \times \frac{760 \text{ torr}}{542 \text{ torr}} \times \frac{283.6 \text{ deg}}{273 \text{ deg}} = 6.38 \text{ g } \ell^{-1}, \text{ answer e.}$$

13.19 The only gases are H_2O and CH_2Cl_2 and their mole fractions can be found as follows:

$$H_2O; \quad \frac{7.70 \text{ g}}{18.0 \text{ g mole}^{-1}} = 0.427 \text{ mole}; \quad \frac{.427}{.506} = .844$$

$$CH_2Cl_2; \quad \frac{6.70 \text{ g}}{84.9 \text{ g mole}^{-1}} = 0.0789 \text{ mole}; \quad \frac{.0789}{.506} = .156$$

$$\text{Total} = 0.506 \text{ mole} \quad \text{Total} = 1.000$$

Partial pressure of CH_2Cl_2 = .156 (124 torr) = 19.3 torr.
The correct answer is (c.)

13.21 To help explain the principle a sketch of the apparatus is shown below:

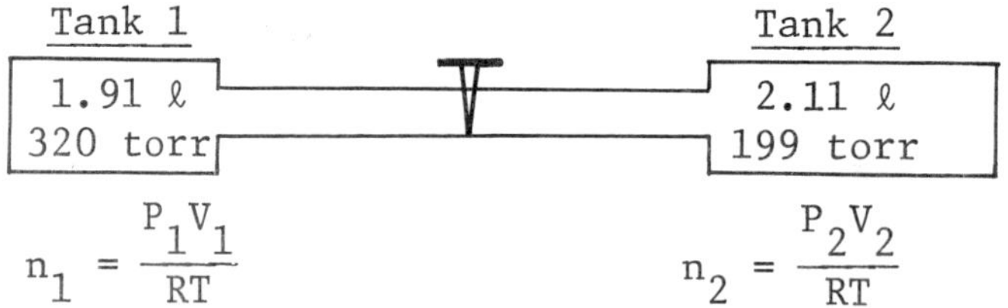

$$n_1 = \frac{P_1 V_1}{RT} \qquad n_2 = \frac{P_2 V_2}{RT}$$

The ideal gas law was used to express the number of the moles of gas in each tank. After they mix, express the total moles, n_T, similarly.

$$n_T = \frac{P_T V_T}{RT} = n_1 + n_2 = \frac{P_1 V_1}{RT} + \frac{P_2 V_2}{RT}$$

$$\frac{P_T V_T}{RT} = \frac{P_1 V_1}{RT} + \frac{P_2 V_2}{RT} \quad \text{or} \quad P_T V_T = P_1 V_1 + P_2 V_2$$

In other words the PV product (which is proportional to moles at constant T) is an additive quantity.

$$P_T(1.91 \text{ } \ell + 2.11 \text{ } \ell) = 320 \text{ torr} \times 1.91 \text{ } \ell + 1.99 \text{ torr} \times 2.11 \text{ } \ell$$

$$P_T = \frac{[(320 \times 1.91) + (199 \times 2.11)] \text{ torr } \ell}{(1.91 + 2.11) \text{ } \ell} = \frac{612 + 420}{4.02} \text{ torr} = \frac{1032}{4.02}$$

= 257 torr, which is answer (c).

13.22 Take advantage of the principle noted in the preceding solution and use the PV product in lieu of moles to find the mole fractions.

$$PV(He) = 3.09 \text{ } \ell \times 795 \text{ torr} = 2460 \text{ } \ell \text{ torr}$$
$$PV(Ar) = 3.98 \text{ } \ell \times 856 \text{ torr} = 3410 \text{ } \ell \text{ torr}$$
$$PV \text{ (Total)} = 5870 \text{ } \ell \text{ torr}$$
$$\text{Mole fraction Ar} = \frac{3410}{5870} = 0.581, \text{ answer (a).}$$

13.23 The relative numbers of moles of the gases are given by their coefficients in the balanced equation. Hence:

$$\text{mole fraction HCl} = \frac{1}{1 + 1} = \frac{1}{2} = 0.50, \text{ and}$$

partial pressure HCl = (0.50)(883 torr) = 442 torr, which is answer e.

13.25 The combustion reaction is

$$C_3H_8(g) + 5\ O_2(g)\ \rightarrow\ 3\ CO_2(g) + 4\ H_2O(g)$$

Since the sum of the coefficients on the right is 7 and that on the left is 6, there is an increase of one mole of gas for every mole of C_3H_8 reacted. The increase in moles, Δn, is:

$$\Delta n = \frac{\Delta(PV)}{RT} = \frac{\Delta P \times V}{RT} = \frac{(6.06 - 5.36)(7.75)}{(273 + 127)(.0821)} = 0.165 \text{ moles.}$$

Hence 0.165 moles of C_3H_8 reacted.

$$g\ C_3H_8 = 0.165 \text{ moles} \times 44.0 \text{ g/mole} = 7.26 \text{ g, answer a.}$$

CHAPTER 14. KINETIC MOLECULAR THEORY

14.1 Graham's Law in this case states:

$$\frac{\text{rate (He)}}{\text{rate (HCl)}} = \sqrt{\frac{\text{MW (HCl)}}{\text{MW (He)}}}$$

where MW stands for molecular weight

$$\frac{\text{rate (He)}}{\text{rate (HCl)}} = \sqrt{\frac{36.5}{4}} = 3.02$$

rate (He) = 3.02 rate (HCl). Answer is c.

14.4 Let ℓ = distance to deposit from 0 in cm., ℓ = distance travelled by A
and $(120 - \ell)$ = distance travelled by B

Both gases were started at the same time so they have travelled for the same length of time. Hence:

$$\frac{\text{rate A}}{\text{rate B}} = \frac{\text{distance A}}{\text{distance B}} = \frac{\ell}{120 - \ell} = \frac{36}{64} = \frac{6}{8} = \frac{3}{4}$$

$$\frac{\ell}{120 - \ell} = \frac{3}{4}; \quad \ell = 51.4 \text{ cm which is answer a.}$$

14.7 a) Since the number of molecules is the same, the number of moles is the same, the mole fraction of each is 0.5, and each exerts a partial pressure equal to half the total. Mark this answer. However, you should check the others.
   b) No; CO_2 molecules are heavier than CO molecules.
   c) No; regardless of the temperature each gas contributes equally to the pressure since their mole fractions remain the same.
   d) Must be wrong since the pressures are equal. The first three phrases are true but CO impacts are less forceful than CO_2 impacts, CO_2 having on the average more momentum.
   e) No; a is correct.

298

14.8 a) No; as long as the vessels are rigid the volume is constant and the
        density must be constant.
     b) No; they contain the same number of molecules (by Avogadro's
        principle) but the He molecules weigh twice as much as the H_2.
     c) It is true that the pressure must be doubled, but you must investigate
        investigate the matter of impacts carefully.
        Originally the H_2 must have made more impacts because the impact rate
        is proportional to velocity.

$$\frac{\text{Impacts } H_2}{\text{Impacts He}} = \frac{v_{H_2}}{v_{He}} = \sqrt{\frac{4}{2}} = \sqrt{2}$$

   or Impacts H_2 = $\sqrt{2}$ x Impacts He = 1.41 x Impacts He.

   If you double the pressure you must double the impact rate of He,
   which is more than 1.41 times, and so the rest of c is true.
   Mark this answer and look at the rest.
     d) No; in fact H_2 will condense at a higher temperature. The conden-
        sation temperature depends not on the density but on the intermolec-
        ular attractions (the Van der Waal's "a") which are weaker in He
        than in any other gas.
     e) No; as discussed under c above.

14.9 a) It is postulated that all collisions of the molecules with the walls
        are elastic, and on the average no energy is transferred since the
        wall and gas are at the same temperature.
     b) On the contrary all the energy is attributed to kinetic energy $mv^2/2$.
     c) Since you may ignore the volume of the molecules their shape and
        compressibility is of no importance.
     d) This is true but you ignore it when you define an "ideal" gas.
     e) Yes, this statement is the basis of the ideal gas temperature scale.

14.10 a) Yes, the vapor pressure is a function of temperature only.
      b) No effect, as long as equilibrium exists between vapor and liquid.
      c) The volume of water and of free space above determines the total mass
         of the water in the vapor phase, but its pressure is the same regard-
         less of the size, as long as you specify that equilibrium prevails.
      d) The ideal gas law pictures all gas molecules moving independently
         of (oblivious to) any other gas molecules sharing their space, so
         other gases don't affect the vapor pressure.

CHAPTER 15. CRYSTALLINE SOLIDS

15.3 The basis of the solution is the unit cell. To find its density, divide
     the mass of the unit cell by its volume. Since there are 4 atoms per
     unit cell in the fcc system,

$$\text{Mass} = \frac{4 \text{ atoms x } 106.4 \text{ g mole}^{-1}}{6.02 \times 10^{23} \text{ atoms mole}^{-1}}$$

$$\text{Volume} = a^3 = (3.87 \times 10^{-8} \text{ cm})^3$$

$$\text{Density} = \frac{\text{Mass}}{\text{Volume}} = \frac{4 \times 106.4}{6.02 \times 10^{23} \times (3.87 \times 10^{-8})^3}$$

$$= 12.1 \text{ g cm}^{-3}, \text{ which is answer (a).}$$

15.6 a) No, in fcc there are only 4 atoms/unit cell:

$$8 \text{ corners } \times 1/8 = 1$$
$$6 \text{ faces } \times 1/2 = 3 \qquad \text{Total 4}$$

b) No, in fcc, atoms touch along a face diagonal so that a $2 = 4$ r.

c) Yes, the second nearest neighbor to the atom in the center of one cell is the atom in the center of the adjacent cell. The answer is c.

d) No, in fcc, which is a closest-packed arrangement, there are 12 nearest neighbors.

e) No, in bcc the atom in the center touches all eight corners, hence 8 nearest neighbors.

The nearest neighbor distance in both cases (d & e) is 2r since the atoms touch, and you always measure from center to center.

15.9 There are 4 cations and 4 anions in the unit cell of the NaCl lattice. The mass of the unit cell is $4 \times 150/6.02 \times 10^{23}$. The volume of unit cell is a^3, where a is the length of an edge in centimeters. Then

$$\rho = 4.14 \text{ g/cm}^3 = \frac{4 \times 150}{6.02 \times 10^{23} \text{ a}^3}$$

solving for a, $a = 6.22 \times 10^{-8}$, or $a = 6.22$ Angstroms, answer (a).

CHAPTER 16. ATOMIC STRUCTURE

16.3 The wavenumber $\bar{\nu}$ is the reciprocal of wavelength λ. It is not a genuine unit of energy but since it is proportional to ν(frequency) which in turn is proportional to energy, the wavenumber is a convenient spectroscopic unit, proportional to energy for purposes of discussion.

Precisely: $E = h\nu = \dfrac{hc}{\lambda} = hc\bar{\nu}$.

$$\lambda = \frac{c}{\nu} = \frac{3.00 \times 10^{10} \text{ cm sec}^{-1}}{5.69 \times 10^{17} \text{ sec}^{-1}} = 5.27 \times 10^{-8} \text{ cm}$$

$$\bar{\nu} = \frac{1}{\lambda} = \frac{1}{5.27 \times 10^{-8} \text{ cm}} = 1.90 \times 10^7 \text{ cm}^{-1} \qquad \text{answer (d)}$$

The answer is also expressed "1.90×10^7 waves per centimeter" or "1.90×10^7 wavenumbers".

16.7 Answer (a) gives the general notation for the outer electrons only, where n is the group number. $n = 1$ is excluded because $1s^2$ is helium.

The alkaline earths, e.g. Mg, Ca, are the group IIA elements and all have just two s electrons in their outermost shell as symbolized ns . (a) is correct.
Following the usual "aufbau" rule, (b) and (c) are also correct.

Cr^+ must have just $24-6 = 18$ electrons and so is just the Ar configuration, no more. (d) is the incorrect one.

300

16.8 $102 \text{ kcal mole}^{-1} \times 6.94 \times 10^{-14} \dfrac{\text{erg photon}^{-1}}{\text{kcal mole}^{-1}} = 7.07 \times 10^{-12} \text{ erg photon}^{-1}$

To break an O–H bond, a photon must possess at least 7.07×10^{-12} erg

$$E = h\nu; \quad \nu = \frac{E}{h}$$

$$\lambda\nu = c; \quad \lambda = \frac{c}{\nu} = \frac{ch}{E}$$

$$\lambda = \frac{3.00 \times 10^{10} \text{ cm sec}^{-1} \times 6.63 \times 10^{-27} \text{ erg sec}}{7.07 \times 10^{-12} \text{ erg}} \times 10^{8} \text{ \AA cm}^{-1}$$

$= 2810 \text{ \AA}$ which is answer (c).

16.9 Calculate the energy of that line in wavenumbers:

$$E_{1-2} = R_H \times \left[\frac{1}{n_1{}^2} - \frac{1}{n_2{}^2} \right]$$

$$= 110,000 \text{ cm}^{-1} \left[\frac{1}{1^2} - \frac{1}{9^2} \right]$$

$$= 110,000 (1.000 - .012) = 109,000 \text{ cm}^{-1}$$

$$= 1.09 \times 10^{5} \text{ cm}^{-1}$$

This falls in the range given by choice d.

16.10 Choice (a) is not correct. Moseley's measurements showed that the atomic number was simply the nuclear charge and did not necessarily follow the order of atomic weight.

16.11 Choice (c) is not correct. Bohr postulated stationary states in which the electron did not radiate energy, in violation of classical electro-magnetic theory.

16.14 The energy of a rapidly moving electron is given by the kinetic energy expression, $\frac{1}{2} mv^2$, where m is the electron mass.

$$E = \frac{1}{2} mv^2 \qquad v = \sqrt{\frac{2E}{m}}$$

$$v = \left[\frac{2 \times 2.46 \times 10^{4} \text{ eV} \times 1.60 \times 10^{-12} \text{ erg(eV)}^{-1}}{9.11 \times 10^{-28} \text{ g}} \right]^{1/2}$$

$v = 9.3 \times 10^{9} \text{ cm sec}^{-1}$ (note: ergs are $\text{g-cm}^2\text{-sec}^{-2}$)

DeBroglie's equation is: $\quad \lambda = \dfrac{h}{mv}$

$$= \frac{6.63 \times 10^{-27} \text{ erg-sec} \times 10^{8} \text{ \AA cm}^{-1}}{9.11 \times 10^{-28} \text{ g} \times 9.3 \times 10^{9} \text{ cm sec}^{-1}} = 0.0783 \text{ \AA}$$

which is answer (d)

Note: According to the theory of relativity, the kinetic energy
expression becomes increasingly more inaccurate as v approaches the
speed of light c. In this problem where v $\sim$ 1/3 c the above value
for v is about 2% too high. Since v can never exceed c the error
gets much worse for higher energy electrons, for which relativistic
calculations are mandatory.

CHAPTER 17. PERIODIC PROPERTIES

17.6 The second electron to be removed is a p electron like the first so the
orbital influence (the distance factor) is about the same. However, it
is removed from a 2+ residue rather than a 1+ residue and so that
electrostatic energy is twice as great. Answer (d) is the closest
to 2 x 22.

17.7 You must compare the configurations:

$$O^+ \quad 1s^2 2s^2\ 2p^3 \quad \uparrow \quad \uparrow \quad \uparrow$$
$$F^+ \quad 1s^2 2s^2\ 2p^4 \quad \uparrow\downarrow \quad \uparrow \quad \uparrow$$

On the basis of size, the F^+ is smaller and would require more energy
to ionize. However, loss of its electron is facilitated because it
leaves behind a stable half-shell, and this compensates for the size
effect. As a result, the net ionization energy is nearly the same.
Select answer (c).

This situation is just like comparing the first ionization energies of
O and N, in which case the configuration effect more than compensates
for the size effect and results in O having a slightly lower ionization
energy than N.

17.11 All elements in the same group as Ga have 3+ as the most stable
oxidation state as do those in the Sc group. However, these are
rather metallic and would have electronegativity less than 1.9.
The least metallic of these is B but it should have a tiny radius
and high ionization energy. The elements in the low end of the
Bi group tend to oxidation number of 3+ rather than 5+ because of
the inertness of the s pair. Since X is just slightly smaller than
Bi (with a consequent slightly larger ionization energy) it is
probably Sb, which is answer (e).

17.13 The percent ionic character is the observed moment divided by the
moment that you would have if a full electronic charge were separated by
the distance.

$$\% \text{ ionic} = \frac{1.29 \text{ D}}{4.80 \times 10^{-10} \text{ esu} \times 1.76 \times 10^{-8} \text{ cm}} \times 10^{-18} \frac{\text{esu-cm}}{\text{D}} \times 100\%$$

$$= 15.3\% \text{ which is answer (a).}$$

17.15 The most electronegative elements have large (endothermic) values for
IE and moderately large (exothermic) values for EA. Therefore,
$\chi \propto$ IE - EA (χ = electronegativity). This gives the following χ values
X 12.4, Y 11.7, Z 19.2. The differences are XY 0.7, XZ 6.8, YZ 7.5.
The smaller this difference the more covalent the bond. Hence sequence
(e) is the answer.

CHAPTER 18. CHEMICAL BONDING

18.3 The wrong answer is (d). The correct Lewis structure and formal
charge analysis is:

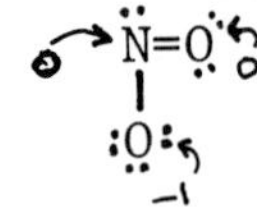

The nitrogen atom has one electron from each of three shared
pairs (3 electrons) plus an unshared pair (2 electrons),
total 5 electrons. The neutral atom has 5 valence electrons
also, so the formal charge is zero. You should confirm the
correctness of the other answers.

18.4 Resonance structures have to be draw in each case, as shown below.
The answer is d.

18.6 There are 18 valence electrons is cyanogen 2 x (4 + 5). Only structure
2 has 18 valence electrons; the answer is b.

18.8 First draw correct Lewis structures for each species. Then note the
following. Choice (a) is correct because B has sigma bonds to four
atoms and no unshared pairs. The four bonding pairs repel each other
at 109.5° putting the four fluorines at the corners of a tetrahedron.
Choice (b) is correct; S has 6 bonding pairs, no unshared pairs.
Choice (c) is wrong (it is the answer); Al has 3 bonding pairs at 120°
forming a trigonal planar structure. (No unshared pair on Al). Choice
(d) is correct; I has 4 shared pairs and 2 unshared pairs; all angles
are 90° and the unshared pairs are at 180° to each other to minimize
the unshared-unshared repulsion. Choice (e) is correct; an unshared
piar on Pb repels the two bonds out of line.

18.13 The molecule with the lowest bond order will have the greatest bond length.

Molecule	# Valence Electrons	MO Configuration	Bond Order
BO	9	$(\sigma_{2s})^2(\sigma^*_{2s})^2(\pi_{2p_{y,x}})^4(\sigma_{2p_z})^1$	2.5
NO^+	10	" " " $(\sigma_{2p_z})^2$	3.0
CO^+	9	" " " $(\sigma_{2p_z})^1$	2.5
O_2	12	" " " $(\sigma_{2p_z})^2(\pi^*_{2p_{y,x}})^2$	2.0

O_2 has the lowest bond order. Therefore, you can predict that O_2 has the greatest bond length. The answer is (d).

CHAPTER 19. ACIDS AND BASES

19.11 In problems like this take advantage of the relationship between pH and pOH.

$$pOH = 14.00 - pH = 14.00 - 3.64 = 11.0 - 0.64$$

$$\log[OH^-] = -pOH = -11.0 + 0.64$$

$$[OH^-] = \text{antilog}(0.64) \times 10^{-11}$$

$$= 4.36 \times 10^{-11} \text{ which is answer (a).}$$

19.21 In the preceeding problems the dissociation of water was ignored. But:

$$H_2O \rightleftharpoons H^+ + OH^- \qquad K_w = [H^+][OH^-] = 1.0 \times 10^{-14}$$

So in neutral aqueous solution, $[H^+] = [OH^-] = 1.00 \times 10^{-7}$, which exceeds the $[H^+]$ obtainable from the 2.80×10^{-8} M $HClO_4$. There are two reactions to consider.

$$HClO_4 \xrightarrow{\ 100\%\ } H^+ + ClO_4^-$$

and the dissociation of water. Since all solutions must contain equal numbers of (+) and (−) charges, we can write a charge balance equation for this system:

$$[H^+] = [OH^-] + [ClO_4^-]$$

Let $x = [H^+]$, then $\dfrac{K_w}{x} = [OH^-]$

Substituting in the charge balance equation:

$$x = \frac{K_w}{x} + 2.80 \times 10^{-8} \text{ or } x^2 - 2.80 \times 10^{-8}\, x - K_w = 0$$

Solving by the quadratic formula,

$$x = \frac{2.80 \times 10^{-8} + \sqrt{(2.80 \times 10^{-8})^2 + 4.0 \times 10^{-14}}}{2}$$

$x = [H^+] = 1.14 \times 10^{-7}$; pH $= 7 - 0.06 = 6.94$, which is answer d.

19.23 These solutes dissolve in and react with water as follows:

1. $NaF \xrightarrow{100\%} Na^+ + F^-$

    $F^- + H_2O \rightleftharpoons HF + OH^-$ (basic solution)

2. $NH_4NO_3 \xrightarrow{100\%} NH_4^+ + NO_3^-$

    $NH_4^+ \rightleftharpoons H^+ + NH_3$ (acidic solution)

3. $Ca_3(PO_4)_2 \xrightarrow{100\%} 3\ Ca^{2+} + 2\ PO_4^{3-}$

    $PO_4^{3-} + H_2O \rightleftharpoons HPO_4^{2-} + OH^-$ (basic solution)

The answer is (c).

19.26 In this problem, the unknown quantity is somewhat unusual but you may solve it by a straightforward application of the equilibrium constant.

$$HNO_2 \rightleftharpoons H^+ + NO_2^- \qquad K_a = \frac{[H^+][NO_2^-]}{[HNO_2]}$$

From the pH we find $[H^+]$ = antilog (-1.71) = 1.95×10^{-2} = $[NO_2^-]$

$$K_a = 4.5 \times 10^{-4} = \frac{(1.95 \times 10^{-2})(1.95 \times 10^{-2})}{[HNO_2]}$$

$[HNO_2]$ = 0.843 <u>at equilibrium</u>

initial $[HNO_2]$ before dissociation = $0.843 + 1.95 \times 10^{-2}$ = .863, answer (d).

19.29 pOH = $14.0 - 10.32 = 4 - .32$; $[OH^-] = [BH^+] = 2.09 \times 10^{-4}$

[B] at equilibrium = $.044 - .000209 = .044$

$$K_b = \frac{(2.09 \times 10^{-4})(2.09 \times 10^{-4})}{(.044)} = 9.9 \times 10^{-7}, \text{ answer (d).}$$

19.34 Some approximations must be made. Assume that the molarity is approximately the same as the molality, a good assumption in dilute aqueous solutions. Assume further that the solution is ideal, i.e. that the freezing point depression accurately reflects the total number of molecules and ions present.

$$\text{Total molality} = \frac{-0.855}{-1.86} = 0.459 = [HA] + [H^+] + [A^-]$$

Let $x = [H^+] = [A^-]$

$.450 - x = [HA]$

$.459 = .450 - x + x + x = .450 + x$

$x = .009 = [H^+] = [A^-]; \quad [HA] = .441$

$$K_a = \frac{(.009)(.009)}{.441} = 2 \times 10^{-4}, \text{ answer (c).}$$

19.37 Dimethylammonium chloride is a strong electrolyte,

$$(CH_3)_2NH_2Cl \xrightarrow{100\%} (CH_3)_2NH_2^+ + Cl^-$$

Dimethylammonium ion is the conjugate acid of dimethylamine,

$$(CH_3)_2NH_2^+ \rightleftharpoons (CH_3)_2NH + H^+ \qquad K_a = K_w/K_b$$

$$0.821 - x \qquad\qquad x \qquad\quad x$$

$$K_a = \frac{1.00 \times 10^{-14}}{5.40 \times 10^{-4}} = \frac{x^2}{0.821-x}$$

Assume $x \ll 0.821$, then $\dfrac{x^2}{.821} \simeq \dfrac{1.00 \times 10^{-14}}{5.00 \times 10^{-4}}$

$$x = 3.90 \times 10^{-6}$$

The assumption $x \ll 0.821$ is clearly valid.

Then $[H^+] = x = 3.90 \times 10^{-6}$

pH = 5.41 which is answer b

Note: Answer a is obtained using $K_a = K_b = 5.40 \times 10^{-4}$

Remember, for acids and conjugate bases: $K_a = K_w/K_b$

19.38 $Al(ClO_4)_3$ is the salt of a weak base and a strong acid, so it hydrolyzes as shown in the problem to give an acid solution.

Let $z = [H^+] = [AlOH^{2+}]$

Then $0.70 - z = [Al^{3+}] \cong 0.70$

$$K_h = 7.9 \times 10^{-6} = \frac{[H^+][AlOH^{2+}]}{[Al^{3+}]} = \frac{(z)(z)}{0.70}$$

$$z^2 = 5.5 \times 10^{-6}$$

$$z = 2.34 \times 10^{-3} = [H^+]$$

$$pH = -.37 + 3.00 = 2.63 \text{ which is answer (c)}$$

Checking our assumption:

$$100\% \times \frac{2.34 \times 10^{-3}}{0.70} = 0.34\%, \text{ which is less than 5\%, OK.}$$

19.42 HNO_3 is a strong acid and HNO_2 is a weak acid.

$$HNO_3 \xrightarrow{100\%} H^+ + NO_3^-$$

$$HNO_2 \rightleftharpoons H^+ + NO_2^- \qquad K_a = 4.6 \times 10^{-4}$$

Assume to a first approximation that the $[H^+]$ from HNO_3 totally suppresses the dissociation of HNO_2. Then $[H^+] = 0.25$ $\underline{M}$, and

$$[NO_2^-] = \frac{K_a[HNO_2]}{[H^+]} = \frac{4.6 \times 10^{-4} \times 0.55}{0.25} = 1.0 \times 10^{-3}$$

The extent of dissociation of HNO_2 is less than 0.2 %. The assumption is justified. The answer is

$$pH = -\log (0.25) = +0.60, \text{ answer (c)}.$$

19.45 Generally, the more dilute the weak acid, the greater the fraction dissociated. In this problem, the concentration is so small you cannot neglect this fraction in the computation and must resort to solving the quadratic equation.

Let $x = [H^+] = [F^-]$

$$(1.94 \times 10^{-3} - x) = [HF]$$

$$\frac{x^2}{1.94 \times 10^{-3} - x} = 6.76 \times 10^{-4}$$

$$x^2 + 6.76 \times 10^{-4} x - 1.31 \times 10^{-6} = 0$$

$$x = \frac{-6.76 \times 10^{-4} \pm \sqrt{4.57 \times 10^{-7} + 5.24 \times 10^{-6}}}{2}$$

$$= -3.38 \times 10^{-4} + 1.19 \times 10^{-3} = 8.5 \times 10^{-4} = [H^+]$$

$$pH = 4 - \log 8.5 = 3.07, \text{ answer c}$$

The percent dissociation is $\dfrac{8.5 \times 10^{-4}}{1.94 \times 10^{-3}} \times 100\% = 43.8\%$ which is far from negligible.

19.47 Note that K_2 is 5 orders of magnitude less than K_1. Therefore, to an excellent approximation the dissociation of HA^- can be ignored relative to the dissociation of H_2A. Considering only the first step:

$$[H^+] = [HA^-] \simeq \sqrt{K_1 \cdot c_a}$$

The second equilibrium expression is

$$K_2 = \frac{[H^+][A^{2-}]}{[HA^-]}$$

But $[H^+] = [HA^-]$

Therefore, $[A^{2-}] = K_2 = 2.61 \times 10^{-12}$ M, which is choice (b)

The above result is general, $[A^{2-}] = K_2$, provided that $K_2 \ll K_1$.

19.50 Assume only the first dissociation contributes significantly to the $[H^+]$. Let x be $[H^+]$,

$$H_3PO_4 \rightleftharpoons H^+ + H_2PO_4^- \qquad K_1 = 7.5 \times 10^{-3}$$
$$0.091-x \qquad x \qquad x$$

$$K_1 = \frac{[H^+][H_2PO_4^-]}{[H_3PO_4]} = \frac{x^2}{0.091 - x} = 7.5 \times 10^{-3}$$

The usual assumption, $x \ll 0.091$, will fail here. Using the quadratic equation, or the method of successive approximations,

$$x = [H^+] = 0.023 \underline{M} = [H_2PO_4^-]$$

Now check the assumption that the second dissociation does not contribute to the $[H^+]$. The additional $[H^+]$ must be equal to the $[HPO_4^{2-}]$ formed.

$$[HPO_4^{2-}] = \frac{K_2[H_2PO_4^-]}{[H^+]} = K_2 = 6.2 \times 10^{-8}$$

The assumption is excellent. pH = $-\log (0.023)$ = 1.65. The answer is (d).

19.52 This problem involves the strong electrolyte ammonium nitrite,

$$NH_4NO_2 \xrightarrow{\ 100\%\ } NH_4^+ + NO_2^-$$

which contains <u>both</u> a conjugate acid of a weak base (NH_4^+-NH_3) and a conjugate base of a weak acid (NO_2^--HNO_2). These species will undergo the following proton transfer reaction,

$$NH_4^+ + NO_2^- \rightleftharpoons NH_3 + HNO_2$$

$$K_{eq} = \frac{[NH_3][HNO_2]}{[NH_4^+][NO_2^-]}$$

If $K_a = \frac{[H^+][NO_2^-]}{[HNO_2]}$ and $K_b = \frac{[NH_4^+][OH^-]}{[NH_3]}$

then it can be seen by inspection that $K_{eq} = \dfrac{K_w}{K_a K_b}$

Let $x = [NH_3] = [HNO_2]$ at equilibrium

then

$$K_{eq} = \frac{x^2}{(.1-x)^2} = \frac{K_w}{K_a K_b} = 1.21 \times 10^{-6}$$

Taking the square root of both sides, $x = [NH_3] = [HNO_2] = 1.10 \times 10^{-4} \underline{M}$

Then $[NH_4^+] = [NO_2^-] = 0.1 - x \approx 0.1 \underline{M}$

The equilibrium pH can be computed using either

$$[H^+] = \frac{K_a[HNO_2]}{[NO_2^-]} = 5.06 \times 10^{-7} \underline{M}, \qquad pH = 6.30$$

or

$$[H^+] = \frac{(K_w/K_b)[NH_4^+]}{[NH_3]} = 5.05 \times 10^{-7} \; \underline{M}, \; pH = 6.30$$

Option (e) is correct.

The initial assumption of the dominant proton transfer reaction is good in this case because the acid and base involved are about equally weak. If, however, one is much stronger than the other, the dominant equilibrium is the hydrolysis of the conjugate of the weaker. For instance, to determine the pH of a solution of NH_4CN, you may consider just the equilibrium

$$CN^- + H_2O \rightleftharpoons HCN + OH^-.$$

19.53 Symbolize the neutral amino acid molecule HA. Then:

$$H_2A^+ \rightleftharpoons HA + H^+ \qquad K_1 = \frac{[HA][H^+]}{[H_2A^+]}$$

$$HA \rightleftharpoons A^- + H^+ \qquad K_2 = \frac{[A^-][H^+]}{[HA]}$$

$$K_1K_2 = \frac{\cancel{[HA]}[H^+]\cancel{[A^-]}[H^+]}{\cancel{[H_2A^+]}\cancel{[HA]}} = [H^+]^2$$

Note that at the isoelectric point $[A^-] = [H_2A^+]$ so these terms cancel.

$$[H^+] = \sqrt{K_1K_2} \qquad or \qquad pH = 1/2(pK_1 + pK_2)$$

This is an exact expression for the pH at the isoelectric point and is truly independent of concentration.

In this problem pH $= \frac{1}{2}(2.34 + 9.10) = 5.72$ which is answer (a).

19.55 Given $\qquad K_1 = \dfrac{[H^+][HA^-]}{[H_2A]} \qquad$ and $\qquad K_2 = \dfrac{[H^+][A^{2-}]}{[HA^-]}$

and $A_{tot} = [H_2A] + [HA^-] + [A^{2-}]$

the following expressions can be derived for the fraction of the three acid species present as a function of $[H^+]$

$$f_{H_2A} = \frac{[H_2A]}{A_{tot}} = \frac{[H^+]^2}{D} \qquad\qquad f_{A^{2-}} = \frac{[A^{2-}]}{A_{tot}} = \frac{K_1K_2}{D}$$

$$f_{HA^-} = \frac{[HA^-]}{A_{tot}} = \frac{K_1[H^+]}{D} \qquad\qquad D = [H^+]^2 + K_1[H^+] + K_1K_2$$

Here $A_{tot} = 1.0 \times 10^{-3}$ M; $[H^+] = 10^{-6.85} = 1.41 \times 10^{-7}$

$D = 1.98 \times 10^{-12}$; $f_{H_2A} = .010$, $[H_2A] = 1.0 \times 10^{-5}$ M,

$f_{HA^-} = .620$, $[HA^-] = 6.2 \times 10^{-4}$; $f_{A^{2-}} = 0.37$, $[A^{2-}] = 3.7 \times 10^{-4}$ M

Answer (a).

This was a complex problem which required some tedious algebra (or a reference to the formulas) for its exact solution. Frequently, however, you can make useful approximations in these problems. In this case, since the pH is much closer to pK_2 than to pK_1 you can assume first that [A] is divided between $[HA^-]$ and $[A^{2-}]$ and that $[H_2A]$ is very small. This reasoning eliminates (c), (d), and (e) as possible answers.

Let $x = [HA^-]$; $0.0010 - x = [A^{2-}]$; $[H^+] = 1.41 \times 10^{-7}$ as above

$$\frac{(1.41 \times 10^{-7})(.0010-x)}{x} = K_2 = 8.40 \times 10^{-8}$$

solving, $x = 6.25 \times 10^{-4} = [HA^-]$; $[A^{2-}] = 3.75 \times 10^{-4}$

At once you can see the answer must be (a). Now you may determine $[H_2A]$ by substituting into K_1.

$$\frac{(1.41 \times 10^{-7})(6.25 \times 10^{-4})}{[H_2A]} = 8.71 \times 10^{-6}$$

$[H_2A] = 1.01 \times 10^{-5}$, as in answer (a).

CHAPTER 20. BUFFERS AND TITRATIONS

20.2 You may consider only that equilibrium involving the two species present in large amounts, HPO_4^{2-} and PO_4^{3-}, that is:

$$K_3 = \frac{[H^+][PO_4^{3-}]}{[HPO_4^{2-}]} = 4.80 \times 10^{-13}$$

Solving for $[H^+]$, $[H^+] = \frac{(4.80 \times 10^{-13})(.928)}{(.586)} = 7.60 \times 10^{-13}$

pH = 12.12, answer (b)

20.3 When NaOH is added, some of the acid HA is converted to its conjugate base A^-, viz.:

$$HA + OH^- \rightleftharpoons H_2O + A^- \tag{1}$$

In one liter of solution initially there are $1.0 \, \ell \times .216 \, \frac{mole}{\ell}$ = 0.216 moles of HA and $1.0 \, \ell \times 0.230 \, \frac{mole}{\ell} = 0.230$ moles of A^-.

As a result of reaction (1) we convert 0.0733 moles of HA was converted to 0.0733 moles of A^-. In the final solution:

moles HA $= 0.216 - 0.0733 = 0.143$ moles; $[HA] = 0.143$

moles A- $= 0.230 + 0.0733 = 0.303$ moles; $[A^-] = 0.303$

$$pK_a = 3.27 = 4.00 - .73$$

310

$$K_a = 5.37 \times 10^{-4}$$

$$K_a = \frac{[H^+][A^-]}{[HA]} \quad ; \quad [H^+] = K_a \frac{[HA]}{[A^-]}$$

$$[H^+] = 5.37 \times 10^{-4} \times \frac{0.143}{0.303} = 2.53 \times 10^{-4}$$

$$pH = 4 - .40 = 3.60 \text{ which is answer (a).}$$

The numerical work in these problems can be simplified by taking logs
of the $[H^+]$ expression above (and changing sign):

$$pH = pK_a - \log \frac{[HA]}{[A^-]}$$

$$pH = pK_a - \log \frac{0.143}{0.303} = pK_a + \log \frac{0.303}{0.143} = 3.27 + 0.33 = 3.60, \text{ answer (a).}$$

20.6 Any solution that contains large quantities of a weak acid and its
conjugate base will serve as a buffer. Examine each choice.
(a) contains H_3A and its conjugate H_2A^-, OK
(b) contains HA^{2-} and its conjugate A^{3-}, OK
(c) after the NaOH has reacted, 0.5 mole of H_2A^- will still remain,
    along with the 0.5 mole of HA^{2-} formed; OK.
(d) After the HCl has reacted, the solution will contain 0.5 mole of
    H_3A plus the 0.5 mole of H_2A^- remaining; OK
(e) the mixture will contain OH^- and A^{3-}, but no conjugate pair; this
    is no buffer, select this answer.

20.7 You must compute the composition of the solution after reaction has
    occurred.

$$\text{moles NH}_3 = \frac{50 \text{ ml}}{1000 \text{ ml } \ell^{-1}} \times \frac{0.40 \text{ mole}}{\ell} = 20 \times 10^{-3} \text{ mole}$$

$$\text{moles HCl} = \text{moles H}^+ = \frac{50 \text{ ml}}{1000 \text{ ml } \ell^{-1}} \times \frac{0.20 \text{ mole}}{\ell} = 10 \times 10^{-3} \text{ mole}$$

When the following reaction is complete:

$$H^+ + NH_3 \rightarrow NH_4^+$$

You will have used all the H^+, formed 10×10^{-3} moles of NH_4^+, and will
have left $20 \times 10^{-3} - 10 \times 10^{-3} = 10 \times 10^{-3}$ moles of NH_3. Since the
total volume is now 100 ml, or 0.100 ℓ, the concentrations are $[NH_4^+]$
= 0.10 M, and $[NH_3] = 0.10$ M.

A mixture of weak base and its conjugate acid is a buffer solution
for which you can write (analogous to the weak acid buffers):

$$[OH^-] = K_b \times \frac{[NH_3]}{[NH_4^+]}$$

$$pOH = pK_b - \log \frac{[NH_3]}{[NH_4^+]}$$

In this problem

$$pOH = 4.75 - \log \frac{0.100}{0.100} = 4.75 - 0 = 4.75$$

$$pH = 14.00 - pOH = 14.00 - 4.75 = 9.25 \text{ which is answer (d).}$$

20.8 To get such a high pH enough NaOH must be added to convert all the H_2A to HA^- and then some of the HA^- to A^{2-}. You must first determine the ratio of HA^- to A^{2-} necessary to obtain the desired pH.

$$pH = pK_2 + \log \frac{[A^{2-}]}{[HA^-]} = 9.95 = 9.65 + \log \frac{[A^{2-}]}{[HA^-]}$$

$$\log \frac{[A^{2-}]}{[HA^-]} = 9.95 - 9.65 = .30$$

$$\frac{[A^{2-}]}{[HA^-]} = \text{antilog } .30 = 2.0 = \frac{\text{moles } A^{2-}}{\text{moles } HA^-}$$

$$\text{Total moles} = \text{orig. moles } H_2A = (1.00)(.6) = 0.600$$

Let y = moles HA^-

$0.600 - y$ = moles A^{2-}

$$\frac{0.600 - y}{y} = 2.0$$

solving for y, $y = 0.200$ moles HA^-

$$.600 - y = 0.400 \text{ moles } A^{2-}$$

Moles NaOH = 0.600 (to convert H_2A to HA^-) + 0.400 (to convert
0.4 moles of HA^- to A^{2-})

= 1.000 moles

g NaOH = 1.00 x 40 = 40 g, which is answer (b).

20.11 At the equivalent point all of the weak base has been converted to the salt $(BH)^+Cl^-$, and one has to solve a typical hydrolysis equilibrium problem:

$$BH^+ \xrightleftharpoons{H_2O} H^+ + B$$

$$K_h = \frac{[H^+][B]}{[BH^+]} = \frac{K_w}{K_b} = \frac{1.0 \times 10^{-14}}{3.47 \times 10^{-6}} = 2.87 \times 10^{-9}$$

$$[H^+] \approx \sqrt{K_h \cdot c_s}$$

To complete the problem, one needs to calculate c_s, the concentration of $(BH)^+Cl^-$ at the equivalence point.

$$\text{volume of HCl} = \frac{(67.0)(0.350)}{(0.250)} = 94.0 \text{ ml.}$$

$$\text{Total volume} = 67.0 + 94.0 = 161.0 \text{ ml.}$$

312

$$\text{Moles (BH)}^+\text{Cl}^- = \text{moles B} = \frac{(67.0)(0.350)}{1000}$$

$$\text{Molarity of (BH)}^+\text{Cl}^- = \frac{(67.0)(0.350)}{1000} \cdot \frac{161.0}{1000}$$

$$= \frac{(67.0)(0.350)}{(161.0)} = 0.146 \ \underline{M}$$

$$[H^+] = \sqrt{K_h \cdot c_a} = \sqrt{2.87 \times 10^{-9} \times 0.146} = 2.05 \times 10^{-5}$$

pH = 4.69, answer (c).

20.12 When a weak acid has been partly neutralized, the solution is a buffer solution, because it contains both the acid and its conjugate base. If you call the acid HA:

$$(1) \qquad HA \rightleftharpoons H^+ + A^-$$

$$(2) \qquad K_a = \frac{[H^+][A^-]}{[HA]}$$

$$(3) \qquad [H^+] = K_a \frac{[HA]}{[A^-]}$$

$$(4) \qquad pH = pK_a - \log\frac{[HA]}{[A^-]}$$

You can use formula (4) to calculate the pH during the course of the titration, or for any buffer solution. In this case, when 57.0% of HA has been converted to A⁻ the ratio:

$$\frac{[HA]}{[A^-]} = \frac{43}{57}$$

and you needn't be concerned about the absolute values of the concentrations.

$$pH = pK_a - \log\frac{43}{57} = -\log(8.87 \times 10^{-8}) - \log\frac{43}{57}$$

$$= -.95 + 8 + .12 = 5.74 - .09$$

$$= 7.17, \text{ which is answer (a).}$$

20.14 First find the pH at the equivalence point. Because the original solution and the titrant were the same concentration, the volume of titrant will also be 50.0 ml and the salt formed Na^+A^- will be diluted to $\frac{0.100}{2} = 0.0500$ M.

$$[OH^-] = \sqrt{K_h \cdot c_s} = \sqrt{\frac{K_w}{K_a} \cdot c_s} = \sqrt{\frac{1.0 \times 10^{-14}}{1.48 \times 10^{-9}} \times 5.00 \times 10^{-2}}$$

$$= \sqrt{3.37 \times 10^{-7}} = \sqrt{33.7 \times 10^{-8}} = 5.8 \times 10^{-4}$$

$$pOH = 4 - 0.76; \quad pH = 14.00 - pH = 10.76$$

An indicator should be chosen such that the midpoint of its range coincides with the pH at the equivalence point. The closest is indicator 11 which is choice (e).

CHAPTER 21. SOLUBILITY

21.4 $Ag_2CO_3 \rightleftharpoons 2\ Ag^+ + CO_3^=$

Let x = molar solubility, then

$$x = [CO_3^=], \text{ and}$$
$$[Ag^+] = 2\ x$$

$$K_{sp} = 8.20 \times 10^{-12} = [Ag^+]^2[CO_3^{2-}] = (2x)^2(x)$$
$$4\ x^3 = 8.20 \times 10^{-12} \qquad x = .00013 \text{ which is answer (c).}$$

21.7 Assume first that the formation of $Tl_2C_2O_4$ is quantitative.

$$2\ Tl^+ \quad + \quad C_2O_4^{2-} \quad \xrightarrow{100\%} \quad Tl_2C_2O_4$$

	$2\ Tl^+$	$C_2O_4^{2-}$	$Tl_2C_2O_4$
Initially (moles)	250×0.040 = 10	750×0.040 = 30	0
After reaction	0	25	5

The concentration of $C_2O_4^{2-}$ after the reaction is 25 mmoles/1000 ml or 0.025 $\underline{M}$. Now let x be the solubility of $Tl_2C_2O_4$ in moles/liter,

$$Tl_2C_2O_4(s) \rightleftharpoons 2\ Tl^+ + C_2O_4^{2-}$$
$$2x \qquad 0.025 + x$$

$$K_{sp} = [Tl^+]^2[C_2O_4^{2-}] = (2x)^2(0.025 + x) = 1.1 \times 10^{-7}$$

Assume x << 0.025. Then

$$x = \left[\frac{1.1 \times 10^{-7}}{4 \times 0.025}\right]^{1/2} , \text{ which is answer (b).}$$

21.9 Let x be the solubility of $Co(OH)_3$. Then

$$Co(OH)_3(s) \rightleftharpoons Co^{3+} + 3\ OH^-$$
$$x \qquad 3x$$

$$K_{sp} = [Co^{2+}][OH^-]^3 = (x)(3x)^3 = 2.5 \times 10^{-43}$$
$$x = 9.8 \times 10^{-12}$$
$$[OH]^- = 3x = 2.9 \times 10^{-11}\ \underline{M}$$

This concentration of hydroxide ion is negligible compared to the $[OH^-]$ from the dissociation of H_2O.

$$H_2O \rightleftharpoons H^+ + OH^-; \quad [OH^-] = [H^+] = 1.0 \times 10^{-7}. \text{ The answer is (e).}$$

21.10 When the ion product exceeds K_{sp} then CaF_2 will precipitate. By
 setting the ion product equal to K_{sp}, you can calculate the maximum
 concentration of fluoride ions that can coexist with the calcium ions
 without precipitation.

$$K_{sp} = 3.9 \times 10^{-11} = [Ca^{++}][F^-]^2 = 5.6 \times 10^{-4} [F^-]^2$$

$$[F^-]^2 = 3.9 \times 10^{-11}/5.6 \times 10^{-4} = 6.96 \times 10^{-8}$$

$$[F^-] = 2.64 \times 10^{-4} = [NaF]$$

mass of $[NaF] = 2.64 \times 10^{-4}$ mole $\ell^{-1} \times 42.0$ g mole$^{-1} = 1.1 \times 10^{-2}$ g ℓ^{-1}

The answer is (e).

Some common errors are:

(a) the student failed to square $[F^-]$

(b) the student used $(2[F^-])^2$ rather than $[F^-]^2$

(c) the student used $2[F^-]$ rather than $[F^-]^2$

(d) the decimal point slipped and the student took the square root of
 69.6 instead of 6.96.

21.11 The concentration of Ba^{2+} required is determined by the relationship:
 $[Ba^{2+}][SO_4^{2-}] = K_{sp} = 1.5 \times 10^{-9}$. The biggest part of the problem is to
 determine $[SO_4^{2-}]$ from the specifications given.

$$\text{moles } SO_4^{2-} = \text{moles } S = \frac{2.6 \times 10^{-5} \text{ g}}{32 \text{ g mole}^{-1}} = 8.13 \times 10^{-7} \text{ mole}$$

$$[SO_4^{2-}] = \frac{8.13 \times 10^{-7} \text{ mole}}{0.250 \ \ell} = 3.25 \times 10^{-6} \text{ M}$$

$$K_{sp} = 1.5 \times 10^{-9} = [Ba^{2+}][3.25 \times 10^{-6}]$$

$$[Ba^{2+}] = 4.6 \times 10^{-4} \text{ , answer (e)}$$

Some common errors are:

(b) the student used the molecular weight of $BaSO_4$ instead of S.
(c) the student used the molecular weight of $SO_4^=$ instead of S.

21.12 The $[Ag^+]$ in equilibrium with the excess CO_3^{2-} will be determined by
 the K_{sp}.

$$[CO_3^{2-}][Ag^+] = 8.2 \times 10^{-12}$$

$$[Ag^+] = \left[\frac{8.2 \times 10^{-12}}{2.5 \times 10^{-2}}\right]^{1/2} = 1.81 \times 10^{-5}$$

The difference between this value and the original $[Ag^+]$ must have
precipitated as Ag_2CO_3.

$$2.65 \times 10^{-5} - 1.81 \times 10^{-5} = 0.84 \times 10^{-5} \text{ M } Ag^+ \text{ precipitated}$$

$$0.84 \times 10^{-5} \text{ mole } \ell^{-1} \times 107.9 \text{ g mole}^{-1} = 0.91 \times 10^{-3} \text{ g per liter,}$$

answer a.

21.13 Since the moles of Ag^+ exceeded the sum of the Cl^- and I^-, they will be practically totally precipitated.

$$\text{moles AgCl} = 0.0045$$
$$\text{moles AgI} = 0.0065$$

Total moles of Ag^+ precipitated = sum of above = 0.011 moles
moles Ag^+ remaining in solution = 0.015 - 0.011 = 0.004

$$[Ag^+] = \frac{0.004 \text{ moles}}{1.0 \text{ liter}} = 0.004 \text{ M} = 4.0 \times 10^{-3} \text{ M}$$

The concentration of the halide ion in solution will be determined by the $[Ag^+]$ above in conjunction with the respective solubility products and will be independent of the quantities of solid halides.

$$[Cl^-] = \frac{1.7 \times 10^{-10}}{4.0 \times 10^{-3}} = 4.3 \times 10^{-8} \text{ which is answer (d).}$$

The errors are:

(a) no, the ratio is determined only by the relative K_{sp}'s.

(b) no, the $[I^-]$ is 2.1×10^{-14} but $[Cl^-]$ is six orders of magnitude greater.

(c) no, the ratio is $\dfrac{1.7 \times 10^{-10}}{8.5 \times 10^{-17}}$ but the quantities of solid do not matter.

(e) no, this is the $[Cl^-]$ in a mixture of pure water and pure AgCl without the common ion, Ag^+, present.

A variation of this problem can be generated in which the quantity of Ag^+ is less than the total iodide and chloride, but more than the iodide. In that case practically all the iodide will precipitate, some of the chloride will precipitate with the remaining Ag^+, and the excess chloride ion will remain in solution. The residual $[Ag^+]$ is then determined by the excess $[Cl^-]$ and the K_{sp} of AgCl. In another variation of the problem, the quantity of Ag^+ is less than the quantity of I^-, so that the excess I^-, and all the Cl^- will remain in solution. In this case, the residual $[Ag^+]$ is determined by the excess $[I^-]$ and the K_{sp} of AgI.

21.14 Since the solution and solids are in equilibrium, the $[Ba^{2+}]$ must be determinable from the K_{sp} for $BaSO_4$.

$$[Ba^{2+}] = K_{sp}/[SO_4^{2-}] = 1.0 \times 10^{-10}/1.0 \times 10^{-2}$$

Also, $[CO_4^{2-}] = K_{sp}/[Ba^{2+}] = 8.0 \times 10^{-9}/1.0 \times 10^{-10}/1.0 \times 10^{-2}$

$$= \frac{8.0 \times 10^{-9} \times 1.0 \times 10^{-2}}{1.0 \times 10^{-10}}$$

Since the above choice is not given the answer is (e).

21.15 The success of the titration depends on the fact that the $[Cl^-]$ must be reduced to a very small value before the red color which signals the end point persists, as shown in this calculation. When the "first bit" of Ag_2CrO_4 forms the $[CrO_4^{2-}]$ is still essentially 0.0050 M.

$$K_{sp} = 1.9 \times 10^{-12} = [Ag^+]^2[CrO_4^{2-}] = [Ag^+]^2 (0.0050)$$

$$[Ag^+]^2 = 3.8 \times 10^{-10}; \quad [Ag^+] = 1.95 \times 10^{-5}$$

$$K_{sp} = 1.7 \times 10^{-10} = [Ag^+][Cl^-] = (1.95 \times 10^{-5})[Cl^-]$$

$$[Cl^-] = 1.7 \times 10^{-10}/1.95 \times 10^{-5} = 8.7 \times 10^{-6}, \text{ answer (e)}$$

21.16 The solubility product for $Fe(OH)_2$ is less than that for $Mn(OH)_2$ so the Fe^{2+} will precipitate first. $Mn(OH)_2$ will precipitate when the product

$$Q = [Mn^{2+}][OH^-]^2 = 4.5 \times 10^{-14}$$

Since $[Mn^{2+}] = 0.020$ $\underline{M}$, precipitation of $Mn(OH)_2$ will begin when

$$[OH^-] = [4.5 \times 10^{-14}/.020]^{1/2} = 1.5 \times 10^{-6} \underline{M}$$

$$pOH = 5.82; \quad pH = 8.18, \text{ answer c.}$$

21.17 Continuing the above problem, the concentration of Fe^{2+} in solution when $[OH^-] = 1.5 \times 10^{-6}$ $\underline{M}$ is:

$$[Fe^{2+}] = 2.0 \times 10^{-15}/(1.5 \times 10^{-6})^2 = 8.9 \times 10^{-4}$$

The answer is option (d).

21.20 The AgCN is much more soluble in a solution buffered on the acid side than in water because the common ion CN^- is removed as HCN. Two equilibria are involved:

$$AgCN(s) \rightleftharpoons Ag^+ + CN^- \qquad K_{sp} = [Ag^+][CN^-] = 1.58 \times 10^{-14}$$

$$HCN \rightleftharpoons H^+ + CN^- \qquad K_a = \frac{[H^+][CN^-]}{[HCN]} = 2.09 \times 10^{-9}$$

Let x = solubility of AgCN

Then $[Ag^+] = x$

and total cyanide $= [CN^-] + [HCN] = x$

but $[HCN] = \dfrac{[H^+][CN^-]}{K_a}$

so $\quad x = [CN^-] + \dfrac{[H^+][CN^-]}{K_a} \qquad\qquad x = [CN^-]\left[1 + \dfrac{[H^+]}{K_a}\right]$

$$[CN^-] = x\left[1 + \frac{[H^+]}{K_a}\right]^{-1}$$

and $K_{sp} = [Ag^+][CN^-] = x^2\left[1 + \dfrac{[H^+]}{K_a}\right]^{-1}$

Solving , $x = \sqrt{K_{sp}\left[1 + \dfrac{[H^+]}{K_a}\right]}$

At pH = 6.48, $[H^+] = 3.32 \times 10^{-7}$

$$x = \sqrt{(1.58 \times 10^{-14})\left(1 + \dfrac{3.32 \times 10^{-7}}{2.09 \times 10^{-9}}\right)} = \sqrt{(1.58 \times 10^{-14})(1 + 159)}$$

x = solubility = 1.6×10^{-6} which is answer (c).

The above solution to the problem is general and can be used at any pH. A rapid approximation can be made for an acid solution, realizing that practically all the cyanide is tied up as HCN, so that $[Ag^+] \cong [HCN]$.

Let x = solubility = $[Ag^+] \cong [HCN]$

$$(x)[CN^-] = K_{sp} = 1.58 \times 10^{-14}; \quad [CN^-] = \dfrac{1.58 \times 10^{-14}}{x}$$

$$K_a = \dfrac{[H^+][CN^-]}{[HCN]} = 2.09 \times 10^{-9} = \dfrac{(3.32 \times 10^{-7})(1.58 \times 10^{-14})}{x^2}$$

$$x = \dfrac{(3.32 \times 10^{-7})(1.58 \times 10^{-14})}{(2.09 \times 10^{-9})} = 1.6 \times 10^{-6}, \text{ answer c.}$$

CHAPTER 22. THERMODYNAMICS

22.4 The first law defines the change in internal energy ΔE in terms of q, the heat absorbed and w, the work done by the system:

$$\Delta E = q - w$$

Since the volume is constant in this experiment no work can be done,   w=0, and q= -176 (heat was evolved, not absorbed, hence the minus sign)

$$\Delta E = -176 \text{ kJ}$$

The enthalpy change, ΔH, is defined:

$$\Delta H = \Delta E + P\Delta V$$

Only the gaseous reactants and products occupy much volume and contribute substantially to ΔV. Assuming they are ideal gases obeying $PV = nRT$

$$P\Delta V = \Delta n\, RT$$

In this reaction, one mole of gaseous reactants produces no gaseous products so $\Delta n = -1$ mole

$$P\Delta V = (-1 \text{ mole})(8.31 \text{ J mole}^{-1}K^{-1})(298 \text{ K})\dfrac{1 \text{ kJ}}{1000 \text{ J}}$$

$$= -2.5 \text{ J}$$

$$\Delta H = -176 - 2.5 = -178, \text{ answer a.}$$

22.7 Since entropy is associated with randomness it will usually be greatest for gases, next for dilute solutions, next for liquids, and least for crystalline solids.

Applying these principles to this problem you should select reaction (e) as probably showing the least change since there is one mole of gas on each side, and the other substances are solids of low entropy. In this problem, since numerical values, are not given, you have to make estimates and cannot be absolutely sure of your answer. Comparing the others:

(a) Since 2 moles of gas disappear, there is probably a large decrease in entropy (offset partly by the $4H^+$).
(b) Since 3 moles of gas appear there is unquestionably a large increase in entropy.
(c) Since 3 moles of gaseous products replace 1 mole of gaseous reactant, there is unquestionably a large increase in entropy.
(d) One mole of liquid appears among the products which probably accounts for an increase in entropy; the solids are low in entropy and there is a net of only one extra mole of solid among reactants.

22.10 For the reaction (left to right)

$$\Delta G^\circ = \Delta H^\circ - T\Delta S^\circ$$
$$= -113.47 - (-114.50) - \frac{298}{1000}(178.29 - 175.03)$$
$$= 1.03 - .2980(3.26) = +.058$$

Since ΔG° is positive, the reaction is <u>not</u> spontaneous. A mixture of standard states will tend to form reactants rather than products. Thus, the reactant, A, is the more stable isomer (reject answers d and e).

In order for an equimolar mixture to be in equilibrium ΔG° must be zero. Assuming ΔH° and ΔS° are constant: $\Delta G^\circ = 0 = 1.03 - (T/1000)(3.26)$.
$$T = \frac{1.03(1000)}{3.26} = 315 \text{ K, answer (c)}$$

22.13 Assuming that the standard heat of vaporization is constant over the temperature range, $86 < T < 148°C$,

$$\log \frac{p_1}{p_2} = \frac{\Delta H^\circ_{vap}}{2.303R} \left[\frac{T_1 - T_2}{T_1 T_2}\right]$$

Here $p_1 = ?$, $p_2 = 760$ torr, $\Delta H^\circ_{vap} = 23.0$ kJ/mole, $R = 8.314$ J/mole-K, $T_2 = 148 + 273$ and T_1 is $86 + 273$ K. This gives $p_1 = 244$ torr, ans. (c).

22.16 If it is assumed that ΔH and ΔS are constant throughout the range, the following explicit formula may be derived:

$$\log \frac{K_1}{K_2} = \frac{\Delta H}{2.303 \ R} \left[\frac{T_1 - T_2}{T_2 T_1} \right]$$

Define conditions 1 and 2 as follows:

$$\text{Condition 1:} \quad K_1 = 12100 = 1.21 \times 10^4 \qquad T_1 = 461 \ K$$
$$\text{Condition 2:} \quad K_2 = K_2 \qquad T_2 = 450 \ K$$

$$\log \frac{1.21 \times 10^4}{K_2} = \frac{-4850 \ J \ mole^{-1}}{(2.303)(8.31 \ cal \ mole^{-1}K^{-1})} \times \frac{11 \ K}{(461)(450)K^2}$$

$$\log \frac{1.21 \times 10^4}{K_2} = -.0134$$

$$\log \frac{K_2}{1.21 \times 10^4} = +.0134$$

$$\frac{K_2}{1.21 \times 10^4} = 1.03 \qquad K_2 = 1.25 \times 10^4, \qquad \text{answer (a)}$$

In a problem like this it is easy to make an error in sign or in arithmetic, so you should consider whether your answer makes sense.

According to Le Chatelier's principle, for an exothermic reaction (ΔH negative) lowering the temperature should drive it to the right, i.e. increase K_{eq}. This is what was calculated. Furthermore, since ΔH was fairly small, the change in K_{eq} was not great.

22.19 $\qquad H_2O \ \rightleftharpoons \ H^+ \ + \ OH^- \qquad\qquad \Delta H = 55.90 \ kJ$

Given $K_{eq} = 1.0 \times 10^{-14}$ at 25°C and assuming ΔH is constant over the temperature range 25–50°C,

$$\log \frac{K_{eq}}{1.0 \times 10^{-14}} = \frac{(55.90)(1000)}{(8.31)(2.303)} \left[\frac{323 - 298}{323 \times 298} \right] = 0.758$$

$$K_{eq} = 5.7 \times 10^{-14}; \ pK_{eq} = 13.24$$

In neutral water $pH = pOH = pK_{eq}/2 = 6.62$

The answer is option (e).

22.20 (a) This statement is true for entropy as it is for every "state function" in thermodynamics.

(b) For an endothermic reaction $\Delta H°$ is positive. $\Delta G° = \Delta H° - T\Delta S°$ If $\Delta S°$ is negative, then $(-T\Delta S°)$ is positive and $\Delta G°$ must be positive at all temperatures. The reaction will never be spontaneous, which requires a negative $\Delta G°$. Mark answer (b). Answers (c) (d) and (e) are all true.

CHAPTER 23.   CHEMICAL EQUILIBRIUM

23.5 Usually one first converts the information in moles and liters to pressures in atmospheres in order to compute K_p. In this particular case the task is simplified because $\Delta n_g = 0$ and $K_c = K_p$.

At equilibrium moles X_2 = moles $Y_2 = \dfrac{4.72 - 4.40}{2} = .165$

$$K_p = K_c = \frac{[X_2][Y_2]}{[XY]^2} = \frac{(.165/2.89)(.165/2.89)}{(4.40/2.89)^2} = \left(\frac{.165}{4.41}\right)^2$$

$$= 1.41 \times 10^{-3} \text{ which is answer (c).}$$

23.6 Let x be the equilibrium concentrations of B in moles per liter, then

$$A(g) \rightleftharpoons B(g) + C(g) \qquad\qquad K_c$$

	A	B	C
Initially (M)	$\dfrac{67.5}{140}$	0	0
At equilibrium	$\dfrac{67.5}{140} - x$	x	x

$$K_c = [B][C]/[A] = x^2/(0.482-x) = 1.00$$

The approximation $x \ll 0.482$ fails here so the quadratic equation must be used.

$$x^2 + x - 0.482 = 0$$

$$x = (-1 + \sqrt{1 + 4 \times 0.482})/2 = 0.356 \text{ M}$$

So $[A] = 0.482 - 0.356 = 0.126$ M
The answer is option (c).

23.8 If there were no decomposition of NH_3, the pressure at 230°C would be, according to the ideal gas laws:

$$P_{NH_3} = 2.10 \times \frac{503}{298} = 3.54 \text{ atm}$$

Let x = the pressure of NH_3 which decomposed, and tabulate the initial and equilibrium pressures.

	$2\ NH_3$	$\rightleftharpoons$	N_2	+	$3\ H_2$
Initial	3.54		0		0
Change	$-x$		$+\dfrac{x}{2}$		$+\dfrac{3x}{2}$
Equilibrium	$3.54 - x$		$\dfrac{x}{2}$		$\dfrac{3x}{2}$

Total pressure $= 4.25 = 3.54 - x + \dfrac{x}{2} + \dfrac{3x}{2} = 3.54 + x$

$$x = 0.71$$

$$P_{NH_3} = 3.54 - x = 2.83; \quad P_{N_2} = \frac{x}{2} = 0.355; \quad P_{H_2} = \frac{3x}{2} = 1.065$$

$$K_p = \frac{(.355)(1.065)^3}{(2.83)^2} = 0.054, \text{ answer (d).}$$

23.9 Set up a table in terms of x, which represents the atmospheres of PCl_5 which decomposed.

$$PCl_5 \rightleftharpoons PCl_3 + Cl_2$$

	PCl_5	PCl_3	Cl_2
Initial	2.75	0	0
Change	$-x$	$+x$	$+x$
Equil.	$2.75-x$	x	x

$$K_p = 1.81 = \frac{x^2}{2.75-x}$$

solving the quadratic, $x = 1.50$ atm $= P_{PCl_3} = P_{Cl_2}$

$$P_{PCl_5} = 2.75 - 1.50 = 1.25$$

Total pressure $= 1.25 + 1.50 + 1.50 = 4.25$, answer (a).

It is wise to check your algebra and numerical work by calculating K_p from your answers.

$$K_p = \frac{(1.50)(1.50)}{(1.25)} = 1.81 \quad OK$$

23.12 Let f be the fraction NO_2 in the N_2O_4-NO_2 mixture. Then the "average molecular weight" of the gas mixture is

$$M_{avg} = f \times 46 + (1-f)\,92$$

and M_{avg} is related to the observed density (assuming ideal behavior) by

$$M_{avg} = \frac{\rho RT}{P} = \frac{(2.507)(0.0821)(323)}{1} = 66.48 \text{ g/mole}$$

Then $66.48 = 46\,f + (1-f)92$

$$\text{or} \quad f = 0.555$$

$$\text{and} \quad 1-f = 0.445$$

Then $P_{NO_2} = fP_{tot} = 0.555$ atm

and $P_{N_2O_4} = (1-f)P_{tot} = 0.445$ atm

so $K_p = \dfrac{P_{NO_2}^2}{P_{N_2O_4}} = \dfrac{(.555)^2}{.445} = 0.692$

The answer is option (b).

23.15 The equilibrium partial pressure of CO in the initial equilibrium system is

$$K_p = \frac{P_{CO}P_{H_2O}}{P_{CO_2}P_{H_2}} = P_{CO} = \frac{(2)(5.75)(6.25)}{5.18} = 13.88 \text{ atm}$$

For the equilibrium partial pressure of H_2 to increase to 10.0 atm, the reaction must be shifted to the left to an extent 10.0–6.25 or 3.75 atm. Then

$$P_{H_2} = 10.0 \text{ atm} \qquad P_{CO_2} = 5.75+3.75 \quad \text{and} \quad P_{CO} = 13.88-3.75$$

Then

$$P_{H_2O} = K_p P_{CO_2} P_{H_2}/P_{CO} = (2)(5.75+3.75)(10.0)/(13.88-3.75) = 18.8 \text{ atm}$$

The answer is (e).

CHAPTER 24. ELECTROCHEMISTRY

24.4 The standard free energy change $\Delta G°$ is the maximum work obtainable from a reaction (all components present in the standard state). It can be shown that this is also the electrical work obtainable when the reaction occurs in a galvanic cell. The overall reaction is written below and $\Delta G°$ is evaluated from the ΔG_f's.

$$CH_3OH + H_2O_2 \rightarrow CH_2O + 2 H_2O$$

$$\Delta G° = 2(-237) - 132 + 175 + 132 = -297, \text{ answer (d)}.$$

Answer (e) is wrong. A negative $\Delta G°$ indicates a spontaneous reaction. The $\Delta H_f°$ and $S°$ values are not required since $\Delta G_f°$'s are given.

24.5 I^- and H_3IO_6 cannot disproportionate being the most highly reduced and most highly oxidized forms, respectively, of the elements iodine.

I_2 and OI^- do disproportionate:

$$I_2 \rightarrow OI^- + I^- \qquad E°_{cell} = 0.535-0.45 \text{ volt} \qquad (>0)$$

$$IO^- \rightarrow IO_3^- + I_2 \qquad E°_{cell} = 0.45-0.14 \text{ volt} \qquad (>0)$$

IO_3^- does not disproportionate:

$$IO_3^- \rightarrow H_3IO_6^{2-} + OI^- \quad E°_{cell} = -0.7+0.14 \text{ volt} \qquad (<0)$$

The answer is (b).

24.7 (a) is wrong. The more powerful oxidizing agents have more positive reduction potentials. The reverse of (a) is true.
   (b) is wrong. M^{2+} is not an oxidizing agent; it is the reduced form of the couple.
   (c) is wrong, as in (b).
   (d) is true, since M^{3+} will not oxidize H_2O (see (a)).
   (e) is wrong. The reverse is true because the better reducing agent has the more negative potential.

24.9 Since the two sides are the same chemically $E°$ is zero and the only voltage arises from the Nernst correction term; its magnitude is

$$\frac{.0592}{2} \log \frac{.45}{.25} = \frac{(.0592)(.255)}{(2)} = 0.0076 \text{ V, answer (a)}.$$

Note that Hg is more likely to dissolve on the left side where there is less Hg^{2+} product present. Thus, the left side is the anode, is the source of electrons, and must be connected to the negative binding post of the voltmeter. The standard notation for writing cells calls for the anode on the left.

24.12 The free energy change is related to the cell potential by:

$$\Delta G = -nFE$$

where n is the no. of electrons transferred in the balanced equation and F is the faraday. For ΔG in kJ and E in volts:

$$\Delta G = -96.5 \text{ nE}$$

If all materials are in the standard states, you obtain the standard free energy change $\Delta G°$.

$$\Delta G° = -96.5 \text{ nE}°$$

If you combine this relation with $\Delta G° = -2.3 \text{ RT} \log K_{eq}$, you obtain (inserting the proper value of R, and T = 298 K)

$$\log K_{eq} = \frac{96.5 \text{ nE}°}{(2.303)(8.314)(298)} = \frac{nE°}{0.0592}$$

The constant $0.0592 = \frac{RT}{F}$ at 298 K is frequently encountered in electrochemistry.

In this problem, $E° = -.758 - (-.248) = -.510$ for the reaction <u>as written</u> (not the direction in which it will go spontaneously). Whence:

$$\log K_{eq} = \frac{(1)(-.510)}{0.0592} = -8.61, \text{ which is answer (e)}.$$

24.13 This is a problem involving the Nernst equation in which the unknown quantity is $[H^+]$. The pH meter uses other electrodes but otherwise operates on this principle. Write the overall reaction following the convention that the left side is the anode.

$$Hg_2Cl_2 + 2 e^- \rightarrow 2 Hg(\ell) + 2 Cl \qquad\qquad +0.24 \text{ V}$$

$$H_2 \rightarrow 2 H^+ + 2 e^- \qquad\qquad +0.00 \text{ V}$$

$$\overline{Hg_2Cl_2 + H_2 \rightarrow 2 Hg(\ell) + 2 H^+ + 2Cl^- \qquad E° = 0.24 \text{ V}}$$

$$E = 0.37 = 0.24 - \frac{.0592}{2} \log \frac{[H^+]^2 [Cl^-]^2}{P_{H_2}}$$

Note: The activities of Hg_2Cl_2 and $Hg(\ell)$, assumed to be pure solid and pure liquid, are unity and were omitted. In this problem P_{H_2} and $[Cl^-]$ are also unity so:

$$0.37 \; = \; 0.24 - \frac{.0592}{2} \log [H^+]^2 = 0.24 - .0592 \log [H^+]$$

$$-\log [H^+] \; = \; pH = \frac{0.37-0.24}{.0592}, \text{ answer a.}$$

24.14 By subtracting the first reaction from the second, the Co and the $3e^-$ are eliminated and the dissociation reaction is obtained. For the overall reaction $E° = -0.13 - 1.26 = -1.39$ Volts

$$\Delta G° \; = \; -96.5 \; nE° \; = \; -2.303 \; RT \log K_{eq}$$

$$\log K_{eq} \; = \; \frac{nE°}{.0592} = \frac{(3)(-1.39)}{.0592} \text{ , answer d}$$

24.16 The balanced equation and the corresponding Nernst equation are

$$Sn^{2+} + Tl^{3+} \; \rightleftharpoons \; Sn^{4+} + Tl^+$$

$$E = E° - \frac{.0592}{n} \log Q, \quad Q = \frac{[Sn^{4+}][Tl^+]}{[Sn^{2+}][Tl^{3+}]}$$

$$= \; 1.25 - 0.15 - \frac{.0592}{2} \log \frac{0.90 \times 0.02}{0.009 \times 0.001}$$

$$= \; 1.10 - .098 \; = \; 1.00 \text{ volt. The answer is (d).}$$

24.17 You'll notice that the desired reaction is the algebraic sum of the two given reactions. This suggests that Hess's Law can be used, but you must be careful to add the correct quantities, namely energies, not voltages. Since $\Delta G = -nFE$ you must multipy the $E°$ values by n in each case to get values proportional to energy which may be added according to Hess' Law.

			$E°$	$nE°$
$M^{3+} + e^-$	$\rightarrow$	M^{2+}	0.108	0.108
$M^{5+} + 2e^-$	$\rightarrow$	M^{3+}	-0.444	-0.888
$M^{5+} + 3e^-$	$\rightarrow$	M^{2+}		-0.780

Since $-0.780 = 3E°$ for the summed reaction, $E° = -0.260$ which is answer (c).

24.18 The Nernst equation is applicable to a half-reaction just as well as to a complete reaction. In neutral solution $[H^+] = 1.0 \times 10^{-7}$, $[H_2O_2]$ in the standard state is 1.0 M, and the activity of water is unity.

$$E = E° - \frac{.0592}{n} \log \frac{1}{[H_2O_2][H^+]^2}$$

$$E = 1.77 - \frac{.0592}{2} \log \frac{1}{(1.0 \times 10^{-7})^2}$$

$$= 1.77 + .0592 \log (1.0 \times 10^{-7})$$

$$= 1.77 - (.0592 \times 7) \qquad \text{answer b.}$$

24.19 The cell reaction, standard cell potential, and equilibrium constant are:

$$5\ Fe^{2+} + 8H^+ + MnO_4^- \rightleftharpoons 5\ Fe^{3+} + Mn^{2+} + 4\ H_2O \qquad E° = 0.75 \text{ volt}$$

$$K_{eq} = \frac{[Fe^{3+}]^5[Mn^{2+}]}{[Fe^{2+}]^5[MnO_4^-][H^+]^3} = \text{antilog}\left(\frac{(5)(.75)}{0.0592}\right) = 2.2 \times 10^{63}$$

Since the equilibrium constant is so large, assume the reaction is quantitative, and then consider the back reaction. Let x be the extent of the back reaction in moles/liter.

$$5\ Fe^{3+} + Mn^{2+} + 4\ H_2O \rightleftharpoons 5\ Fe^{2+} + 8\ H^+ + MnO_4^-$$

$$0.8\text{-}5x \qquad \frac{0.8}{5}-x \qquad - \qquad\qquad 5x \qquad\qquad - \qquad x$$

$$K'_{eq} = \frac{1}{K_{eq}} = \frac{(5x)^5(x)(1)^8}{(0.8-5x)^5(\frac{0.8}{5}-x)} = 4.5 \times 10^{-64}$$

The $[Mn^{2+}]$ is $\frac{0.8}{5}-x$ because a stoichiometric quantity of MnO_4^- was added to the initial Fe^{2+} solution. Since K'_{eq} is very small, assume $x \ll 0.8/5$. Then

$$\frac{5^6x^6}{(0.8)^6} = 4.5 \times 10^{-64}$$

$$\frac{5x}{0.8} = [4.5 \times 10^{-64}]^{1/6} = 2.8 \times 10^{-11}$$

$$5x = [Fe^{2+}] = 0.8 \times 2.8 \times 10^{-11} = 2.2 \times 10^{-11}\ \underline{M}$$

The assumption is excellent. The answer is (b).

The following is an alternative solution.
At the equivalence point, the solution will contain 0.80 M Fe^{3+} and a very small concentration ω of Fe^{2+}. The concentration of Mn^{2+} formed will be 0.80/5 = 0.16 M and there will be a very small residual concentration of MnO_4^-, stoichiometrically equivalent to the residual Fe^{2+}, that is $\omega/5$. In terms of these quantities, you can calculate the standard potential of the mixture from the point of view of either iron or manganese; you must get the same value either way assuming the mixture is in equilibrium.

$$E = 0.77 - \frac{.0592}{1} \log \frac{[Fe^{2+}]}{[Fe^{3+}]} = 1.52 - \frac{0.592}{5} \log \frac{[Mn^{2+}]}{[MnO_4^-][H^+]^8}$$

Substituting the values of the concentrations at equivalence (and $[H^+] = 1.0$):

$$0.77 - .0592 \log \frac{\omega}{.80} = 1.52 - \frac{.0592}{5} \log \frac{.16 \times 5}{\omega}$$

$$-0.592 \log \frac{\omega}{.80} - \frac{.0592}{5} \log \frac{\omega}{.80} = 1.52 - 0.77 = 0.75$$

$$\frac{6(.0592)}{5} \log \frac{.80}{\omega} = 0.75$$

$$\frac{.80}{\omega} = \text{antilog } 10.55 = 3.5 \times 10^{10}$$

$$\omega = 2.3 \times 10^{-11} \quad \text{answer (b)}$$

One can derive a useful result from the above:

$$E = 0.77 - .0592 \log \frac{\omega}{.80}$$

$$5E = 5(1.52) - .0592 \log \frac{.80}{\omega} = 5(1.52) + .0592 \log \frac{\omega}{.80}$$

Adding
$$6E = 0.77 + 5(1.52)$$
$$E = \frac{0.77 + 5(1.52)}{6} = 1.39 \text{ V}$$

The above result can be generalized: in a redox titration, the voltage at equivalence is the weighted average of the potentials, the weighting factors being the respective n's.

24.20 The objective here is to use the five observations to rank the six half reactions in increasing (or decreasing) order of oxidation (or reduction) potential. In this example, the half reactions will be ordered according to decreasing reduction potential since the half-reactions are written as reductions. From observation 1:

$$HNO_2 \rightarrow N_2O$$

$$H_3PO_3 \rightarrow H_3PO_2$$

That is the reduction potential for the HNO_2-N_2O couple will be greater than the reduction potential of the H_3PO_3-H_3PO_2 couple. From observation 2)

$$HNO_2 \rightarrow N_2O$$

$$H_3PO_3 \rightarrow H_3PO_2$$

$$U^{4+} \rightarrow U^{3+}$$

The final tabulation is

$$Mn^{3+} \rightarrow Mn^{2+}$$
$$HNO_2 \rightarrow N_2O$$
$$Ag^+ \rightarrow Ag$$
$$Pb^{2+} \rightarrow Pb$$
$$H_3PO_3 \rightarrow H_3PO_2$$
$$U^{4+} \rightarrow U^{3+}$$

Mn^{3+} is the strongest oxidizing agent and U^{3+} is the strongest reducing agent. The correct order of decreasing strength of the metallic species as reducing agents is

$$U^{3+} > Pb > Ag > Mn^{2+}$$

The answer is option (b).

CHAPTER 25. KINETICS

25.5 It can be shown that in the case of a first order reaction:
$$\log \frac{[AB]_o}{[AB]} = \frac{kt}{2.3}$$
where $[AB]_o$ is the initial concentration of AB when time $t = 0$, and $[AB]$ is the concentration at time $t = t$. When 72.8% of AB has decomposed 27.2% remains, and the ratio $[AB]_o/[AB]$ must be $100/27.2$. Solving for t

$$t = \frac{2.3}{k} \log \frac{100}{27.2} = \frac{(2.3)(.566)}{58.7 \text{ sec}^{-1}} = 2.22 \times 10^{-2} \text{ sec, answer (b)}$$

25.7 The Arrhenius Law applied to kinetic data at two temperatures has the form:
$$\log \frac{k_1}{k_2} = \frac{E_a}{2.3 \, R} \left[\frac{T_1 - T_2}{T_1 T_2} \right]$$

If you let $T_1 = 474$, $T_2 = 438$, then $k_1 = 24 \, k_2$,

$$\log 24 = \frac{E_a \, (474-438) \text{ K}}{(2.3)(8.31 \text{ cal mole}^{-1} \text{ K}^{-1})(474)(438)\text{K}^2}$$

$$E_a = \frac{(1.38)(2.3)(8.31)(474)(438)}{(36)}$$

$$= 152,100 \text{ J mole}^{-1} \text{ or } 152.1 \text{ kJ mole}^{-1}, \text{ which is answer (a).}$$

25.8 In this problem the same relationship is used as in the previous one, but the unknown quantity is different.

Let $T_1 = $ unknown, $T_2 = 619$

then $k_1 = 3.00 \, k_2$

$$\log 3.00 = \frac{(68,600)(T_1 - 619)}{(2.3)(8.31)(619 \, T_1)}$$

solving, $T_1 = 675$, which is answer (b).

25.9 It is assumed that the equation, $k = Ae^{-E_a/RT}$ holds over this
temperature range. Then,

$$\log \frac{k_1}{k_2} = \frac{E_a}{2.303 \ R} \left[\frac{T_1-T_2}{T_1 T_2} \right]$$

$$E_a = \frac{(2.303)(8.31)(273+45)(273+100) \ \log\left(\frac{2}{6.2 \times 10^{-4}}\right)}{(273+100) - (273+45)} = 103{,}800 \text{ J}$$

$$E_a = 103.8 \text{ kJ}$$

If $k = Ae^{-E_a/RT}$ then $A = ke^{E_a/RT} = k\exp\left(\frac{E_a}{RT}\right)$

Using $k = 6.2 \times 10^{-4}$ sec^{-1} and $T = 45 + 273 = 318$ K

$$A = 6.2 \times 10^{-4} \ \exp\left(\frac{103{,}800}{8.31 \ \times \ 318}\right)$$

$$= 6.4 \times 10^{13} \ sec^{-1}$$

Alternatively, $A = 0.2 \ \exp\left(\frac{103{,}800}{8.31 \ \times \ 373}\right)$

$$= 6.4 \times 10^{13} \ sec^{-1}$$

The answer is (a). Some numberical discrepancy arises from round-off
errors and too few significant figures, expecially with exponentials.

25.10 The rate law for the reaction

$$H_2O_2 + 2 \ H^+ + 3 \ I^- \rightarrow I_2 + 2 \ H_2O$$

is $\quad$ Rate $= k[H_2O_2][I^-]$

Therefore, the rate-determining step of a mechanism consistent with
this rate law should have a rate proportional to $[H_2O_2]$ and $[I^-]$.
For option (a) to be correct, the rate law would have to be
Rate $= k_a[H_2O_2][H^+]^2[I^-]^2$. Option (b) is correct because the slow
(rate-determining step) has the rate law $\quad$ Rate $= k_b[H_2O_2][I^-]$. This
is the same as the observed rate law. Mechanisms (c) and (d) do not
give this rate law.

CHAPTER 26.   COORDINATION CHEMISTRY

26.6 The bidentate ligand AA must go on cis. It is then possible to have
trans B_2, trans C_2, or neither B_2 nor C_2 trans. These three
possibilities are pictured below. The answer is (b).

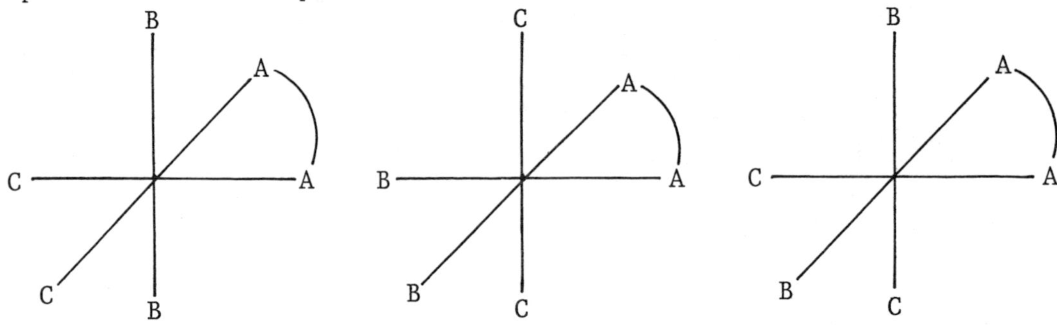

26.7 A complex with one or more planes of symmetry does <u>not</u> exhibit optical isomerism. Compounds one and 2 have several planes of symmetry, assuming the puckering of the ethylene groups can be ignored. Compound 3 does not contain a plane of symmetry. A sketch of compound 3 and its mirror image will show that the two are not superimposable. Compound 3 is optically active. The answer is (b).

26.10 The solution is acidic because of the hydrolysis of the Ni^{2+}.

$$Ni^{2+} + H_2O \rightleftharpoons NiOH^+ + H^+ \qquad\qquad K_h$$

$$[H^+] = \text{antilog } (-5.0) = 1.0 \times 10^{-5} = [NiOH^+]$$

$$K_h = \frac{[NiOH^+][H^+]}{[Ni^{2+}]} = \frac{(10^{-5})(10^{-5})}{0.1-10^{-5}} = 1.0 \times 10^{-9}$$

The formation constant, K_f, is related to K_h as follows:

$$Ni^{2+} + OH^- \rightleftharpoons NiOH^+ \qquad\qquad K_f$$

$$K_f = \frac{[NiOH^+]}{[Ni^{2+}][OH^-]} \times \frac{[H^+]}{[H^+]} = \frac{K_h}{K_w} = \frac{1.0 \times 10^{-9}}{1.0 \times 10^{-14}} = 1.0 \times 10^5$$

answer (b).

26.11 The dissociation constant is defined as:

$$K_d = \frac{[M^{3+}][X^-]^6}{[MX_6{}^{3-}]}$$

Orig. $[MX_6{}^{3-}] = \dfrac{1 \times 10^{-3} \text{ mole}}{50 \times 10^{-3} \text{ } \ell} = 0.020 \text{ M}$

At Equil: $[M^{3+}] = 6.0 \times 10^{-5}$

$$[MX_6{}^{3-}] = 0.020 - 6.0 \times 10^{-5} \simeq 0.020$$

$$[X^-] = 6 \times [M^{3+}] = 3.6 \times 10^{-4}$$

$$K_d = \frac{6.0 \times 10^{-5} \times (3.6 \times 10^{-4})^6}{0.020} = 6.5 \times 10^{-24} \text{ which is answer (e)}.$$

26.12 The formation constant is defined as:

$$Zn^{2+} + 4\ OH^- \rightleftharpoons Zn(OH)_4{}^{2-} \qquad K_f = \frac{[Zn(OH_4{}^{2-}]}{[Zn^{2+}][OH^-]^4}$$

Since K_f is a very large number, assume that very little of the complex has dissociated and consequently the $[Zn(OH)_4{}^{2-}]$ and $[OH^-]$ remain essentially unchanged.

$$K_f = 5.00 \times 10^{14} = \frac{0.050}{[Zn^{2+}](.450)^4}$$

$$[Zn^{2+}] = 2.44 \times 10^{-15} \text{ which is answer (d)}.$$

You will note that your assumption is justified by this result.

26.16 Always start by writing down the appropriate reaction and equilibrium expression.

$$Ni^{2+} + CN^- \rightleftharpoons NiCN^+ \qquad K_f = \frac{[NiCN^+]}{[Ni^{2+}][CN^-]} = 1.00 \times 10^{30}$$

Since K_f is so large you expect that practically all the reactants that can react will react to form the complex. Since CN^- is in excess, the solution at equilibrium will contain:

$$[Ni^{2+}] = \text{very little}$$

$$[NiCN^+] = \text{same as original } [Ni^{2+}] = 0.500$$

$$\text{excess } [CN^-] = 0.550 - 0.500 = 0.050$$

$$1.00 \times 10^{30} = \frac{0.500}{[Ni^{2+}](0.050)}$$

$$[Ni^{2+}] = 1.00 \times 10^{-29} \text{ which is answer (d)}.$$

26.18 The dissociation reaction is

$$Zn(OH)_4^{2-} \rightleftharpoons Zn^{2+} + 4\ OH^- \qquad K_d = 1/K_f$$

Let x be $[Zn^{2+}]$, then $[OH^-] = 4x$ and $[Zn(OH)_4^{2-}] = 0.1 - x$

$$K_d = \frac{[Zn^{2+}][OH^-]^4}{[Zn(OH)_4^{2-}]} = \frac{(x)(4x)^4}{0.1-x} = \frac{1}{5 \times 10^{14}}$$

Assuming $x \ll 0.1$, $\quad x \simeq \left[\dfrac{0.1}{4^4 \times 5 \times 10^{14}}\right]^{1/5} = 2.4 \times 10^{-4}$

The assumption is justified.

Then $[OH^-] = 4x = 9.6 \times 10^{-4}\ \underline{M}$

and $pH = 14 - pOH = 14 - 4 + \log 9.6$

$pH = 10.98$

The answer is (d).

26.19 Let x be the solubility of AgI in moles/liter.

In pure H_2O: $\qquad AgI_{(s)} \rightleftharpoons Ag^+ + I^-$

Let $x = [Ag^+] = [I^-]$; $\quad x^2 = K_{sp}$; $\quad x = \sqrt{K_{sp}}$

In $1.0\ \underline{M}\ NH_3$: $\qquad AgI_{(s)} + 2\ NH_3 \rightleftharpoons Ag(NH_3)_2^+ + I^- \qquad K_{eq}$

$$Ag^+ + 2\ NH_3 \rightleftharpoons Ag(NH_3)_2^+; \quad K_f = \frac{[Ag(NH_3)_2^+]}{[Ag^+][NH_3]^2}$$

$$K_{sp}K_f = \frac{[Ag^+][I^-][Ag(NH_3)_2^+]}{[Ag^+][NH_3]^2} = K_{eq}$$

Let $x = [Ag(NH_3)_2^+] = [I^-]$; $\quad [NH_3] = 1.0 - 2x$

$$K_{eq} = \frac{[Ag(NH_3)_2^+][I^-]}{[NH_3]^2} = K_{sp}K_f = \frac{x^2}{1.0-2x}$$

Assume $2x \ll 1.0$ Then $x = \sqrt{K_{sp}K_f}$

The answer is (a).

26.21 If L has a greater ligand field strength, there will be a larger gap
between energy levels and a higher energy quantum will be absorbed.
Higher energy corresponds to higher frequency, _i.e._, the absorption
band will approach the ultraviolet and the transmitted light will
include a broader mixture of colors, becoming less red and even colorless (colorless means all colors are equally transmitted). Select
answer (b). Answers (a), (d), and (e) imply a shift to lower frequency
absorption; and (c) implies an absorption in the red, which is low
frequency.

26.23 a) NH_3 > F^- means that NH_3 causes a greater splitting. (a) is OK.
   b) When Cl^- replaces NH_3 the splitting is less; hence less energy, lower
      frequency, and longer wavelength absorbed. (b) is OK.
   c) When H_2O replaces NH_3, the splitting is less and (as in b) a longer
      wavelength (toward the red) is absorbed. When the absorption moves
      toward the red, the transmitted light moves toward the blue,
      so (c) is OK.
   d) The diamagnetism of this d^6 complex implies a strong field. Since en
      produces an even stronger field than NH_3, the $Co(en)_3^{3+}$ is also
      diamagnetic. (d) is correct.
   e) Outer orbital complexes are more likely with lower field strength
      ligands, hence are more likely with Cl^- than with NO_2^-. Select this
      choice; (e)is the wrong statement.

26.24 Calculate the number of moles of metal and each of the ligands.

moles of M^{2+} = molarity x liters = .0090 x 1.0 = 9.0 x 10^{-3}

$$\text{moles of } NH_3 = \text{moles of } H^+ \text{ titrant} = \frac{180 \text{ ml}}{1000 \text{ ml } \ell^{-1}} = 0.10 \frac{\text{mole}}{\ell}$$

$$= 18.0 \times 10^{-3}$$

$$\text{moles of } I^- = \text{moles of } AgI_{(s)} = \frac{8.46 \text{ g}}{235 \text{ g mole}^{-1}} = 36.0 \times 10^{-3}$$

The ratio, moles of I^-: moles NH_3: moles M^{2+}

is 36.0 x 10^{-3}; 18.0 x 10^{-3}: 9.0 x 10^{-3} = 4:2:1

Hence the formula is $[M(NH_3)_2I_4]^{2-}$, answer (c).

26.25 There are 10 Dq units of splitting in an octahedral field, the t_{2g}
levels are 4 Dq below the average and the e_g are 6 Dq above. Mn(VI) is
a d^1 ion, which is diagrammed:

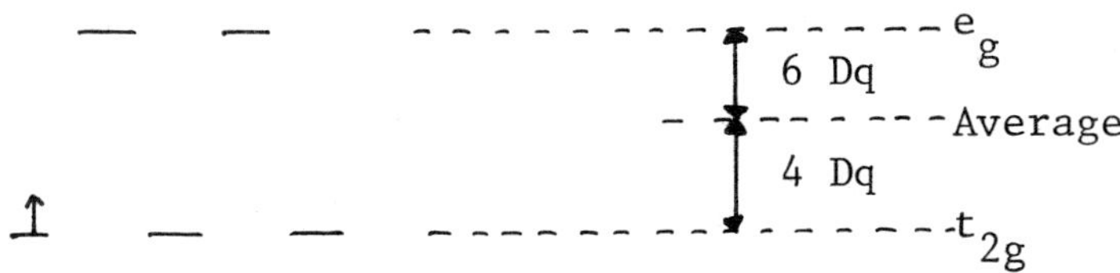

Since there is one electron below the average, the ion is stabilized by
4 Dq units, answer d.

CHAPTER 27.   NUCLEAR CHEMISTRY

27.3 The normal ratio of neutrons to protons in stable isotopes is given by the "stability line". It varies from 1 to 1 for low Z elements (except ^1_1H) to about 1.6 to 1 for high Z elements. Unstable isotopes will decay in such a fashion as to produce isotopes with a ratio closer to the normal.

(a) ^8_5B is neutron-poor. The indicated decay is expected because the product ^8_4Be has the normal 1:1 ratio of neutrons to protons.

(b) $^{16}_6C$ is neutron-rich and is expected to decay as shown toward the less neutron-rich $^{16}_7N$.

(c) ^7_4Be is neutron-poor and so is expected to decay to the stable and abundant ^7_3Li.

Answer (d).

27.7 You must first determine the decrease in mass. The reaction given is a <u>nuclear</u> reaction. If you add $17e^-$ to each side you will have the equivalent of one chlorine atom on the left and one argon atom (one nucleus and $18e^-$) on the right. Recall, a β^- is an electron. The decrease in mass then is the difference between a Cl-36 atom and an Ar-36 atom. $35.96831 - 35.96755 = 0.00076$ amu. The energy release $= 0.00076 \times 931$ MeV/amu $= 0.708$ MeV, answer (e).

Note: The energy released is largely in the form of kinetic energy of the β^- particle which is emitted from the decaying nucleus with great velocity.

27.8 First consider the nuclear process. The energy released in the formation of one mole of ^4_2He is

$$\frac{\#kJ}{mole} = \frac{17.6 \text{ MeV}}{molecule} \times \frac{10^6 eV}{MeV} \times \frac{96.5 \text{ kJ/mole}}{1 \text{ eV/molecule}} = 1.70 \times 10^9 \text{ kJ/mole}$$

$$\frac{\#kJ}{1b \text{ of He}} = \frac{1.70 \times 10^9 \text{ kJ}}{mole} \times \frac{1 \text{ mole}}{4 \text{ g}} \times \frac{454 \text{ g}}{1b} = 1.93 \times 10^{11} \text{ kJ/1b}$$

Now consider the chemical process.

$$\frac{\#kJ}{ton \text{ of } C} = \frac{394 \text{ kJ}}{mole} \times \frac{1 \text{ mole}}{12 \text{ g}} \times \frac{2000 \times 454 \text{ g}}{ton} = 2.98 \times 10^7 \text{ kJ/ton of C}$$

Then

$$\frac{\# \text{ tons of C}}{1 \text{ 1b He}} = \frac{1 \text{ ton of C}}{2.98 \times 10^7 \text{ kJ}} \times \frac{1.93 \times 10^{11} \text{ kJ}}{1 \text{ 1b He}} = 6.47 \times 10^3$$

The answer is (b).

27.13 The first order nuclear decay equation can be most conveniently used
in the form:
$$2.3 \log \frac{A_o}{A} = kt$$

where A_O is the initial specific activity, A, the activity at time t,
and k the decay constant. In place of the ratio A_O/A you could also
use the ratio m_o/m where m is the mass of radioactive isotope. In
first order _chemical_ decay you must use a ratio of concentrations
c_o/c. The half-life $t_{1/2}$ is related to the decay constant by:

$$t_{1/2} = \frac{.693}{k}$$

In radiocarbon dating, time zero is the time the object ceased living;
its specific activity was then 15.3. Time t is the number of years
from then until the present. Substituting the data from the problem:

$$k = \frac{.693}{5770}$$

$$2.3 \log \frac{15.3}{10.2} = \frac{.693}{5770} \text{ x } t$$

$$t = \frac{(2.3)(.176)(5770)}{(.693)} = 3370 \text{ years,}$$

which is answer (c).

27.14 You must first determine the fraction of original thorium remaining for
use in the decay formula.

$$\text{Moles thorium remaining} = \frac{1.29}{232} = 5.56 \text{ x } 10^{-3}$$

$$\text{Moles thorium decayed} = \text{moles } {}^{208}\text{Pb} = \frac{1.97}{208} = 9.46 \text{ x } 10^{-3}$$

$$\text{Orig. moles thorium} = 5.56 \text{ x } 10^{-3} + 9.46 \text{ x } 10^{-3} = 15.02 \text{ x } 10^{-3}$$

$$2.3 \log\left(\frac{15.02 \text{ x } 10^{-3}}{5.56 \text{ x } 10^{-3}}\right) = kt = \frac{.693 \, t}{1.39 \text{ x } 10^{10}}$$

$$t = \frac{(2.3)(.432)}{(.693)} \text{ x } 1.39 \text{ x } 10^{10} = 2.00 \text{ x } 10^{10}, \text{ which is answer (d).}$$

A common error made by students is to use a mass ratio instead of a mole
ratio, giving answer (a).

CHAPTER 28. ORGANIC CHEMISTRY

28.3 (a) is a planar molecule and the C-Cl dipoles (as well as the smaller
C-H dipoles) point in opposite directions. No dipole moment.
(b) H-C≡C-C≡N is linear and the large C≡N dipole is not compensated;
in fact the H-C dipole points in the same direction. Answer is (b).
(c) In this compound, the three C-I dipoles are at 120° to each other in
a plane and cancel out. No dipole moment.

28.5 The structural formulas, hybridization of the underlined atoms, and expected bond angles are as follows:

a) $\underline{C}H_2O$ (H₂C=O structure) sp^2 $<HCO \approx 120°$ OK

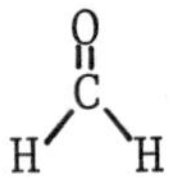

b) $\underline{C}H_3Cl$ (CH₃Cl structure) sp^3 $<HCCl \approx 109.5°$ OK

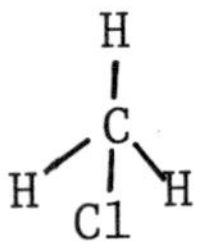

c) $C_6H_{13}\underline{N}H_2$ $H_3C-CH_2-CH_2-CH_2-CH_2-\ddot{N}\langle^H_H$ sp^3 $<HNH \approx 109.5°$

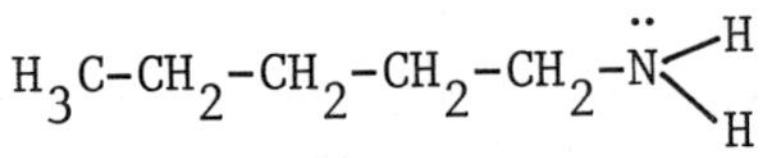

The listed angle is wrong. Select answer c.

d) $CH_3\underline{C}OOH$ $H_3C-C\overset{\ddot{O}:}{\diagup}OH$ sp^2 $<OCO \approx 120°$ OK

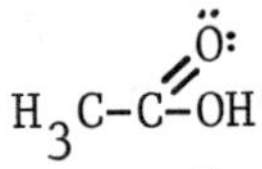

e) $CH_3CO\underline{O}H$ $H_3C-C\overset{\ddot{O}:}{\diagup}\ddot{O}-H$ 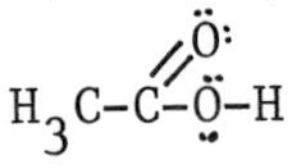 sp^3 $<COH \approx 109.5°$ OK

28.6 Let $x = [H_2O] = [RCOOR]$

$$(0.065-x) = [ROH] = [RCOOH]$$

$$\frac{x^2}{(.065-x)^2} = 5.2 \times 10^{-3}$$

$$\frac{x}{.065-x} = 7.2 \times 10^{-2}; \quad x = 4.4 \times 10^{-3}, \text{ answer (b).}$$

Note that in non-aqueous media the concentration of H_2O enters into the equilibrium constant expressions.

28.7 The only possibilities are (b) and (c) which are the only sp^3 (tetrahedral) carbons. Atom (c) has 2 hydrogens and so is not asymmetric. Atom (b) is the answer; it has four different groups attached to it.

28.9 All three have a planar group at atoms around the C=C bond. In II there are two Cl atoms on the same carbon so a 180° rotation around the C=C bond axis makes no change, it has no geometric isomers; the others do, mark answer (b).

CHAPTER 29. DESCRIPTIVE CHEMISTRY

29.1 This question covers many aspects of chemistry and each choice must
be individually analyzed.

a) At the <u>cathode</u> reduction occurs, producing H_2 gas. Heavy water
would produce D_2 gas, which is twice as dense as H_2. (D is the
symbol for deuterium)

b) The reaction is $Ca + 2\ H_2O \rightarrow Ca(OH)_2 + H_2$. Heavy water would
produce D_2 instead of H_2 which would weigh twice as much.

c) The reaction $2\ H_2O \rightarrow 2\ H_2 + O_2$ requires 4 Faradays. One Faraday
decomposes 1/2 mole of either light or heavy water. This is 9.0
g of light water, but 10.0 g of heavy water.

d) At the <u>anode</u> oxidation occurs to produce O_2 gas, regardless of
whether it is heavy or light water. This is the answer to the
question.

e) The reaction is $2\ Li + 2\ H_2O \rightarrow 2\ LiOH + H_2$. With heavy water
one would get D_2, twice as dense. Note however, that the <u>volume</u>
of gas would be the same since the same number of moles of D_2 and
H_2 would be generated by a given amount of Li.

29.2 The law of Dulong and Petit relates the atomic weight of a solid
element to its heat capacity, or specific heat. The answer is e.

29.3 Each statement must be discussed separately.

a) The existence of a very small dense nucleus was discovered by
Rutherford in 1910 and provided the first clue to our modern
understanding of the structure of the atom.

b) These are the fundamental particles of modern theory.

c) This finding caused the abandoning of the Bohr orbit in favor of
the orbital concept of modern atomic theory.

d) It is known now that this is not so, but it was a postulate of
classical theory, and is the answer to this question. Classical
chemists did not realize that there were isotopes because all
samples of an element had virtually the same isotope composition.
Ordinary stoichiometry calculations are still made as if all atoms
of an element have the same mass (the average mass).

e) Atomic number used to mean just the position of an element in a
list or table. Modern theory identifies it with the number of
protons, i.e., the positive charge on the nucleus.

29.7 Items (a) and (e) are chemical properties, and items (b) and (c) are
properties of individual molecules without regard to interactions. The
answer is (d). To dissolve starch and cellulose requires that there
be a strong interaction between solute and solvent to compensate for
the crystal energy of the solute; this is provided by hydrogen bonding
between the many −OH groups of the solute and water.

29.9 The best solute in a polar solvent would be the most polar solute.
(Like dissolves like.) Methanol (d) is by far the most polar because
of its C-O and H-O bonds.

29.11 (a), (b), and (e) are ionic salts in the solid state and so dissolve
as ions. HI, item (c), is molecular but ionizes 100% because I^- is
such a weak base compared to H_2O; $HI + H_2O \rightarrow H_2O^+ + I^-$. The
answer is (d) phenol in which the bonds are primarily covalent. It
does act as a very weak acid; $C_6H_5OH + H_2O \rightleftharpoons C_6H_5O^- + H_3O^+$.

29.12 Unlike many other pollutants, phosphates are not harmful biologically
but on the contrary are important nutrients. The answer is (c). In
most lakes, growth of algae is limited by lack of phosphate. Phosphate
pollution permits algae to proliferate. The decay of the algae consumes
all the oxygen killing off higher forms of life. This whole process is
termed "eutrofication".

29.13 Coal contains a great deal of sulfur which upon exposure to air and
bacteria is oxidized to SO_3, which is converted by water to H_2SO_4.
Answer is (c).

29.14 The filler gas must be chemically inert and so argon is used. Argon
is a noble gas, which is present to the extent of about 1% and is
easily distilled from liquid air.   Answer is (d).

29.15 (a) is reversed, all soaps are carboxylates and most anionic
detergents are sulfonates.
(b) is correct, soaps are derived from natural fatty acids which are
straight chains and biodegradable.
(c) and (e) are each reversed with respect to soaps & detergents.
(d) Most detergents are ionic, some are not. However, all soaps are
ionic salts with organic anions.

29.17 The answer is (a), all entries correct.
In (b), Si is not a metal; it is a semiconductor and an example of a
covalent network solid.
In (c), HCl(s) is polar molecular, not ionic
In (d), $POCl_3$ is polar molecular, not non-polar
In (e), both SrS and $NaHCO_3$ are ionic.

APPENDIX A.   SAMPLE EXAMS

There are four 20-item sample exams in this Appendix. The exams cover chapters 2-12, 13-16, 17-18 and 19-29 respectively. Allow three to four minutes for each question. The answer keys are given in Appendix C.

<u>EXAM 1</u>

1. A neutral atom which has 53 electrons and an atomic mass of 111 has

    a) no other isotopes.
    b) an atomic number of 58.
    c) a nucleus containing 58 neutrons.
    d) a nucleus containing 53 neutrons.
    e) a nucleus containing 58 protons.

2. The 1970 standard established by the United States Government for carbon monoxide emission for automobiles limits exhaust to 23.0 grams CO per vehicle mile. Assume that in a typical metropolitan area of 565000 people there are 90400 automobiles, driven an average of 15.6 miles per 24 hour period. How much CO (in tons) is (legally) discharged into the area's atmosphere per day?   (1 ton = 2000 x 454 gr.)

    a) 0.0677 tons/day
    b) 35.8 tons/day
    c) 4.38×10^{-9} tons/day
    d) 0.146 tons/day
    e) 2.95×10^{13} tons/day

3. The following nitrogen compounds could possibly be added to the soil as fertilizers. Which compound has the greatest percent nitrogen by weight?

    a) ammonium sulfite, $(NH_4)_2SO_3$ MW = 116 g/mole
    b) magnesium nitrate, $Mg(NO_3)_2$ MW = 148 g/mole
    c) iron(II) azide, $Fe(N_3)_2$ MW = 140 g/mole
    d) iron(II) nitride, Fe_3N_2 MW = 195 g/mole
    e) More information is required.

337

4. A 22.9 gram sample of an iron oxide was heated with a stream of hydrogen
    gas converting it completely to metallic iron. If the resultant iron
    weighed 17.8 grams, what is the empirical formula of the iron oxide?

    a) Fe_2O_3
    b) Fe_3O_2
    c) FeO
    d) Fe_3O_4
    e) Fe_4O_3

5. Which one of the following measurements or techniques can be used to
    find the atomic weight, or molecular weight of a substance?

    a) heat of combustion
    b) boiling point elevation
    c) boiling point of pure liquid
    d) electrical conductivity
    e) density of liquid state

6. How many grams of arsenic must be burned in an atmosphere of oxygen to
    form 43.4 grams of As_2O_3?

    a) 0.00585
    b) 32.8
    c) 8.21
    d) 0.0304
    e) 39.4

7. How many grams of HCl are required to react completely with 2.17 grams
    of Zn?

    a) 4.85
    b) 9.70
    c) 1.21
    d) 0.412
    e) 2.43

8. The Duquesne Light Company burns approximately 504 tons of coal per day.
    If the sulfur content of the coal is 2.47% by weight, how many tons of
    SO_2 are dumped into the Pittsburgh atmosphere each day? (1 ton = 908 kg)

    a) 24.8
    b) 12.4
    c) 24800
    d) 0.161
    e) 6.22

9. What is the sum of the coefficients of all the species in the balanced
    equation:
$$PbO_2 + NO \xrightarrow{H^+} Pb^{2+} + HNO_2$$

    a) 20
    b) 18
    c) 13
    d) 8
    e) 6

10. What weight of zinc (AW = 65.4 g/mole) will be deposited at an inert
    electrode on passage of a current of 28.5 amps through a solution of
    zinc chloride for 4.87 hours? The reduction equation is:
$$Zn^{2+} + 2\ e^- \rightarrow Zn$$

    a) $\dfrac{(28.5)(4.87)(3600)}{(96500)(65.4)}$ grams
    b) $\dfrac{(65.4)(28.5)(4.87)}{(96500)(2)}$ grams

    c) $\dfrac{(65.4)(4.87)(3600)}{(96500)(28.5)}$ grams
    d) $\dfrac{(65.4)(28.5)(4.87)(3600)}{96500}$ grams

    e) None of the above

11. The same quantity of electricity that caused the deposition of 143 grams
    of silver from a $AgNO_3$ solution was passed through a solution of a
    titanium salt containing Ti cations of unknown charge. 15.9 grams of
    Ti were deposited. What is the charge of the Ti cation?
    (AW's: Ag = 108, Ti = 47.9)

    a) 3
    b) 4
    c) 2
    d) 1
    e) None of the above

12. Calculate the boiling point of a .163 molal solution of a non-
    electrolyte in acetic acid. The boiling point and boiling point
    elevation constant for acetic acid are 118°C and 3.07°C per molal
    respectively.

    a) 118 – 3.07
    b) 118 + (3.07 x 0.163)
    c) 118
    d) 118 – (3.07 x 0.163)
    e) None of the above

13. Given the following information:

Compound	Standard heat of formation (kJ/mole)
$CO_2{}_{(g)}$	-394
$H_2O_{(\ell)}$	-286
$C_2H_6{}_{(g)}$	84.5

$$C_2H_{6(g)} + 7/2\ O_{2(g)} \rightarrow 2\ CO_{2(g)} + 3\ H_2O_{(\ell)}$$

Calculate the enthalpy change for the combustion of C_2H_6.

a) 594 kJ/mole
b) -84.5 kJ/mole
c) 594 kJ/mole
d) -1560 kJ/mole
e) 1560 kJ/mole

14. Given that the standard heats of formation of SO_2 and SO_3 are -297 and -396 kJ /mole respectively, find the heat of reaction for the following reaction: $$SO_2 + \frac{1}{2}\ O_2 \rightarrow SO_3$$

a) -396 - (-297)
b) -396
c) 396 - (-297)
d) 297 - (-396)
e) None of the above

15. What is the molarity of a solution containing 31.9 grams of HCl in 5.05 liters of solution?   (MW of HCl = 36.5)

a) $\dfrac{31.9}{(36.5)(5.05)}$

b) $\dfrac{(1000)(31.9)}{(36.5)(5.05)}$

c) $\dfrac{(31.9)(36.5)}{5.05}$

d) $\dfrac{31.9}{5.05}$

e) None of the above

16. If 37.0 ml of 1.64 molar NaOH is mixed with 42.9 ml of .588 molar HNO_3 the final solution will be:

a) 1.08 molar in OH^-

b) 0.445 molar in H^+

c) 1.08 molar in H^+

d) 0.860 molar in H^+

e) 0.445 molar in OH^-

17. A standard solution of 0.198 molar HCl is being used to determine the concentration of an unknown NaOH solution. If 35.0 milliliters of the acid are required to neutralize 35.7 milliliters of the base, what is the molarity of the base?

a) $\dfrac{(35.7 + 35.0)}{0.198}$

b) $\dfrac{0.198}{(35.7 + 35.0)}$

c) $\dfrac{(0.198)(35.0)}{35.7}$

d) $\dfrac{35.0}{(0.198)(35.7)}$

e) None of the above

18. Suppose 0.40 moles of Zn metal is added to 400 ml of a solution which is 0.10 molar in Cu^{2+} and 6.0 molar in acid. The first reaction is:

$$Zn_{(s)} + Cu^{2+}_{(aq)} \rightarrow Zn^{2+}_{(aq)} + Cu_{(s)}$$

Then the excess Zn reacts with the acid:

$$Zn_{(s)} + 2 H^+_{(aq)} \rightarrow Zn^{2+}_{(aq)} + H_{2(g)}$$

(Assume both reactions go to completion.) After the zinc reacts, which one computation below is correct?

a) 0.40 mole of Cu metal is produced.
b) 6.35 g of Cu metal is produced.
c) 2.02 ℓ of H_2 is produced (at STP).
d) The solution contains 26.2 g of Zn^{2+}.
e) The concentration of Zn^{2+} is 0.10 M.

342

19. What mass (in grams) of Cu should be added to an excess of concentrated H_2SO_4 to generate as much $SO_{2(g)}$ as 7.94 grams of sodium sulfite, Na_2SO_3? The reactions are:

$$Cu \ + \ 2 \ H_2SO_4 \ \rightarrow \ CuSO_4 \ + \ SO_2 \ + \ 2 \ H_2O$$

$$Na_2SO_3 \ + \ H_2SO_4 \ \rightarrow \ Na_2SO_4 \ + \ SO_2 \ + \ H_2O$$

a) 2.00 grams
b) 4.01 grams
c) 0.00283 grams
d) 31.5 grams
e) 15.7 grams

20. A freshman chemist can make an estimate of Avogadro's number by simply adding a few drops of an alcoholic solution of oleic acid to a water surface dusted with lycopodium powder. To calculate the result, some numerical values are necessary and some assumptions have to be made, such as listed below. Select the one item which does not belong (either incorrect, unnecessary, or irrelevant).

a) assume all the alcohol disappears by evaporation or solution.
b) the charge on the electron.
c) the density of oleic acid
d) the concentration of oleic acid solution
e) assume a geometric shape for the oleic acid molecule

EXAM 2

1. What is the volume occupied by .277 mole of an ideal gas at 764 torr and 14.1°C?

   a) 0.154 liters
   b) 0.319 liters
   c) 3.13 liters
   d) 6.49 liters
   e) 0.00854 liters

2. The pressure of a gas in a 25.0 liter tank is 709 torr. If the gas is transferred to a 43.9 liter container at the same temperature, what is the final pressure of the gas (in torr)?

   a) 709 torr
   b) 0.00495 torr
   c) 404 torr
   d) 0.531 torr
   e) More information is required

3. What is the pressure on 2.85 grams of an ideal gas at 10.8°C occupying 7.71 liters? (The molecular weight of the hypothetical gas is 20.1 grams per mole.)

   a) 326 torr
   b) 2.33 torr
   c) 0.0162 torr
   d) 0.429 torr
   e) 61.6 torr

4. The Haber process for preparing ammonia involves the direct conversion of hydrogen and nitrogen gases at high temperature and pressure using a catalyst:

$$N_{2(g)} + 3\ H_{2(g)} \rightleftharpoons 2\ NH_{3(g)}$$

   How many liters of ammonia can be prepared from 11.2 liters of N_2 and 49.2 liters of H_2, assuming complete conversion at a constant temperature of 1000°K and constant external pressure of 500 atm?

   a) (.333)(49.2) liters
   b) (2)(11.2) liters
   c) (1/2)(11.2) liters
   d) (.667)(49.2) liters
   e) None of the above.

5. What volume of HI can be made if 14.2 liters of H_2 and 23.5 liters of I_2 react at STP?

$$H_2 \;+\; I_2 \;\rightarrow\; 2\;HI$$

   a) 28.4
   b) 47.0
   c) 23.5
   d) 14.2
   e) 51.9

6. Determine the density of a gas given the following data: MW = 53.2 grams mole^{-1}; pressure = 782 torr; temperature = 42.5°C.

   a) 15.7 g/liter
   b) .473 g/liter
   c) 2.11 g/liter
   d) 1.61 x 10^3 g/liter
   e) More information is required.

7. According to Graham's Law of gas effusion, helium gas will effuse X times as fast as SO_2. What is the value of X?

   a) 4.00
   b) 16.0
   c) 0.250
   d) 256
   e) 0.0624

8. If 90.0 liters of H_2 effuse through a porous partition in 30.0 min., what volume of C_2F_4 will effuse through the same partition at the same temperature and pressure in 20.0 min? (MW's: H_2 = 2.02, C_2F_4 = 100)

   a) 1.21
   b) 8.53
   c) 422
   d) 2970
   e) 46.9

9. Given:

Gas	MW (grams/mole)
C_2H_2	26.0
H_2	2.02

What is the rate of effusion of C_2H_2 compared to H_2?

a) 3.59
b) 0.0777
c) 12.9
d) 0.279
e) 0.928

10. A 7.20 gram sample of pure $Cu(NO_3)_2$ was placed in an evacuated container and heated until it completely decomposed as follows:

$$2\ Cu(NO_3)_2 \rightarrow 2\ CuO_{(s)} + O_{2(g)} + 4\ NO_{2(g)}$$

If the total pressure in the container is now 930 torr, what is the partial pressure of O_2?

a) 35.7 torr
b) 186 torr
c) 133 torr
d) 159 torr
e) 310 torr

11. A gas mixture contains an equal number of CO molecules and CO_2 molecules, and no others. Assuming ideal gas behavior, which of the following statements is true?

a) The total mass of CO_2 is the same as the CO.
b) The average energy is the same for CO and CO_2.
c) Since CO_2 molecules have more mass, they have more inertia and contribute more to the pressure.
d) The average velocity is the same for CO and CO_2 .
e) None of the above.

12. Absolute zero is defined as the temperature which:

a) a straight line graph of P vs. T for a confined sample of an ideal gas intersects the T axis.
b) the density of a liquid equals that of its vapor.
c) mercury freezes solid.
d) a straight line graph of V vs. 1/P intersects the 1/P axis.
e) None of the above.

13. Select the proper set of names for the following four compounds:

    1. $Cu(CN)_2$ 2. $Sn(ClO_3)_4$ 3. $Fe_2(SO_4)_3$ 4. $AgCl_2$

a) 1. copper(II) cyanide   2. tin(IV) chlorate
   3. iron(III) sulfate   4. silver(II) chloride

b) 1. curium(II) cyanide   2. tin(IV) chlorate
   3. iron(III) sulfite   4. gold(III) chloride

c) 1. copper(II) cyanide   2. tin(IV) perchlorate
   3. iron(III) sulfate   4. silver(II) chloride

d) 1. curium(II) cyanide   2. tin(IV) chlorate
   3. iron(III) sulfate   4. silver(II) chloride

e) None of the above

14. Which of the following four compounds is <u>incorrectly</u> named?

    a. $TiSe_2$ b. $Na_2C_2O_4$ c. $V(ClO_3)_3$ d. $AlPO_4$

a) titanium(IV) selenide
b) sodium acetate
c) vanadium(III) chlorate
d) aluminum phosphate
e) None of the above

15. Cu metal crystallizes in a face centered cubic unit cell with the length
of an edge equal to 3.62 Angstroms. Assuming the atoms are identical
hard spheres, calculate the atomic radius of the metal. (HINT: The
square roots of 2 and 3 are 1.414 and 1.732 respectively.)

a) $\dfrac{3.62}{2}$

b) $\dfrac{(1.414)(3.62)}{4}$

c) $\dfrac{(1.414)(3.62)}{2}$

d) $\dfrac{(1.732)(3.62)}{2}$

e) None of the above

16. The length of a unit cell of Fe metal is 2.91 Angstroms and its density
is 7.53 grams/cm^3. How many atoms of Fe are present in the unit cell?
(AW of Fe = 55.8). The cell is cubic.

a) 14
b) 9
c) 5
d) 4
e) 2

17. TlBr crystallizes into the CsCl unit cell with the length of an edge equal to 4.74 Angstroms. What is the density of TlBr? (MW of TlBr = 284)

    a) 17.7 g/cc
    b) 398 g/cc
    c) 99.5 g/cc
    d) 4.43 g/cc
    e) 0.0603 g/cc

18. A quantum of electromagnetic radiation has frequency equal to 2.11×10^{14} sec^{-1}. What is the wavelength of this radiation in centimeters? ($c = 3.00 \times 10^{10}$ cm/sec, $h = 6.63 \times 10^{-27}$ erg sec.)

    a) 9.82×10^{-9}
    b) 1.40×10^{-12}
    c) 1.42×10^{-4}
    d) 7.15×10^{11}
    e) 7.03×10^{3}

19. Which of the following combinations of quantum numbers do <u>not</u> represent permissible solutions of the Schroedinger wave equation for the hydrogen atom?

	n	ℓ	m	s
1.	9	0	0	-5/2
2.	2	1	0	-1/2
3.	1	3	3	1/2

    a) 1
    b) 3
    c) 1 and 2
    d) 1 and 3
    e) None of them is permissible

20. The Rydberg equation for the hydrogen atom in frequency units is

$$\nu = 3.29 \times 10^{15} \times (1/N_1^2 - 1/N_2^2) \text{ sec}^{-1}$$

What is the wavelength in Angstroms of a quantum of electromagnetic radiation that corresponds to the transition from the ground state ($N_1 = 1$) to the first excited state ($N_2 = 2$)?

    a) 6.84×10^{2}
    b) 1.82×10^{3}
    c) 7.40×10^{33}
    d) 3.04×10^{17}
    e) 1.22×10^{3}

EXAM 3

1. Which of the following electron configurations is _incorrect_?

    a) Halide ions ns^2np^6 $n = 2,3$
                       $ns^2(n-1)d^{10}np^6$ $n = 4,5$
    b) Se   [Ar] $5s^24d^{10}5p^4$
    c) Be^{2+} $1s^2$
    d) Co^{3+} [Ar] $3d^6$
    e) All of the above configurations are correct.

2. Electromagnetic radiation of sufficient energy can be used to break
    chemical bonds. Calculate the wavelength in Angstroms of photons with
    enough energy to break H-Cl single bonds. The H-Cl single bond energy
    is 103 kcal/mole. (1 kcal/mole = 6.94 x 10^{-14} erg/photon)

    a) 3.59 x 10^{12} Angstroms
    b) 2.78 x 10^3 Angstroms
    c) 7.15 x 10^{-4} Angstroms
    d) 3.59 x 10^4 Angstroms
    e) 2.78 x 10^{-5} Angstroms

3. A typical electron diffraction experiment is conducted with electrons
    accelerated through 40900 volts, or with 40900 eV of energy. What is
    the wavelength of the electrons in Angstroms? (1 electron volt =
    1.60 x 10^{-12} erg; electron mass = 9.11 x 10^{-28} g)

    a) 0.0486 Angstroms
    b) 7.29 x 10^{-10} Angstroms
    c) 6.07 x 10^{-10} Angstroms
    d) 0.0607 Angstroms
    e) 4.86 x 10^{-10} Angstroms

4. Atom A has 3 valence electron(s) and atom B has 7 valence electrons.
    The formula expected for an ionic compound of A and B is:

    a) AB
    b) AB_3
    c) AB_2
    d) A_2B
    e) A_3B

5. Which one of the following trends in atomic and ionic radii is _not_
    correct?

    a) N < Ge < Sr < Cs
    b) $Li^+ < Na^+ < K^+ < Rb^+$
    c) $Ga^{3+} < K^+ < Cl^- < P^{3-}$
    d) O < S < Se < Te
    e) K < Ca < Ga < As

6. Which of the following trends in first ionization potential is <u>not</u> correct?

    a) Rb < Sb < Cl < Ne
    b) Na < Mg < P < Cl
    c) Bi < Sb < P < N
    d) F < N < B < Li
    e) Te < Se < S < O

7. The first ionization potential of As is 231 kcal/mole. Which of the values below would your predict for the first ionization potential of Br?

    a) 231 kcal/mole
    b) 273 kcal/mole
    c) 190 kcal/mole
    d) 546 kcal/mole
    e) 116 kcal/mole

8. Given the following table of Pauling electronegativity values

Atom	Be	In	Pb	Bi	B	H	Se	C	Br	N	O	F
Electro-negativity	1.5	1.7	1.8	1.9	2.0	2.1	2.4	2.5	2.8	3.0	3.5	4.0

    Which one of the following sequences would you predict to have decreasing ionic character of the bond indicated?

    a) Pb–O > Pb–Br > Pb–Se > Pb–C
    b) Pb–O > Pb–C > Pb–Se > Pb–Br
    c) Pb–O > Pb–Br > Pb–C > Pb–Se
    d) Pb–Se > Pb–C > Pb–Br > Pb–O
    e) Pb–O > Pb–Se > Pb–C > Pb–Br

9. In which of the following molecules is the octet rule violated?

    a) BeH_2
    b) $PbCl_4$
    c) F_2
    d) None of them
    e) All of them

10. Which one of the following oxidation states would not be expected?

    a) Ar(0)
    b) Pb(IV)
    c) Nb(V)
    d) N(VI)
    e) All of these oxidation states would be expected

11. Indicate whether the bonding in the following compounds is primarily ionic (I), polar covalent (PC), or non-polar covalent (NC).

	KF	Cl_2	BaO	HI	NH_3
a)	I	NC	I	PC	PC
b)	I	NC	PC	PC	I
c)	PC	PC	I	NC	PC
d)	PC	NC	PC	PC	PC
e)	I	I	I	NC	NC

12. How many unpaired electrons are there in Mn^{2+}?

a) 1
b) 2
c) 4
d) 5
e) None, Mn^{2+} is diamagnetic

13. Which one of the following gives the correct hybridization for the central atom in the molecule?

	$AlCl_3$	N_3^-	CS_2
a)	sp^3	sp^2	sp^2
b)	sp^3	sp	sp
c)	sp	sp^3	sp
d)	sp^2	sp	sp
e)	None of the above		

14. Use molecular orbital theory to predict which one of the following diatomic molecules has the greatest bond length.

a) O_2^-
b) BF
c) OF^-
d) O_2
e) All of the above species have the same bond length.

15. How many sigma and pi bonds are there in Br_2?

	σ	π
a)	1	1
b)	1	0
c)	1	2
d)	0	1
e)	None of the above.	

16. Which of the following molecules is (are) diamagnetic?

    a) O_2
    b) NO^+
    c) B_2
    d) N_2^+
    e) All are paramagnetic.

17. Which of the following molecules would be expected to have a dipole moment?

    1. NO   2. $GaCl_3$ 3. HF   4. $COCl_2$

    a) 1
    b) 2 and 4
    c) 1, 3 and 4
    d) All of them
    e) None of them

18. The observed dipole moment of ICl is 0.65 debyes, and the ICl bond distance is 2.32 Angstroms. Estimate the percent ionic character in the ICl bond. (Electron charge, $e = 4.80 \times 10^{-10}$ esu; 1 debye = 1.0×10^{-18} esu cm.)

    a) 35.7%
    b) 74.4%
    c) 5.84%
    d) 4.83%
    e) 13.5%

19. Which one of the following is <u>not</u> an expected description of the geometry of the indicated molecule or ion based on the valence shell electron pair repulsion theory?

    a) $SnCl_2$ – planar, bent
    b) NH_3 – pyramidal
    c) CoF_6^{3-} – octahedral
    d) ClO_4^- – distorted tetrahedron
    e) XeF_4 – square planar

20. Which one of the following best describes the F-S-F bond angle in F_2SO?

    a) 180
    b) 90
    c) 109.5
    d) 109.5 < angle < 120
    e) 90 < angle < 109.5

<u>EXAM 4</u>

1. The $[H^+]$ of the aqueous solution is 2.34×10^{-6}. What is the pOH?

    a) 7.00
    b) 4.27×10^{-9}
    c) 5.63
    d) 8.37
    e) 2.34×10^{-6}

2. What is the pH of a 0.0100 molar HNO_3 aqueous solution?

    a) 7.00
    b) 12.00
    c) 2.00
    d) 2.76
    e) 4.60

3. What is the pH of a .875 molar hydrazine solution? (K_b for N_2H_4 is 9.80×10^{-7})

    a) 6.01
    b) 9.80×10^{-7}
    c) 10.97
    d) 11.97
    e) 3.03

4. What is the pH of a 1.42 molar sodium hypobromite solution? (K_a for hypobromous acid, HOBr, is 2.09×10^{-9})

    a) 9.74
    b) 8.68
    c) 11.42
    d) 7.00
    e) 4.26

5. 40.0 milliliters of a .250 molar weak acid are titrated with 0.250 molar NaOH. What is the pH at 60.0% of the titration? (The dissociation constant of the weak acid is 1.16×10^{-8})

   a) 7.00
   b) 4.27
   c) 5.66
   d) 7.93
   e) 8.12

6. What is the pH of a 1.08×10^{-3} molar cyanic solution? K_a for HOCN is 2.19×10^{-4}? (Assume water dissociation is negligible.)

   a) 3.41
   b) 10.59
   c) 11.03
   d) 7.00
   e) 2.97

7. The numerical value of K_b for hydrazine is 9.80×10^{-7}. What percentage of a .772 molar solution of N_2H_4 is ionized?

   a) 9.80×10^{-7} %
   b) 0.113%
   c) 7.56×10^{-5} %
   d) 0.00113%
   e) 9.80×10^{-5} %

8. Calculate the pH of a solution that is .841 molar in HPO_4^{2-} and .660 in PO_4^{3-}. For phosphoric acid, H_3PO_4, K_1, K_2, and K_3 are: 7.50×10^{-3}, 6.20×10^{-8}, and 4.80×10^{-13} respectively.

   a) 1.79
   b) 12.32
   c) 12.21
   d) 6.12×10^{-13}
   e) 12.42

9. Calculate the pH of a solution made by mixing 32.0 ml of 0.243 molar ammonia with 50.0 ml of 0.20 molar HCl. K_b for ammonia is 1.79×10^{-5}.

a) 0.43
b) 11.57
c) 11.43
d) 1.57
e) 12.43

10. The solubility product for $AgIO_3$ is 3.10×10^{-8}. What is the molar solubility of $AgIO_3$ in pure H_2O? (Neglect hydrolysis.)

a) 7.51
b) 3.10×10^{-8}
c) 0.00314
d) 0.000176
e) 1.55×10^{-8}

11. Calculate the molar solubility of Ag_2CO_3 in a solution of .863 molar $AgNO_3$. (K_{sp} for $Ag_2CO_3 = 8.20 \times 10^{-12}$; neglect hydrolysis.)

a) 8.20×10^{-12}
b) 0.000127
c) 1.10×10^{-11}
d) 3.32×10^{-6}
e) 9.50×10^{-12}

12. Given the standard enthalpies of formation and third-law entropies tabulated below, calculate the standard free energy change (in kcal) at 298°K for the reaction:

$$2\ PbO_2{(s)} \rightarrow 2\ PbO_{(s)} + O_2{(g)}$$

Compound	O_2	PbO_2	PbO
ΔH_f°, (kJ/mole)	0.0	−277	−219
S° (J/mole·K)	205	76.6	67.8

a) 60.1
b) 49.0
c) 123
d) −123
e) −112

13. Complete the statement below correctly. (ΔH = enthalpy change, ΔS = entropy change, Δn = change in no. of moles.)

   The enthalpy change in a chemical reaction:

   a) can be measured directly in a bomb calorimeter.
   b) is the heat of reaction under conditions of constant volume.
   c) is the criterion for spontaneity of reaction.
   d) can be predicted precisely from heats of formation.
   e) is the heat obtained upon combustion in pure oxygen.

14. Given the following reaction: $\quad\quad\quad Y + 2\,W \rightleftharpoons 3\,Z$

   What concentration of Z will be in equilibrium with .865 moles/liter of Y and .505 moles/liter of W if the equilibrium constant = 4.48×10^4?

   a) $\left[\dfrac{(.505)^2(.865)}{4.48 \times 10^4} \right]^{1/3}$

   b) $\dfrac{(.505)(.865)}{4.48 \times 10^4}$

   c) $(4.48 \times 10^4)(.505)(.865)$

   d) $\dfrac{4.48 \times 10^4}{(.505)(.865)}$

   e) None of the above.

15. Given the following hypothetical half-reactions:

Reaction	E°
$M^{2+} + e^- \rightleftharpoons M^+$	$-.579$
$M^{3+} + e^- \rightleftharpoons M^{2+}$	0.735

   Which of the following is true?

   a) M^{2+} will not disproportionate, $E^\circ = -1.31$ volts
   b) M^{2+} will not disproportionate, $E^\circ = 0.156$ volts
   c) M^{3+} will disproportionate, $E^\circ = 1.31$ volts.
   d) M^{2+} will disproportionate, $E^\circ = -1.31$ volts.
   e) More information is required.

16. Given the following standard electrode potentials:

Reaction	$E°$
$Cr^{3+} + 3\ e^- \rightleftharpoons Cr$	-0.774
$Au^{3+} + 3\ e^- \rightleftharpoons Au$	1.50

Concerning the cell:

$$Cr\,|\,Cr^{3+}\,||\,Au^{3+}\,|\,Au$$

Which one of the statements below is <u>incorrect</u>?
(Assume all substances are in the standard state.)

a) Au^{3+} ions will drift away from Au.
b) The negative terminal of the voltmeter should be attached to Cr and the positive to Au.
c) The concentration of Cr^{3+} will increase.
d) If the standard hydrogen half cell were substituted for the Au^{3+} half cell the voltage would be 0.774.
e) The anode is Cr.

17. If 27.0 percent of a radioactive isotope decays in 169 days, what is its half-life?

a) $\dfrac{.693}{2.303} \times 169 \times \log(100-27.0)$ days

b) $\dfrac{.693}{2.303} \times (169)/\log \dfrac{100}{73}$ days

c) $\dfrac{2.303}{.693} \times (169)/\log \dfrac{100}{73}$ days

d) $.5 \times \dfrac{2.303}{.693} \times \dfrac{27.0}{169}$ days

e) None of the above

18. In each of the gas phase experiments listed below the initial rate
     was measured for the reaction

$$2\ NO\ +\ Cl_2\ \rightarrow\ 2\ NOCl$$

Exp. #	P_{NO}	P_{Cl_2}	Rate of pressure drop
1	0.50 atm	0.50 atm	2.3×10^{-4} atm/sec
2	1.00 atm	1.00 atm	1.84×10^{-3} atm/sec
3	0.50 atm	1.00 atm	4.6×10^{-4} atm/sec

Based on the above, select the one correct statement.

a) Knowing the rate law we may calculate the rate constant from
   experiment 1, 2 or 3.
b) The order of reaction with regard to each reactant is given by
   the coefficients in the balanced equation.
c) The reaction is second order with respect to NOCl.
d) The reaction is second order with respect to Cl_2.
e) To find the order with respect to NO we compare experiments 1 and 2.

19. Calculate the amount of energy in MeV liberated in the following nuclear
     fission reaction, given the information below:

$$^{239}Pu\ +\ neutron\ \rightarrow\ ^{90}Rb\ +\ ^{146}Ba\ +\ 4\ neutrons$$

Conversion factor for $E = mc^2$ is 931 MeV/amu

     mass of neutron   = 1.01 amu
     mass of ^{239}Pu = 239.05
     mass of ^{90}Rb = 89.91
     mass of ^{146}Ba = 145.92

a) 1770
b) 177
c) 58.20
d) 582
e) 7630

20. In a present day nuclear reactor a typical reaction is:

$$^{235}U\ +\ neutron\ \rightarrow\ ^{93}Kr\ +\ ^{140}Ba\ +\ 3\ neutrons$$

Select the one correct statement.

a) The products have exceptionally low neutron/proton ratios.
b) The above reaction is endothermic (ΔH is plus).
c) Neutrons which escape the reactor will accumulate in the
   atmosphere and eventually threaten our environment.
d) The above reaction could not become explosive under any conditions.
e) The products will probably decay by electron (β ray) emission.

APPENDIX B.   CONSTANTS AND FORMULAS

Chapter 4. <u>Atoms and Molecules</u>

Avogadro's number $N = 6.02 \times 10^{23}$ mole^{-1}

Chapter 5. <u>Chemical Reactions</u>

Molar volume at STP (T = 273 K, P = 1.00 atm)

$\bar{V}$ = 22.4 liters/mole

Chapter 7. <u>Faraday's Laws</u>

Faraday's constant   F = Ne = 96500 coulombs/equivalent

current (amps) = charge(coulombs)/time(seconds)

Chapter 8. <u>Solution Stoichiometry</u>

molarity = # moles solute/# liters solution
molality = # moles solute/# kilograms solvent
normality = # equivalents solute/# liters solution

Chapter 10. <u>Colligative Properties</u>

Boiling point elevation constant for H_2O = 0.52°C/mole-kg^{-1}
Freezing point depression constant for H_2O = 1.86°C/mole-kg^{-1}

Chapter 13. <u>The Gas Laws</u>

Gas constant   R = 0.082 liter-atm/mole-deg K

Boyle's Law $P_1V_1 = P_2V_2$

Charles' Law $\dfrac{V_1}{T_1} = \dfrac{V_2}{T_2}$

Ideal Gas Law $PV = nRT$

Dalton's Law $P_{tot} = P_1 + P_2 + P_3 + \ldots$

$P_i = X_iP_{tot}$ (X_i = mole fraction)

Chapter 14. <u>Kinetic Molecular Theory</u>

Graham's Law
$$\frac{R_1}{R_2} = \sqrt{\frac{\rho_2}{\rho_1}} = \sqrt{\frac{M_2}{M_1}}$$

where R = rate of effusion, ρ = density and
M = molecular weight

$$V_{rms} = \sqrt{\frac{3RT}{M}} \qquad R = 8.317 \times 10^7 \text{ ergs/mole-deg K}$$

Chapter 15. <u>Crystalline Solids</u>

$$\rho = \frac{nM}{N\ell^3} \quad \text{where} \quad \rho = \text{density (g/cm}^3)$$

n = number of atoms or ion-pairs per unit cell,
M = atomic or molecular weight (g/mole),
N = Avogadro's number, and
ℓ = length of the unit cell in cm.

Chapter 16. <u>Atomic Structure</u>

Bohr radius $\qquad a_o = 0.529 \text{ Å} = 0.529 \times 10^{-8} \text{ cm}$

Planck's constant $\quad h = 6.63 \times 10^{-27} \text{ erg sec}$

Rydberg formula $\qquad \nu = R_H \left[\frac{1}{n_1^2} - \frac{1}{n_2^2} \right]$

Where ν = frequency (sec^{-1}), R = Rydberg constant in
frequency units, and n_1 and n_2 are positive integers

$$R_H = 3.29 \times 10^{15} \text{ sec}^{-1}$$

Wave equation $\qquad \nu = c/\lambda \quad$ (all electromagnetic radiation)

where ν = frequency in sec^{-1}; c = speed of light,
3.00 x 10^{10} cm/sec; and λ = wavelength in cm.

Wave number $\quad \bar{\nu} = 1/\lambda$

where $\bar{\nu}$ = wave number (cm^{-1}) and λ = wavelength (cm)

de Broglie's hypothesis $\qquad \lambda = \frac{h}{mv}$

where λ = wavelength (cm), m = mass (grams) and
v = velocity (cm/sec).

$$1 \text{ eV/molecule} = 1.60 \times 10^{-12} \text{ erg/molecule}$$
$$= 23.06 \text{ kcal/mole}$$
$$= 96.48 \text{ kJ/mole}$$

Chapter 19. <u>Acids and Bases</u>

$$pH = -\log_{10} [H^+]$$

$$pOH = -\log_{10} [OH^-]$$

$$pH + pOH = 14.00 \text{ at } 25°C$$

Chapter 22. <u>Thermodynamics</u>

$$\Delta G° = \Delta H° - T\Delta S° = -2.303 \, RT \log K_{eq}$$

where $\Delta G°$ = standard free energy change,
$\quad \Delta H°$ = standard enthalpy change,
$\qquad T$ = absolute temperature,
$\quad \Delta S°$ = standard entropy change
$\qquad R = 8.314$ J/mole-deg K
and $\quad K_{eq}$ = thermodynamic equilibrium constant.

$$\Delta H = \Delta E + P\Delta V \simeq \Delta E + \Delta n_g RT$$

$$\log \frac{K_1}{K_2} = \frac{\Delta H°}{2.303 \, R} \left[\frac{T_1 - T_2}{T_1 T_2} \right]$$

where K_1 and K_2 are equilibrium constants at absolute temperatures T_1 and T_2 respectively and $\Delta H°$ is the standard enthalpy change for the reaction.

Chapter 23. <u>Chemical Equilibrium</u>

$$K_p = K_c (RT)^{\Delta n_g}$$

Chapter 24. <u>Electrochemistry</u>

$$\Delta G° = -nFE° = -2.303 \, RT \log K$$

$$\Delta G° = -96.48 \, nE° \qquad \text{(kilojoules)}$$

$$\text{at } T = 298°K, \quad \log K = \frac{nE°}{0.0592}$$

$$\text{Nernst equation} \quad E = E° - \frac{0.0592}{n} \log Q \text{ at } 298°K$$

Chapter 25. <u>Kinetics</u>

First order kinetics: $k = 0.693/t_{1/2}$

$$\log \frac{C_o}{C} = \frac{kt}{2.303}$$

where k = first order decay constant; $t_{1/2}$ = half life; C_o and C are concentration and t is time.

Arrhenius relationship $k = Ae^{-E_a/RT}$

$$\log \frac{k_1}{k_2} = \frac{E_a}{2.303\ R} \left[\frac{T_1-T_2}{T_1 T_2}\right]$$

where k = rate constant, A = a constant, and E_a = activation energy.

Chapter 27. <u>Nuclear Chemistry</u>

Atomic mass unit   1 amu = 1.66×10^{-24} gram

1 amu = 931 MeV (million electron volts)

APPENDIX C.   ANSWERS TO QUESTIONS

Chapter 2

2.1 - a
2.2 - c
2.3 - a
2.4 - e
2.5 - e
2.6 - a
2.7 - c
2.8 - b*
2.9 - c
2.10 - e
2.11 - c
2.12 - b
2.13 - d
2.14 - b
2.15 - e
2.16 - a
2.17 - b
2.18 - a*
2.19 - c
2.20 - a
2.21 - d
2.22 - d
2.23 - c
2.24 - b
2.25 - c*
2.26 - b
2.27 - c
2.28 - d
2.29 - c
2.30 - d*

Chapter 3

3.1 - d
3.2 - b
3.3 - d
3.4 - e
3.5 - c
3.6 - c

Chapter 3

3.7 - a
3.8 - b
3.9 - d
3.10 - b

Chapter 4

4.1 - d
4.2 - e
4.3 - d
4.4 - c
4.5 - c
4.6 - d
4.7 - c*
4.8 - c
4.9 - a
4.10 - e
4.11 - d
4.12 - e
4.13 - e
4.14 - a*
4.15 - e
4.16 - e*
4.17 - e*
4.18 - a
4.19 - a
4.20 - c*
4.21 - b
4.22 - b*
4.23 - d
4.24 - c*
4.25 - a

Chapter 5

5.1 - d
5.2 - d
5.3 - e
5.4 - d

Chapter 5

5.5 - b
5.6 - c*
5.7 - e
5.8 - c
5.9 - c
5.10 - a*
5.11 - d
5.12 - c*
5.13 - e
5.14 - a
5.15 - d*
5.16 - d
5.17 - b
5.18 - c
5.19 - c
5.20 - c
5.21 - c
5.22 - b
5.23 - b
5.24 - b
5.25 - e
5.26 - c
5.27 - b
5.28 - e*
5.29 - b
5.30 - c

Chapter 6

6.1 - b
6.2 - d
6.3 - a
6.4 - e
6.5 - b*
6.6 - c*
6.7 - e
6.8 - d
6.9 - c
6.10 - a*

Chapter 7

7.1 - a
7.2 - b
7.3 - c
7.4 - d
7.5 - b
7.6 - c
7.7 - c
7.8 - b*
7.9 - d*
7.10 - b*

Chapter 8

8.1 - b
8.2 - d
8.3 - b
8.4 - b
8.5 - c
8.6 - c
8.7 - d
8.8 - c*
8.9 - d
8.10 - a
8.11 - b*
8.12 - d
8.13 - b
8.14 - c*
8.15 - b
8.16 - c
8.17 - a*
8.18 - a*
8.19 - b
8.20 - d
8.21 - c*
8.22 - b*
8.23 - e
8.24 - c*
8.25 - c*

Chapter 9

9.1 - c
9.2 - b
9.3 - a*
9.4 - b
9.5 - c
9.6 - c*
9.7 - b
9.8 - a
9.9 - b*
9.10 - b*
9.11 - c
9.12 - c
9.13 - b*
9.14 - b*
9.15 - b*

Chapter 10

10.1 - d
10.2 - d
10.3 - d*
10.4 - b
10.5 - d
10.6 - d
10.7 - a
10.8 - b*
10.9 - e
10.10 - d*

Chapter 11

11.1 - b*
11.2 - a*
11.3 - e*
11.4 - c*
11.5 - b*
11.6 - a
11.7 - d*
11.8 - d
11.9 - b
11.10 - d
11.11 - c
11.12 - c
11.13 - b*
11.14 - d*
11.15 - a*

Chapter 12

12.1 - e*
12.2 - e*
12.3 - a*
12.4 - b*
12.5 - d*

Chapter 13

13.1 - c
13.2 - d
13.3 - d
13.4 - a*
13.5 - e
13.6 - a
13.7 - b
13.8 - b*
13.9 - d
13.10 - c
13.11 - e
13.12 - a
13.13 - e*
13.14 - d
13.15 - e*
13.16 - e
13.17 - d
13.18 - a
13.19 - c*
13.20 - a
13.21 - c*
13.22 - a*
13.23 - e*
13.24 - d
13.25 - a*

Chapter 14

14.1 - c*
14.2 - c
14.3 - d
14.4 - a*
14.5 - d
14.6 - d
14.7 - a*
14.8 - c*
14.9 - e*
14.10 - a*

Chapter 15

15.1 - b
15.2 - c
15.3 - a
15.4 - a
15.5 - c*
15.6 - c*
15.7 - d
15.8 - e
15.9 - a*
15.10 - b

Chapter 16

16.1 - d
16.2 - c
16.3 - d*
16.4 - c
16.5 - c
16.6 - c
16.7 - d*
16.8 - c*
16.9 - d*
16.10 - a*
16.11 - c*
16.12 - e
16.13 - a
16.14 - d*
16.15 - b

Chapter 17

17.1 - b
17.2 - a
17.3 - e
17.4 - e
17.5 - d
17.6 - d*
17.7 - c*
17.8 - c
17.9 - a
17.10 - b
17.11 - e*
17.12 - b
17.13 - a*
17.14 - d
17.15 - e*

Chapter 18

18.1 - a
18.2 - d
18.3 - d*
18.4 - d*
18.5 - d
18.6 - b*
18.7 - e
18.8 - c*
18.9 - e
18.10 - a
18.11 - a
18.12 - a
18.13 - d*
18.14 - a
18.15 - e

Chapter 19

19.1 - a
19.2 - d
19.3 - a
19.4 - e
19.5 - e
19.6 - b
19.7 - a
19.8 - c
19.9 - c
19.10 - e
19.11 - a*
19.12 - e
19.13 - b
19.14 - d
19.15 - d
19.16 - c
19.17 - a
19.18 - b
19.19 - c
19.20 - d
19.21 - d*
19.22 - a
19.23 - c*
19.24 - b
19.25 - d
19 26 - d*
19.27 - d
19.28 - c
19.29 - d*
19.30 - a
19.31 - c
19.32 - b

Chapter 19

19.33 - c
19.34 - c*
19.35 - c.
19.36 - d
19.37 - b*
19.38 - c*
19.39 - a
19.40 - d
19.41 - c
19.42 - c*
19.43 - e
19.44 - c
19.45 - c*
19.46 - b
19.47 - b*
19.48 - d
19.49 - b
19.50 - d*
19.51 - c
19.52 - e*
19.53 - a*
19.54 - c
19.55 - a*

Chapter 20

20.1 - a
20.2 - b*
20.3 - a*
20.4 - d
20.5 - b
20.6 - e*
20.7 - d*
20.8 - b*
20.9 - c
20.10 - d
20.11 - c*
20.12 - a*
20.13 - b
20.14 - e*
20.15 - e

Chapter 21

21.1 - d
21.2 - a
21.3 - c
21.4 - c*
21.5 - e

Chapter 21

21.6 - a
21.7 - b*
21.8 - d
21.9 - e*
21.10 - e*
21.11 - e*
21.12 - a*
21.13 - d*
21.14 - e*
21.15 - e*
12.16 - c*
21.17 - d*
21.18 - e
21.19 - e
21.20 - c*

Chapter 22

22.1 - c
22.2 - d
22.3 - a
22.4 - a*
22.5 - d
22.6 - a
22.7 - e*
22.8 - b
22.9 - d
22.10 - c*
22.11 - e
22.12 - a
22.13 - c*
22.14 - b
22.15 - b
22.16 - a*
22.17 - e
22.18 - c
22.19 - e*
22.20 - b*

Chapter 23

23.1 - b
23.2 - a
23.3 - c
23.4 - d
23.5 - c*
23.6 - c*
23.7 - b
23.8 - d*

Chapter 23

23.9 - a*
23.10 - c
23.11 - b
23.12 - b*
23.13 - a
23.14 - c
23.15 - e*

Chapter 24

24.1 - a
24.2 - a
24.3 - c
24.4 - d*
24.5 - b*
24.6 - a
24.7 - d*
24.8 - c
24.9 - a*
24.10 - d
24.11 - c
24.17 - e*
24.13 - a*
24.14 - d*
24.15 - e
24.16 - d*
24.17 - c*
24.18 - b*
24.19 - b*
24.20 - b*

Chapter 25

25.1 - d
25.2 - c
25.3 - b
25.4 - d
25.5 - b*
25.6 - e
25.7 - a*
25.8 - b*
25.9 - a*
25.10 - b*

Chapter 26

26.1 - b
26.2 - e
26.3 - e

Chapter 26

26.4 - a
26.5 - c
26.6 - b*
26.7 - b*
26.8 - b
26.9 - b
26.10 - b*
26.11 - e*
26.21 - d*
26.13 - e
26.14 - a
26.15 - e
26.16 - d*
26.17 - b
26.18 - d*
26.19 - a*
26.20 - c
26.21 - b*
26.22 - c
26.23 - e*
26.24 - c*
26.25 - d*

Chapter 27

27.1 - b
27.2 - c
27.3 - d*
27.4 - e
27.5 - b
27.6 - b
27.7 - e*
27.8 - b*
27.9 - c
27.10 - b
27.11 - d
27.12 - d
27.13 - c*
27.14 - d*
27.15 - b

Chapter 28

28.1 - b
28.2 - e
28.3 - b*
28.4 - d
28.5 - c*

<u>Chapter 28</u>

28.6 - b*
28.7 - b*
28.8 - c
28.9 - b*
28.10 - c

<u>Chapter 29</u>

29.1 - d*
29.2 - e*
29.3 - d*
29.4 - b
29.5 - b
29.6 - d
29.7 - d*
29.8 - d
29.9 - d*
29.10 - a
29.11 - d*
29.12 - c*
29.13 - c*
29.14 - d*
29.15 - b*
29.16 - d
29.17 - a*
29.18 - d
29.19 - b
29.20 - a
29.21 - a
29.22 - c
29.23 - e
29.24 - d
29.25 - a

<u>Sample Exams</u>

<u>Exam 1</u>		<u>Exam 2</u>		<u>Exam 3</u>		<u>Exam 4</u>	
1	c	1	d	1	b	1	d
2	b	2	c	2	b	2	c
3	c	3	a	3	d	3	c
4	c	4	b	4	b	4	c
5	b	5	a	5	e	5	e
6	b	6	c	6	d	6	a
7	e	7	a	7	b	7	b
8	a	8	b	8	c	8	c
9	d	9	d	9	a	9	d
10	e	10	b	10	d	10	d
11	b	11	b	11	a	11	c
12	b	12	a	12	d	12	a
13	d	13	a	13	d	13	d
14	a	14	b	14	c	14	e
15	a	15	b	15	b	15	a
16	e	16	e	16	b	16	a
17	c	17	d	17	c	17	b
18	d	18	c	18	c	18	a
19	b	19	d	19	d	19	b
20	b	20	e	20	e	20	e

Periodic Table of the Elements

PERIOD	IA	IIA	IIIA	IVA	VA	VIA	VIIA	VIII	VIII	VIII	IB	IIB	IIIB	IVB	VB	VIB	VIIB	NOBLE GASES
1	1 H 1.0079†																1 H 1.0079†	2 He 4.00260
2	3 Li 6.941†	4 Be 9.01218											5 B 10.81	6 C 12.011	7 N 14.0067	8 O 15.9994†	9 F 18.99840	10 Ne 20.179†
3	11 Na 22.98977	12 Mg 24.305											13 Al 26.98154	14 Si 28.086†	15 P 30.97376	16 S 32.06	17 Cl 35.453	18 Ar 39.948†
4	19 K 39.098†	20 Ca 40.08	21 Sc 44.9559	22 Ti 47.90†	23 V 50.9414†	24 Cr 51.996	25 Mn 54.9380	26 Fe 55.847†	27 Co 58.9332	28 Ni 58.71†	29 Cu 63.546†	30 Zn 65.38	31 Ga 69.72	32 Ge 72.59†	33 As 74.9216	34 Se 78.96†	35 Br 79.904	36 Kr 83.80
5	37 Rb 85.4678†	38 Sr 87.62	39 Y 88.9059	40 Zr 91.22	41 Nb 92.9064	42 Mo 95.94†	43 Tc 98.9062	44 Ru 101.07†	45 Rh 102.9055	46 Pd 106.4	47 Ag 107.868	48 Cd 112.40	49 In 114.82	50 Sn 118.69†	51 Sb 121.75†	52 Te 127.60†	53 I 126.9045	54 Xe 131.30
6	55 Cs 132.9054	56 Ba 137.34†	57 *La 138.9055†	72 Hf 178.49†	73 Ta 180.9479†	74 W 183.85†	75 Re 186.2	76 Os 190.2	77 Ir 192.22†	78 Pt 195.09†	79 Au 196.9665	80 Hg 200.59†	81 Tl 204.37†	82 Pb 207.2	83 Bi 208.9804	84 Po (210)	85 At (210)	86 Rn (222)
7	87 Fr (223)	88 Ra 226.0254	89 ʼAc (227)	104 § (260)	105 § (260)													

*LANTHANOID SERIES

58 Ce 140.12	59 Pr 140.9077	60 Nd 144.24†	61 Pm (147)	62 Sm 150.4	63 Eu 151.96	64 Gd 157.25†	65 Tb 158.9254	66 Dy 162.50†	67 Ho 164.9304	68 Er 167.26†	69 Tm 168.9342	70 Yb 173.04†	71 Lu 174.97

ʼACTINOID SERIES

90 Th 232.0381	91 Pa 231.0359	92 U 238.029	93 Np 237.0482	94 Pu (244)	95 Am (243)	96 Cm (247)	97 Bk (247)	98 Cf (251)	99 Es (254)	100 Fm (257)	101 Md (256)	102 No (255)	103 Lr (256)

§The International Union for Pure and Applied Chemistry has not adopted official names or symbols for these elements.

†These weights are considered reliable to ±3 in the last place. Other weights are reliable to ±1 in the last place.

Atomic weights corrected to conform to the 1971 values of the Commission on Atomic Weights.